VOLUME ONE HUNDRED AND FIFTEEN

ADVANCES IN COMPUTERS

Role of Blockchain Technology in IoT Applications

VOLUME ONE HUNDRED AND FIFTEEN

ADVANCES IN COMPUTERS

Role of Blockchain Technology in IoT Applications

Edited by

SHIHO KIM
*Yonsei Institute of Convergence Technology,
Yonsei University, Seoul, South Korea*

GANESH CHANDRA DEKA
*Deputy Director (Training),
Directorate General of Training Ministry of Skill
Development & Entrepreneurship,
Government of India, New Delhi, India*

PENG ZHANG
*Department of Biomedical Informatics,
Vanderbilt University Medical Center,
Nashville, TN, United States*

Academic Press is an imprint of Elsevier
50 Hampshire Street, 5th Floor, Cambridge, MA 02139, United States
525 B Street, Suite 1650, San Diego, CA 92101, United States
The Boulevard, Langford Lane, Kidlington, Oxford OX5 1GB, United Kingdom
125 London Wall, London, EC2Y 5AS, United Kingdom

First edition 2019

ISBN: 978-0-12-817189-9
ISSN: 0065-2458

For information on all Academic Press publications
visit our website at https://www.elsevier.com/books-and-journals

Publisher: Zoe Kruze
Acquisition Editor: Zoe Kruze
Editorial Project Manager: Peter Llewellyn
Production Project Manager: James Selvam
Cover Designer: Greg Harris

Typeset by SPi Global, India

Contents

Preface

Blockchain technology is a decentralized technology for storing immutable records of transactions grouped into blocks, which are secured by cryptography. Blockchain technology will be revolutionizing the way we interact and exchange information. It offers data security and integrity and eliminates the role of trusted third-party valuators.

There are prospects of applying blockchain technology in healthcare, supply chain management, banking, to list a few examples. As forecasted by BR SofTech (https://www.brsoftech.com/), the market of blockchain technology in healthcare will be worth $829 million by 2023. Blockchain has also permeated into the increasingly popular Internet of Things technology. According to "Bain & Company" estimates, the B2B IoT market will surpass $300 billion by 2020. This edited book with 10 chapters will deliberate upon the various aspects of blockchain technology and its prospects.

Chapter 1 presents the technical aspects of blockchain and IoT, including various applications and challenges of IoT-based systems. IoT and blockchain are two emerging technologies that are expected to have an immense impact in the society around the world. Chapter 2 discusses integrated platforms for blockchain enablement. Chapter 3 provides a discussion of the intersection between IoT and distributed ledger technology (DLT), which is a family of different types of blockchain-like systems. It first provides an overview of the DLT by highlighting its main components, benefits, and challenges and then summarizes centralized IoT systems and their limitations. The integration of blockchain with IoT and integration benefits is presented along with potential application and challenges of the integration. Chapter 4 explores the potential of blockchain technology for decentralizing the autonomous organizations. Chapter 5 examines how blockchain technology will revolutionize the healthcare system. Chapter 6 describes the design and testing methodologies and tools to test IoT and blockchain applications. Chapter 7 surveys commonly used consensus mechanisms and information security technologies used in DLT (distributed ledger technology). Chapter 8 presents strategies to address privacy and security issues in IoT. Chapter 9 deliberates upon prospects of blockchain technology in cost savings in supply chain management system. Chapter 10 highlights the research challenges and future research directions in the integrated blockchain-based IoT with homomorphic encryption.

We hope the readers of the book will be benefited by the wide coverage of the various aspects of IoT in the context of blockchain technology and its application.

Prof. Shiho Kim
Ganesh Chandra Deka
Dr. Peng Zhang

CHAPTER ONE

Technical aspects of blockchain and IoT

Hany F. Atlam[a,b], **Gary B. Wills**[a]

[a]Electronic and Computer Science Department, University of Southampton, Southampton, United Kingdom
[b]Computer Science and Engineering Department, Faculty of Electronic Engineering, Menoufia University, Menoufia, Egypt

Contents

Abstract

Blockchain technology is getting a growing attention from various organizations and researchers as it provides magical solutions to the problems associated with the classical centralized architecture. Blockchain, whether public or private, is a distributed ledger with the capability of maintaining the integrity of transactions by decentralizing the ledger among participating users. On the other hand, the Internet of Things

Advances in Computers, Volume 115
ISSN 0065-2458
https://doi.org/10.1016/bs.adcom.2018.10.006

(IoT) represents a revolution of the Internet which can connect nearly all environment devices over the Internet to share their data to create novel services and applications for improving our quality of life. Although the centralized IoT system provides countless benefits, it raises several challenges. Resolving these challenges can be done by integrating IoT with blockchain technology. To be prepared for the integration process, this chapter provides an overview of technical aspects of the blockchain and IoT. It started by reviewing blockchain technology and its main structure. Applications and challenges of the blockchain are also presented. This is followed by reviewing the IoT system by highlighting common architecture and essential characteristics. Various applications and challenges of the IoT system are also discussed.

1. Introduction

Before the invention of the blockchain, managing various activities and actions over the Internet was achieved through a centralized server to guarantee non-repudiation of data. A group of distributed entities could not verify transactions without using the centralized authority [1]. There was no trust between communicating parties, so a third party was needed to build the required trust and manage the communication process. This problem was known as the Byzantine Generals Problem (BGP) [2]. This problem supposed that there were three sections of the Byzantine army waiting outside an enemy city and planning to attack it. The army general of each section was independent; however, a common course of action should be achieved to be able to conquer the city. The army generals' capabilities to interconnect with one another were only allowed through a messenger service, and there was a traitor who corrupts generals' actions to make sure there is no chance to make a united attack [3].

To find a solution to the problem of Byzantine generals, the blockchain increases transparency and reliability by using a probabilistic approach to distribute data among several users of the network. Generally, blockchain is a distributed database/ledger of transactions used to manage a constantly increasing set of records. It provides an efficient way to maintain security and data integrity in which a transaction must be verified by the majority of participating users in the blockchain network to be eligible to add in the ledger [3,4]. The blockchain does not use a third party to store information instead, every participating user in the blockchain network holds a genuine copy of the ledger. So, if a user has breached and added a malicious transaction, the system will discard it, as it should be verified by all other network users. Also, there is a multi-signature protection to validate each transaction, which adds another layer of security [5]. Therefore, injecting the distributed ledger with false or malicious data by an intruder or malicious user is significantly reduced. It is possible to happen but only if the intruder uses more computational power than the complete blockchain network, which is very rare to occur [6].

Another technology added significant developments to our community by having the capability to connect environment devices over the Internet to share their data and create new applications and services for improving our quality of life. This technology is known as the Internet of Things (IoT). The IoT is considered as an evolution of the Internet which involves both virtual and physical things of our environment, which are in billions [7]. Using a set of cheap sensors, the IoT enables several advantages to users by collecting relevant information that will ultimately change their lifestyles and improve their quality of life [8].

The next phase of developments is to merge the IoT with blockchain technology. Although the centralized IoT architecture provides various benefits, it raises severe challenges regarding costs, scalability and security. The blockchain provides a decentralized model that can process billions of operations between various IoT devices. This is, in turn, will reduce the costs associated with building and maintaining large centralized data centers. Moreover, in the absence of a third party, the security issues regarding the single point of failure will be eliminated [6].

To be prepared for the integration of IoT with blockchain, this chapter presents technical aspects of blockchain and IoT. It started by providing a discussion of the blockchain technology and its main components. Current applications and challenges of the blockchain are also presented. This is followed by providing an overview of the IoT system including its common architecture and essential characteristics. Various applications and challenges of the IoT system are also discussed.

This chapter is structured as follows; Section 2 provides the technical aspects of blockchain technology by discussing its history, structure, main characteristics, applications and challenges; Section 3 presents an overview of the IoT system by providing its definitions, architecture, essential characteristics, applications and challenges; and Section 4 is the conclusion.

2. Blockchain technology

This section provides a discussion of the technical aspects of blockchain technology. It discusses various definitions of blockchain with highlighting its essential characteristic and different types of the blockchain. Various applications and challenges of the blockchain are also presented.

2.1 An overview of blockchain

Blockchain is a distributed and decentralized ledger of transactions used to manage a constantly increasing set of records. To store a transaction in the ledger, the majority of participating users in the blockchain network should agree and record their consent. A set of transactions are grouped together and allocate a block in the ledger, which is chained ofblocks. To link the blocks together, each block encompasses a timestamp and hash function to the previous block. The hash function validates the integrity and non-repudiation of the data inside the block. Moreover, to keep all participating users of the blockchain network updated, each user holds a copy of the original ledger and all users are synchronized and updated with newly change [4].

Blockchain has defined by many organizations from different perspectives. For instance, Coinbase, the world's largest cryptocurrency exchange, defined blockchain as "*a distributed, public ledger that contains the history of every bitcoin transaction*" [9]. This definition explains the blockchain from the cryptocurrency's perspective which does not consider the fact that the blockchain can be used in various applications independently. Whereas Oxford dictionary provides a more common definition for the blockchain. It stated "*a digital ledger in which transactions made in bitcoin or another cryptocurrency are recorded chronologically and publicly*" [10].

In addition, a broader definition for the blockchain is provided by Webopedia. It stated "*a type of data structure that enables identifying and tracking transactions digitally and sharing this information across a distributed network of computers, creating in a sense a distributed trust network. The distributed ledger technology offered by blockchain provides a transparent and secure means for tracking the ownership and transfer of assets*" [11]. This definition provides a more detailed description of the blockchain by highlighting its essential features with confirming that the blockchain is not only a distributed technology but also a decentralized environment.

Moreover, highlighting the main elements of blockchain technology, Sultan et al. [12] provides a general definition for the blockchain. It stated *"a decentralized database containing sequential, cryptographically linked blocks of digitally signed asset transactions, governed by a consensus model."*

2.2 History of blockchain

The history of blockchain technology has their roots in the 1980s and 1990s in the 20th century. However, it became widely acknowledged in 2008 after the discovery of the Bitcoin. According to Pilkington [13], the first notion of digital currency was invented based on a centralized server to avoid double-spending, which is the process of using the same bitcoins more than once. However, this perception failed to provide a solution for double-spending, anonymity and centralization problems.

The world remains several years to utilize the centralized architecture which use a third party to control and maintain the trust between communication parties until Szabo at the end of 1990 invented a decentralized digital currency which was called bit gold. After about 10 years, Bitcoin cryptocurrency was presented. Blockchain became broadly popular after the legendary paper of Nakamoto [6]. He proposed substituting the classical centralized architecture with a new technique based on a consensus mechanism. Initially, the technology was named as blockchain as two words "block" and "chain"; however, by 2016, two words are combined into one word to be what we all know now blockchain [14].

During the period from 2011 to 2013, blockchain has widely used in cryptocurrencies especially in currency transfer and digital payment. Nowadays, blockchain technology has emerged in various applications and services to make use of decentralization and immutability features. Fig. 1 depicts the history of blockchain technology from 1990 till now.

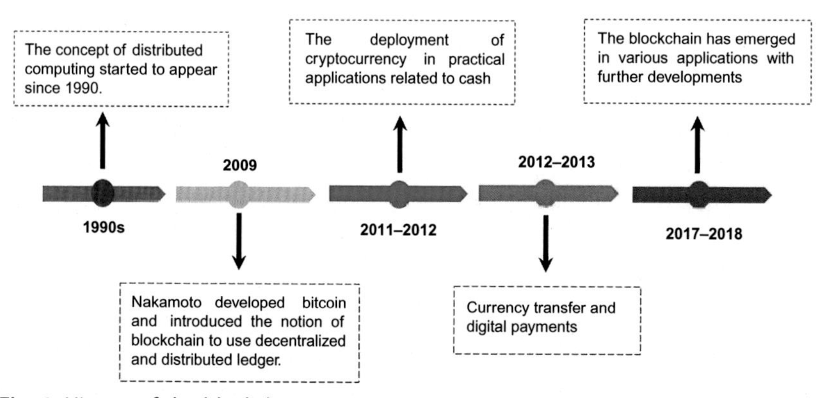

Fig. 1 History of the blockchain.

2.3 Building blocks of blockchain

Blockchain technology has the potential to deliver an effective way of storing transactions in the ledger with ensuring transparency, security and auditability. Although the blockchain still in the early stage of approval, the commercial community should adopt it for industries and businesses to avoid disruptive surprises or wasted chances.

The next section provides a brief discussion of significant building blocks of blockchain technology.

2.3.1 Database

A classical database is a data structure used for storing information. It uses a relational model to provide more composite ways of querying and collecting data by linking information from multiple databases. The information stored in databases can be organized using a DataBase Management System (DBMS). A simple database is stored in data elements called a table which contains fields. Each field contains columns to describe the field and rows to define a record stored in the database [15].

One of the main elements of the blockchain is the database. However, this is not a normal database containing rows and columns; instead, it is a ledger of all previous transactions for all participating users within the blockchain network. This type of databases is characterized by having a high-throughput, decentralized control, low latency, immutable data storage and built-in security.

2.3.2 Block

Block is the key storage element in the blockchain. It contains and perseveres data related to multiple transactions. The blocks are chained together by storing the hash of the previous block in the current block, which makes blocks chained as a circle for enclosure in the public ledger.

Blocks are typically divided into two segments, header and a group of transactions. The header contains the block metadata which is used to contain all details about the block in the ledger [16]. Fig. 2 shows the structure of a block and illustrates how blocks chained together with the block header information described as follows:

- *Version number*: 4 bytes to indicate the version number of the *block*.
- *Previous block hash*: 32 bytes to describe the hash of the previous block of the blockchain. It acts as a pointer between the current block and the previous block in the ledger.

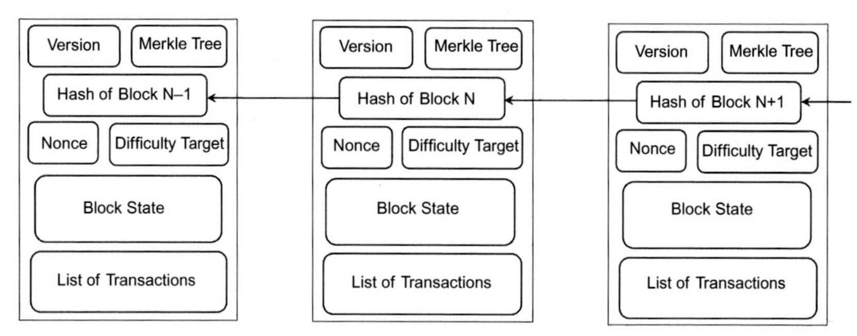

Fig. 2 Structure of block showing how blocks chained together with header information.

- *Timestamp*: 4bytes to record the time at which the block has been created.
- *Merkle tree*: 32bytes which are a hash (SHA-256) of all transactions that are related to this block.
- *Difficulty target*: 4bytes to identify the difficulty target of the block.
- *Nonce*: 4bytes to create the block and compute different hashes.

2.3.3 Hash

The hash function is a complex mathematical problem which the miners have to solve in order to find a block. The notion of hash function is used as a way to search for data in a database. Hash functions are collision-free, which means it is very difficult to find two identical hashes for two different messages. Hence, the blocks are identified through their hash, serving two purposes; identification and integrity verification [17].

For linking blocks together, each block encompasses the hash of its parent inside its own header which places a chain going all the way back to the first block which creates a sequence of hashes. The hash values are kept in a hash table which is a well-organized indexing mechanism to increase the performance of the search operations [18].

2.3.4 Minor

A CPU that tries to solve a computationally intense mathematical problem to discover a novel block is known as a miner. The miners can work either alone or in pools to try to find out the solution of the mathematical problem.

The process of finding a new block is started by broadcasting new transactions to all the users of the blockchain network. Each user collects new transactions into a block and works to find the block's proof-of-work.

If a user finds it, the block will be broadcasted to all the users to verify it. The block will be verified only if all inside transactions are valid. The block can be considered as accepted from all the participating users in the blockchain network when they start working on generating the next block in the chain using the hash of the accepted block as the previous hash. In some cases, the transactions with the highest costs are first selected from minors, since the minor who finds the block earns the costs or fees of all the transactions in that block [17].

2.3.5 Transaction

A blockchain transaction can be defined as a small unit of task that is stored in public records. These records are implemented, executed and stored in the blockchain only after being verified by the majority of users involved in the blockchain network. Each previous transaction can be reviewed at any time but cannot be updated. The size of the transaction is significant for miners since larger transactions need more space in the block and consume more power, while smaller transactions are easier to validate and consume less power [18].

2.3.6 Consensus mechanism

Blockchain is a type of distributed ledger used to store a record of all previous transactions. It called distributed since it is stored across multiple computers over the network worldwide. The main operation of a distributed ledger is to guarantee that the entire network approves the contents of the ledger, which is done by using the consensus mechanism.

There are a number of consensus mechanisms. However, the most common blockchain consensus mechanisms are Proof of Stake (PoS) and Proof of Work (PoW). The key difference between various consensus mechanisms is the way they delegate and reward the verification of transactions [15].

PoW is a popular consensus mechanism used by the most widespread cryptocurrency networks like Bitcoin and Litecoin. The participant-user in the blockchain network is required to prove the work was done to qualify them to obtain the ability to add new blocks to the ledger. However, the mining process requires high energy consumption and processing time. PoS is another public consensus mechanism to provide a low-cost, low-energy consumption in comparison with the PoW mechanism. Also, it allocates the responsibility to the participant users in proportion to the number of virtual currency tokens held by it. However, this derives a downside as it encourages crypto-coin saving, instead of spending it [16].

2.4 Characteristics of blockchain

Blockchain can be considered as a decentralized architecture with built-in security to increase the trust and integrity of transactions. This section aims to provide a discussion of common characteristics associated with the blockchain. These features, as summarized in Fig. 3, include:

- *Decentralization*: In contrast to the centralized architecture which presents several issues including single point of failure and scalability, the blockchain uses a decentralized and distributed ledger to utilize the processing capabilities of all the participating users in the blockchain network, which reduce latency and eliminate the single point of failure.
- *Immutability*: One of the essential features of the blockchain is the ability to ensure the integrity of transactions by creating immutable ledgers. In traditional centralized architectures, databases can be altered and a trust with a third party needs to be created to guarantee information integrity. While in blockchain technology, since each block in the distributed ledger relates to the previous block constituting a chain of blocks, the blocks are permanently saved and never changed as long as the participating user continue to maintain the network [15].

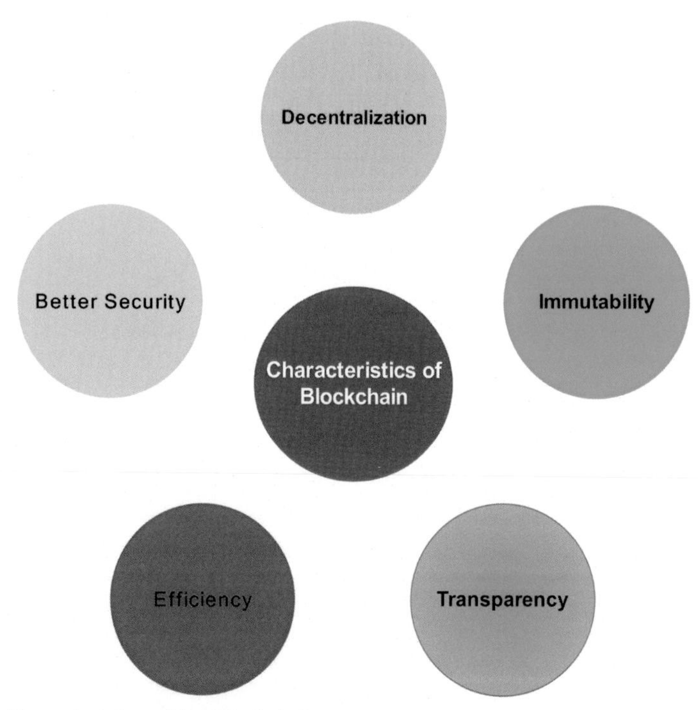

Fig. 3 Characteristics of the blockchain.

- *Transparency*: Blockchain delivers a high level of transparency by sharing transaction details between all participants users involved in those transactions. In a blockchain environment, no need for a third party which improve business friendliness and guarantees a trusted workflow.
- *Better Security*: Although security represents an essential issue for most new technologies, blockchain provides better security since it uses public key infrastructure that protects against malicious actions to change data. The participating users of the blockchain network place their trust in the integrity and security features of the consensus mechanism. In addition, blockchain eliminates the single point of failure which affects the entire system [12].
- *Efficiency*: Blockchain improves the classical centralized architecture by distributing database records between various users involved in the blockchain network. The distribution of transactions makes it more transparent to verify all records stored in the database. Blockchain is more efficient than the classical centralized architecture in terms of cost, settlement speed and risk management [19].

2.5 Types of blockchain

There are three types of blockchain; public, private and federated, as shown in Fig. 4.

- *Public Blockchain*: It is a blockchain that allows any anonymous user to be added to the blockchain network, sends a new transaction, verifies newly added blocks and reads the content of the blockchain. Public blockchain

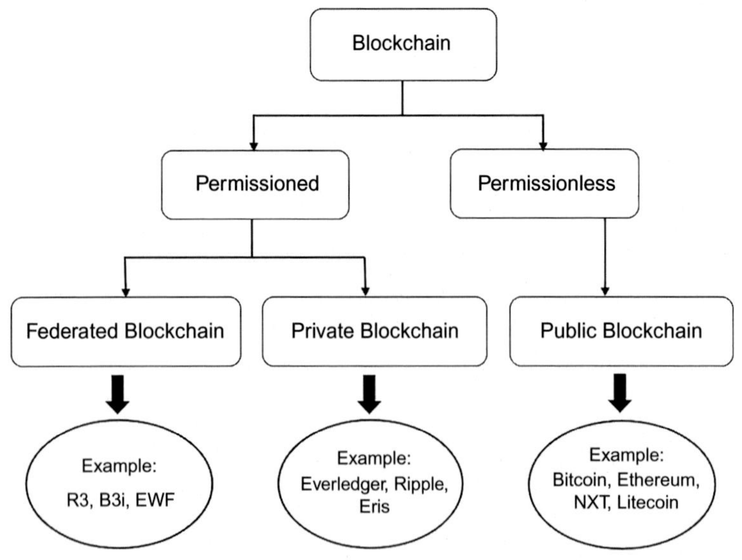

Fig. 4 Types of blockchains.

is open for all types of entities to participate in the network. Securing the public blockchain is done using cryptoeconomics which is a mixture of cryptographic verification and economic incentives using consensus mechanisms such as PoW or PoS. There are many examples of public blockchains; however, the most common are Ethereum, Bitcoin and NXT [20].

- *Private Blockchain*: In this type of blockchains, only a specific organization has the authority to join the blockchain network, send a new transaction, and participate in the consensus mechanism. Users willing to participate have to gain their permissions from the organization before joining the blockchain network. Likely applications that used private blockchain include database management and auditing. Common instances of private blockchain are Ripple, Everledger and Eris [21]. In comparison with public blockchain, the private blockchain is easier since the number of participating users are small so that verifying the new blocks does not take huge processing power and time. Also, the private blockchain provides a better privacy as only users identified within the blockchain network can read the transactions.
- *Federated Blockchain*: It is considered as partly private blockchain. It is operated under the authority of a group of companies or organizations. So, it is a private blockchain for a specific set of organizations. Unlike public blockchain, federated blockchain is faster and delivers better scalability and privacy. Examples of federated blockchains are R3, EWF, and B3i [13].

Table 1 provides a comparison between public, private and federated blockchains in terms of access permission, speed of transaction execution, efficiency, security, immutability, consensus mechanism, network and asset.

Table 1 Differences between public, private and federated blockchain.

Item	Public	Private	Federated
Access	Read/write for anyone	Read/write for a single organization	Read/write for multiple selected organizations
Speed	Slower	Lighter and faster	Lighter and faster
Efficiency	Low	High	High
Security	Proof of work, proof of stake, and other consensus mechanisms	Pre-approved participants and voting/multi-party consensus	Pre-approved participants and voting/multi-party consensus
Immutability	Nearly impossible to tamper	Could be tampered	Could be tampered
Consensus process	Permissionless and anonymous	Permissioned and known identities	Permissioned and known identities
Network	Decentralized	Partially decentralized	Partially decentralized
Asset	Native Asset	Any Asset	Any Asset

2.6 How does blockchain work?

As said earlier, the blockchain is a decentralized and distributed ledger for maintaining the integrity of transactions. Before discussing how the blockchain works, let us talk about the classical ledger or centralized architecture. For a long time, ledgers were used as means for bankers and governments to store various transactions regarding land possession and other activities that require maintaining a record of the transaction. Maintaining and building a trust relationship between parties of a certain transaction were the major problem, so the bank or government office was used as a central authority to accomplish the required changes in the transactions and design contracts to define who possess what. Therefore, distinguishing between genuine and fake transactions are only done by the central authority.

The ledger manager (bank or government office) built the required trust, so people can sell and buy without having to worry since the centralized manager controls the access to their information on the ledger. These ledgers are totally centralized in which a third-party person or organization is trusted by all users and has a full control over transactions management. Also, these ledgers are black-boxed since the ledger contents are only visible to the ledger manager.

On the other hand, blockchain provides similar functions in terms of storing and maintaining transactions but no third-party (ledger manager) is required. It solves the problem of the central authority that verifies transactions by decentralizing the ledger in which each participating user within the blockchain network holds a copy of the original ledger. In addition, any participating user can request to add a transaction; however, the transaction is added to the block only if the majority of participating users in the blockchain network verify it. An automatic checking is reliably done for each user to generate a fast and protected ledger that is significantly tamper-proof the transactions and blocks [22].

Once a transaction is verified, it will be added and linked with other transactions in a block, which is linked with previous blocks in the ledger through a timestamp and hash function. This forms chains of blocks, which create what is known as blockchain. Once the block is generated, all participating users in the blockchain network start to look for the next block by trying to solve the complex mathematical function and generate a genuine encrypted block of transactions to add it to the ledger. This process is called mining, in which all users (minors) compete to generate the new block. The first minor to generate a genuine block and add it to the ledger is rewarded with the sum of fees for its transactions. Fees are applied to each transaction. Since blocks involve a large number of transactions which are added repeatedly, minors could collect multiple fees.

The ledger held by all participating users in the network is updated once a novel block is added. If the newly added block has been verified by all participating users and all its transactions are genuine, the block will be added and remains permanently in the ledger as a public record. If a conflict is discovered, the block will be discarded. Corrupting a classical ledger needs an attack on the third party (centralized manager). While the blockchain is immutable, so if there is a malicious attempt to alter the contact of any transaction, this will need repeated computations of PoW for the involved block and all other blocks afterward. These calculations are very difficult to accomplish unless most of the users in the blockchain network are malicious. Also, the possibility of having a fake ledger does not exist since all participating users have their own genuine copy of the ledger to compare with [23].

Fig. 5 shows the flow process of a typical financial transaction using the blockchain when a User A wants to send money to User B. The flow starts when User A requests to add a block to the ledger which contains information regarding his financial transfer transaction. After creating the block, it

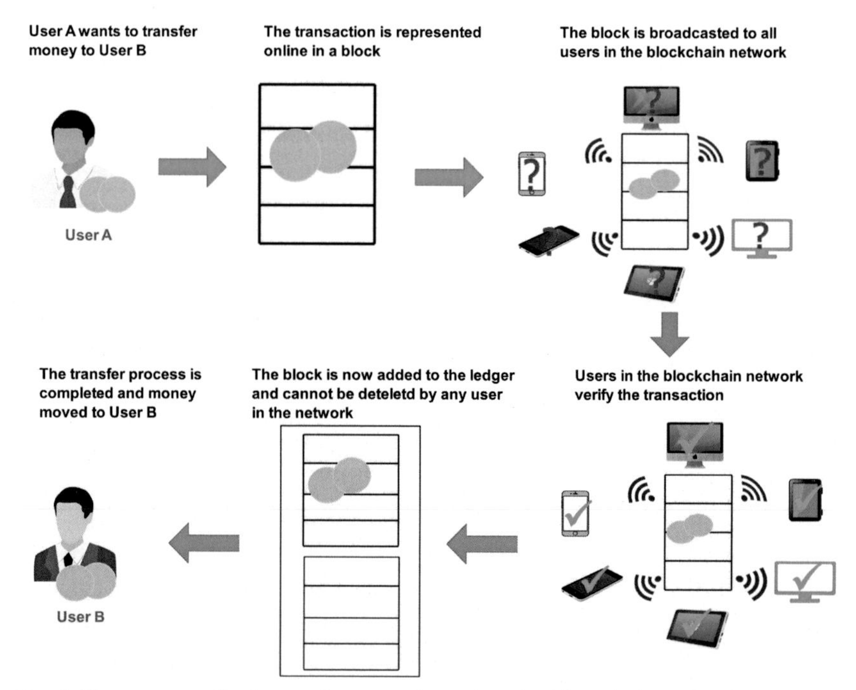

Fig. 5 The process of a financial transaction using the blockchain technology.

broadcasted between all participating users in the blockchain network to verify it. When the new block is verified by all participating users in the network, the block will be added to the ledger and the transfer operation will be completed. At last User B can receive the money.

2.7 Applications of blockchain

There are several applications that can benefit from various capabilities of blockchain technology. These applications include:

2.7.1 Music industry

Due to the evolution of the Internet and ease of accessibility to various streaming facilities over the Internet, the music industry has become one of the applications that can benefit from enormous benefits provided by blockchain technology. The music industry involves a variety of entities such as publishers, songwriters, artists, labels and streaming service providers. The music ownership has changed and become more difficult due to the growth of the Internet. There is a need for transparency in the copyrights and ownership payments for songwriters and artists [24].

Integrating the music industry with blockchain technology can solve several issues regarding transparency and ownership payment. The blockchain can be used to create a precise distributed database to protect information of music rights in a ledger. Also, smart contracts can be used to provide a digital and secure contract for the music industry.

2.7.2 Education

Education is one of the applications that started to adopt the blockchain in interesting and innovative applications such as management of credentials and transcripts, proof of learning, management of reputation and management of student records. The blockchain can be used as a decentralized database to store different types of education information permanently. This, in turn, can help universities to adopt cryptographical-signed and confirmable certificate on the blockchain which allow both employers and students to access it easily [25].

Integrating the blockchain with learning societies can create innovative educational applications which build a new learning model where the exchange of ideas and concepts coupled with a tracking system for evaluating the learning results. The blockchain can be used in the regulation of contracts and payments to assess learning and record academic progress such as paying tuition fees by peer-teaching with other students.

2.7.3 Public services

Data generated by governmental organizations are internally fragmented and opaque to citizens and businesses. While with the use of blockchain technology, data records can be created and verified quickly with ensuring security and transparency of data. Blockchain features such as digital signatures and time-stamping are predicted to provide countless advantages in public services to allow citizens to handle transactions and generate accounts independently without the need for lawyers, government officials and other third parties.

Several governments started to adopt blockchain technology to support various public services to their citizens. For instance, the Estonian government has utilized blockchain technology to allow citizens to perform several tasks using their ID cards such as voting, register for their businesses, order medical prescriptions and pay taxes. In addition, the UK work and pensions department has started to adopt the blockchain in welfare payments. Also, Sweden has conducted tests to put real estate transactions on the blockchain [22].

2.7.4 Healthcare

Blockchain has great potentials to resolve interoperability problems of the existing healthcare systems. It can be utilized to enable healthcare objects and researchers to share their Electronic Health Record (EHR) in a safe and protected way. Also, it allows for improving medical care and doctor endorsement.

Managing the healthcare data whether by storing or analyzing is not an easy operation especially regarding data privacy. To provide a secure environment for the healthcare sector with blockchain, Healthcare Data Gateway (HDG) can be used to manage and control data storage and sharing easily. Also, improving privacy can be ensured by adopting the private blockchain which allows only specific persons to store or modify the medical information [26].

2.7.5 Cybersecurity

Security is one of the major issues for all current and new technologies. Popular companies faced many security problems. For instance, more than 50 million Facebook profiles have been breached by Cambridge Analytica to target them with personalized political advertisements which affected the US voters on their final decision on the presidential election. Also, in 2016, Yahoo, the famous search engine, faced a major attack and around one billion Yahoo accounts were compromised. When security companies did their research about common security vulnerabilities, they found that 65% of the data breaches were occurred because of weak, default, or stolen passwords. Also, they found phishing emails steal sensitive data such as username, password, and financial records [27].

Blockchain has several benefits that can be used to solve the cybersecurity issue. First, blockchain is a trustless system where people trust does not exist. It assumes that any insider or outsider can attack the system, so it is completely independent of human ethics. Second, blockchain is immutable, so anyone can store data and secure it with different cryptographic features such as hashing and digital signatures. As soon as data formed as a block in the blockchain, it cannot be altered or deleted. Third, blockchain involves multiple users in the network, so changing or adding a block needs to be verified by the majority of users which make the attack very difficult to achieve [16].

2.7.6 Voting

Voting is an important tool for any democratic government. It is a must process for all people; however, the traditional paper ballot system of voting faces several issues. For instance, the system cannot be automated, and people have to physically go to the venues where the ballot boxes are kept which make them wait in lines for long times. Also, counting the votes takes a long time and the election can be breached by inserting bogus ballot papers. Also, the cost and amount of papers wasted in the operation are very high [28].

With transparency and immutability features of the blockchain, the voting process can be much simpler and save the huge cost wasted in the classical paper ballot voting system. The voter can create a block, which is their vote, so once the vote is done, anyone can verify whether the signature is valid or not and make sure that none of the votes has been tampered with.

2.8 Challenges of blockchain

Blockchain is not straightforward. It raises several challenges with existing technologies that need to be resolved. These challenges are summarized in Fig. 6 and described as follows.

- *Scalability*: As discussed earlier, every transaction is stored in the distributed ledger. These transactions are increasing every day. To validate a transaction, each user has to store it on the ledger to examine the source of the existing transaction. Moreover, creating a block faces several constraints regarding time and size. For example, Bitcoin can only create almost seven transactions per second, which cannot realize the processing needs of

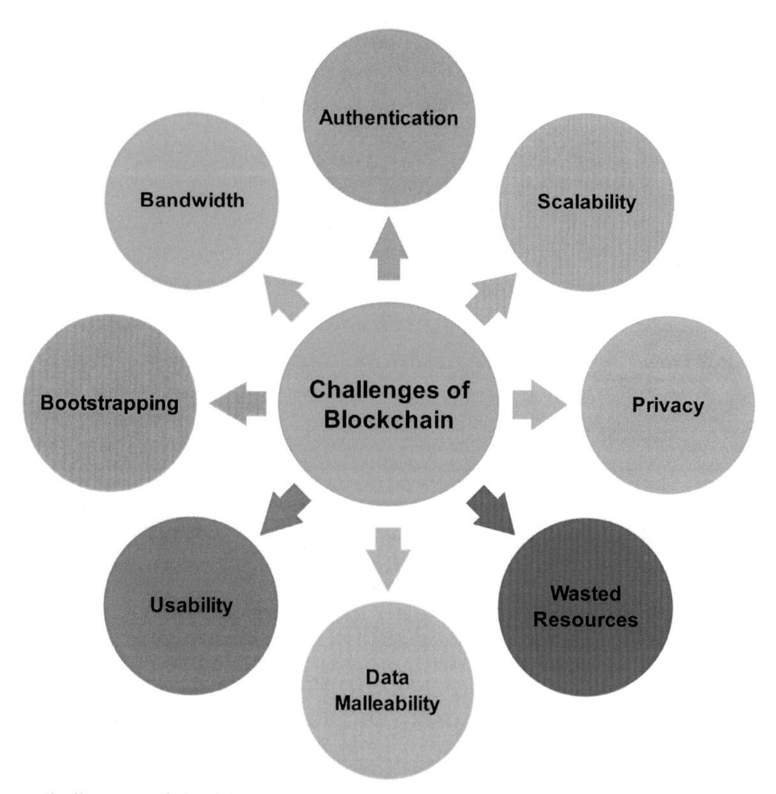

Fig. 6 Challenges of the blockchain technology.

billions of transactions in real-time applications [21]. Also, the size of a transaction plays an important role in the execution order since minors prefer to generate blocks with large transaction size and high transaction fees. This leads to more latency for small transactions. Some research studies suggested solutions to the scalability issue. For instance, Bruce [29] proposed a storage optimization method for the blockchain to delete old transaction records from the ledger. However, more research is needed to address this challenge.

- *Privacy*: A certain amount of privacy can be protected through the blockchain technology. The user uses anonymous identity to create and verify transactions using their private and public key. However, since all participating users in the blockchain network can view values of all transactions, the blockchain cannot guarantee transactions' privacy. Also, J. Barcelo indicated that the user's transaction can be linked to disclose user's personal information. Therefore, the privacy issue of blockchain technology needs more research to increase the adoption rate of blockchain in various applications.

- *Wasted Resources*: Till the moment, energy efficiency is one of the significant challenges in computer engineering that needs to be resolved. Regarding blockchain technology, the mining process needs a huge volume of computation power to compute and verify transactions in a secure manner. However, it is essential to reduce wasted resources in the mining process. Some researchers have proposed several solutions to resolve this problem. For example, Janish [30] has proposed a scheme to speed the mining process by using simultaneous Central Processing Units (CPUs) and Graphics Processing Units (GPUs) in individual machines in mining pools.
- *Data Malleability*: Maintaining the integrity of data is one of the critical aspects in the blockchain. Data should not be altered or tampered with when transmitted or verified. Malleability attack on data integrity indicates that the signature of transactions used to verify the possession of Bitcoin does not deliver any integrity assurance for signatures themselves. Consequently, an intruder can capture, alter, and rebroadcast a transaction which causes the transaction's creator to think that the transaction was not verified [31].
- *Usability*: The usability issue refers to the fact that the blockchain Application Programming Interface (API) is hard and difficult to use. The main target of all new technologies should involve providing usable and easy-to-use interfaces for both users and developers. Blockchain usability from the perspective of the cryptocurrency domain should allow users to analyze the blockchain. In the blockchain environment, blocks are generated continuously and validated by the participating users, which generate an exciting atmosphere of transaction flows. Therefore, it is critical to improve the blockchain usability by providing the required tools to allow users to analyze the entire blockchain network [32].
- *Bootstrapping*: Transferring the present business documents, contracts or frameworks to the novel blockchain based technology introduces multiple migration responsibilities which require to be performed. For instance, in the event of a land ownership, the existing forms require to be migrated and formatted to be equivalent for the blockchain form, which take time and cost.
- *Bandwidth*: The block size in blockchains determines the number of transactions needed for each block. To keep equality and give all users of the blockchain network equal chances to become a leader in the next round, all users should be informed about newly blocks at the same time. The block size is generally restricted to the uplink bandwidth of users. For instance, the current block size in Bitcoin is 1 MB, which is around

1000 transactions [15]. Therefore, the block size and user's bandwidth should be considered when creating new blocks.

- *Authentication*: In the existing blockchain, once a user identity is created, there is no guarantee that the user requesting the identity is the correct owner of that identity and not a malicious one. There are some problems in the authentication of Bitcoin. For instance, there is a famous incident in Mt. Gox, where private keys of their users were breached. So, maintaining a strong authentication scheme is one of the fundamental priorities for the blockchain technology which needs more research and efforts to develop [32].

3. Internet of Things

The IoT allows different devices/objects around us in the environment to be addressable, recognizable and locatable via sensor devices. Also, it allows these devices to be controllable over the Internet using either wired or wireless communication networks. Everyday objects involve not only normal electronic devices or technological development products like vehicles, phones, etc., but also other objects such as food, animals, clothes, trees, etc. The key purpose of the IoT system is to allow various objects to be connected in anyplace, anytime by anyone ideally using any path/network and any service [33].

This section provides a discussion of technical aspects of the IoT system. It started by discussing the history of IoT and how it developed. This followed by presenting various definitions of the IoT which are suggested by various organizations and researchers. Architecture, essential characteristics, applications and challenges of the IoT system are also presented.

3.1 History of IoT

The concept of IoT is not new, it passed through several phases until reaches to what is known now. The IoT notion starts in 1982 when four students from Carnegie Mellon University invented the ARPANET-connected coke machine to indicate whether drinks contained in the coke machine are cold or not. Their main idea was to count how many coke bottles had remained in each row and for how long. If the loaded bottle is left for a long time in the machine, it is labeled "cold." All this data was then remotely available to customers via a finger interface. This experiment had inspired a lot of inventors all over the world to create their own connected appliance [34].

In the early 1990s, IBM scientists presented and patented an Ultra-High Frequency (UHF) Radio Frequency Identification (RFID) that covers wider distance and provides faster data transfer. Although IBM performed few pilot experiments, it never commercialized this new technology. In the mid-1990s, IBM has suffered from tough financial problems which make them sell their patent to Intermec, a barcode system provider. Several applications are built using Intermec RFID systems, but due to the high cost of this technology at this time and low capacities of sales, this technology did not spread as was expected [35].

In 1999, Auto-IDentification Centre at the Massachusetts Institute of Technology (MIT) has funded through various organizations to involve UHF RFID in connecting various objects together. This occurred when two professors, David Brock and Sanjay Sarma, have proposed using RFID tags to track products through the supply chain. Their proposal was essentially to use only the tag's serial number to track products to save costs, since producing a more complicated chip with a large memory storage will be more expensive. Data linked with the RFID tag was kept in a database that can be accessed over the Internet [36].

Several research and publications confirmed that the term "Internet of Things" was first presented by Ashton, who is the executive director of MIT ID Centre in 1999 [37]. Ashton has said, "*The Internet of Things has the potential to change the world, just as the Internet did. Maybe even more so*" [37]. However, other researchers argued that Neil Gershenfeld is the first one who spoke about the idea of IoT in his book titled "*When Things Start to Think*" [38]. Table 2 presents the development phases of the IoT system starting from 1982 till 2021.

Table 2 Summary of IoT development phases starting from 1982 till 2021.

Year	Contribution
1982	Four students from Carnegie Mellon University invented the ARPANET-connected coke machine
1989	Tim Berners-Lee proposed the World Wide Web
1990	John Romkey introduced a toaster connected to the Internet
1999	Neil Gershenfeld talked about foundations of the IoT in his book titled "*When Things Start to Think*"
1999	Kevin Ashton introduced the concept of "Internet of Things" for the first time
2000	LG announced the world's first Internet-enabled refrigerator
2004	The concept of IoT becomes more popular. There were enormous publications in newspapers and magazines about the IoT

Table 2 Summary of IoT development phases starting from 1982 till 2021.—cont'd

Year	Contribution
2005	The Internet-connected device Nabaztag appeared. It was a small robot for consumer use, manufactured to connect to Wi-Fi networks to gather weather information, news and stock market changes, and read them aloud to the owner
2008	The first international conference on the IoT which took place in Zurich, Switzerland
2008	The IoT has been notified as one of the "Disruptive Civil Technologies" by US national intelligence council with possible effects on US interests out to 2025
2008	Cisco reported that the IoT was born since there were more connected devices than people population
2009	Google starts testing self-driving cars. Using sensor-enabled devices on the car deck, Toyota Prius was able to detect pedestrians, cyclists, road work, and other valuable objects
2010	China picks the IoT as a key industry. Chinese Premier Wen Jiabao considered the IoT as a significant industry domain for China
2011	IPv6 public launch. Several organizations have motivated the Internet providers to be prepared for the transition from IPv4 to IPv6
2013	Google announced smart glasses. It featured a display that had the ability to show information hands-free, and a natural language voice recognition module to connect to the Internet via spoken commands
2015	Mattel produces IoT-enabled toys. They produced a Barbie with an embedded Wi-Fi module and a toy house with built-in interactive features such as voice-controlled light bulbs and a toy oven with fire
2016	Apple introduced products home kit to provide the developers with comprehensive tools for developing smart home appliances' software
2016	Google releases Google Home which allows for integrating third-party services to enable users with a wide field of interaction
2017	Microsoft launches Azure IoT edge that enables small devices to utilize cloud services even if they are not connected to the cloud
2017	Google releases Cloud IoT Core that allows devices to connect to the cloud more easily
2018	Governments started to think about the security of IoT devices and encourage manufacturers to adopt security by design
2020	According to Cisco, there will be around 50 billion connected devices
2021	BMW, Ford, Volvo say that there will be fully autonomous cars

3.2 Definitions of IoT

The IoT concept describes the capacity of network connectivity of various types of objects in the environment, not just computers. These objects can act intelligently and exchange data with other devices with negligible human involvement. Although the popularity of the IoT system and high acceptance of this new technology globally, a precise definition does not exist. There are various definitions that focus on a specific view of the IoT. We will try to provide common definitions of the IoT from different perspectives [39].

The International Telecommunication Union (ITU) in 2012 has provided a common definition for the IoT which has been adopted by several researchers. It stated "*a global infrastructure for the information society, enabling advanced services by interconnecting (physical and virtual) things based on, existing and evolving, interoperable information and communication technologies*" [40]. While the Internet Architecture Board (IAB) describes the IoT as "*Internet of Things denotes a trend where a large number of embedded devices employ communication services offered by the Internet protocols. Many of these devices, often called 'smart objects,' are not directly operated by humans, but exist as components in buildings or vehicles, or are spread out in the environment*" [41].

There were other views from various researchers. For instance, Atzori et al. [42] have suggested the fundamental idea of the IoT is the universal existence of diversity of things such as sensors, RFID tags, actuators and mobile phones which can interact with each other to achieve a common goal. In addition, Ma [43] has proposed a definition for the IoT which stated as "*The IoT can enable the interconnection and integration of the physical world and the cyberspace; representing the trend of future networking while leading the third wave of the IT industry revolution.*"

Similarly, Gubbi et al. [44] have defined the IoT system from the perspective of a smart environment. It described the IoT as "*interconnection of sensing and actuating devices providing the ability to share information across platforms through a unified framework, developing a common operating picture for enabling innovative applications.*" Also, Guillemin and Friess [45] have defined the IoT in simple terms, as shown in Fig. 7. It stated: "*The Internet of Things allows people and things to be connected Anytime, Anyplace, with anything and anyone, ideally using any path/network and any service.*"

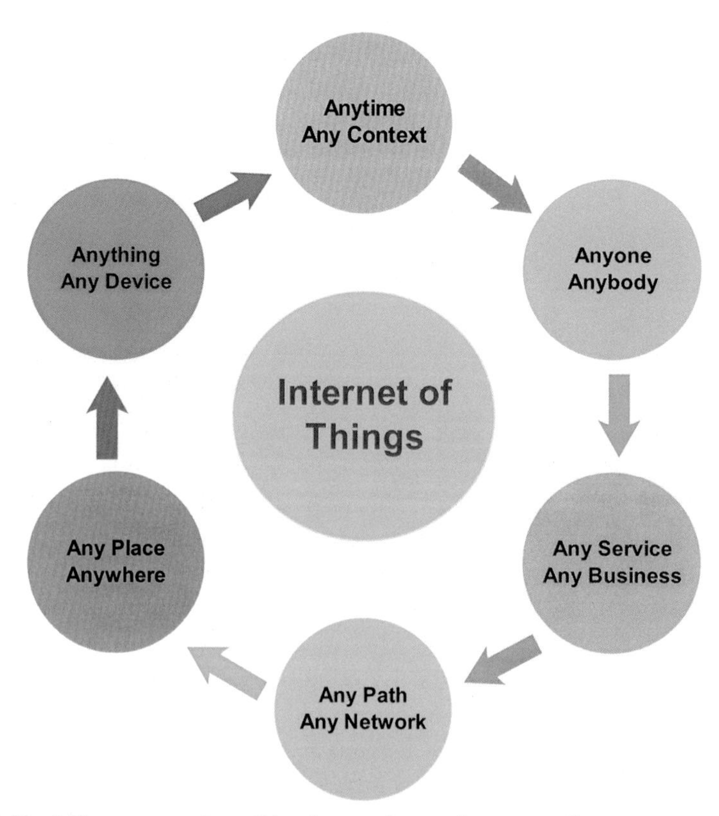

Fig. 7 The IoT can connect anything in anywhere using any path.

3.3 IoT expansion

The IoT refers to a huge network of devices and sensors that able to capture and share data with one another. These devices involve both physical and virtual objects that are interconnected together over the Internet. There are huge technological developments that extended the IoT to include other technologies such as Cloud computing and Wireless Sensor Networks (WSNs) [45]. The IoT has the capability to primarily modify business models and value chains in different organizations. It is not just a smart oven connected to the Internet. In some stage, all products will have the ability to connect to the Internet in an economic way.

The number of connected devices exceeds the population worldwide from 2008 and with unlimited capabilities of the IoT system, novel applications

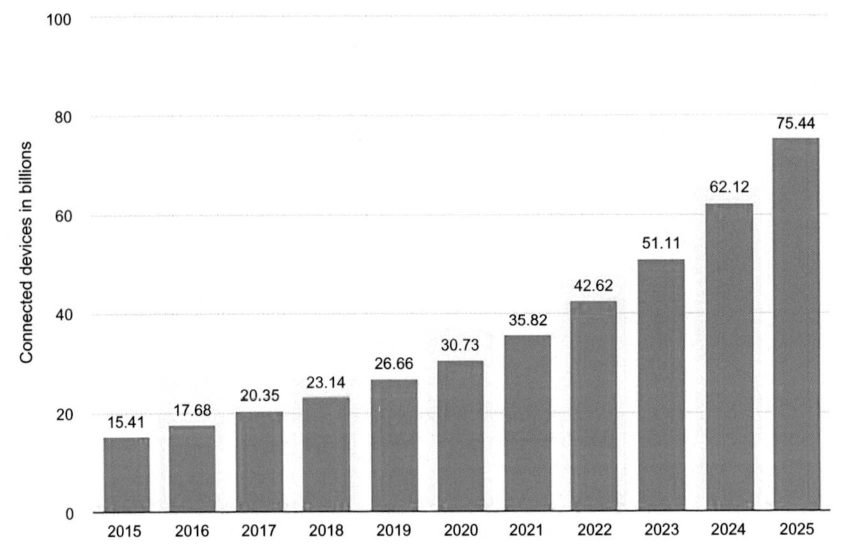

Fig. 8 Expectation of IoT growth from 2015 to 2025 [47].

and services can be created every day using this fascinating technology. The number of IoT devices is increasing every day. Regarding Statista, the number of IoT devices is expected to reach about 31 billion worldwide at the end of 2020. This number will significantly increase to about 75 billion devices at the end of 2025 [46], as shown in Fig. 8. In addition, the IoT has an expected revenue estimated to reach about $1.8 trillion by 2026.

One of the major issues standing as a barrier to adopting various IoT products is the security and privacy challenges. The growth of IoT devices creates new services and applications, but at the same time, it creates several security vulnerabilities that became more apparent. Manufacturers of IoT devices are not considering security in their priorities [48]. With low public awareness about security and privacy, IoT devices could lead to severe problems that could literally lead to losing our lives. The governments should encourage manufacturers of IoT devices to adopt new security measures in their products. Also, manufacturers should employ the concept of security by design to implement built-in security algorithms within their products to ensure minimum security and safety for various consumers.

3.4 Architecture of IoT

There are different architectures for the IoT system that represent various perspectives about the IoT and its functions. However, the most common architecture for the IoT is the one made by IoT World Forum (IWF) architecture committee in October 2014 [49]. This reference model provides a common framework to allow deploying the IoT easily and quickly in the industry. Similar to Open System Interconnection (OSI) reference model of the network, the IoT reference model is divided into seven layers to promote the association and expansion of IoT deployment models, as shown in Fig. 9. It identifies where various kinds of processing are operated through different layers of the IoT reference model and enables various manufacturers to produce compatible IoT products working with each other smoothly and efficiently. Also, this architecture model converts the IoT from a conceptual model into a real and approachable system [50].

Layer 1 is the physical layer. It is the hardware layer which collects data from the physical world and transfers it to the upper layer. This layer involves physical objects and sensors. Essentially, the purpose of this layer is to identify different objects and collect information about the surrounding environment such as temperature, humidity, pressure, water quality, motion detection, amount of dust in the air, etc. [51].

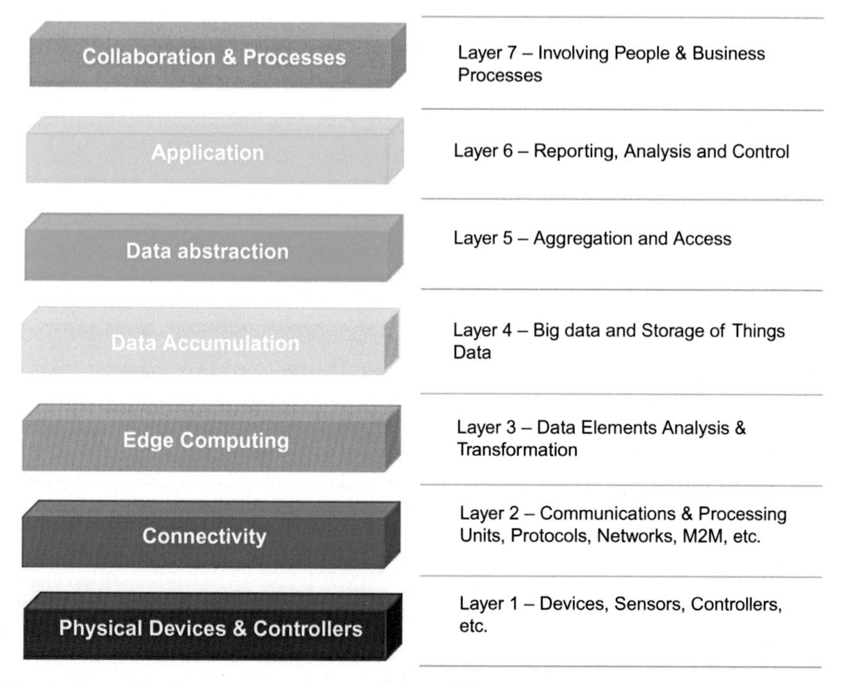

Fig. 9 The IoT reference model according to IWF.

Layer 2 is connectivity. This layer is used to interconnect different IoT things with each other using interconnection devices such as switches, gateway and router. It also transfers gathered data securely from sensors to the upper layer for processing. Layer 3 is edge computing. This layer takes data coming from the connectivity layer and converts it into information appropriate for storage and higher-level processing. At this layer, the processing components work with a huge amount of data which could execute some data transformation to reduce the size of data.

Data accumulation occurs in layer 4. The main function of this layer is to store data coming from layer 3. It absorbs a huge amount of data and places them in storages to be accessible by upper layers. So, it simply changes event-based data to query-based processing information for upper layers. Layer 5 is data abstraction. This layer combines data coming from different sources and converts stored data into the appropriate format for applications in a manageable and efficient manner [47].

Layer 6 is the application layer. This layer is concerned with the information interpretation of various IoT applications. It includes various IoT applications such as healthcare, smart city, smart grid, smart home, connected car, smart agriculture, etc. [49]. Layer 7 is collaboration and processes. This layer identifies individuals who can communicate and collaborate to make use of the IoT data efficiently. It provides other functionalities like creating graphs and business models and other based on data retrieved from the application layer. It also assists managers to make precise choices about their business based on their data analysis [52].

3.5 Characteristics of IoT

The basic notion of the IoT is to provide an autonomous system capable of sharing useful information between uniquely identifiable real-world objects using RFID tags and WSNs. The IoT system shows common characteristics. This section provides common characteristics that better describe the IoT system, as summarized in Fig. 10.

- *Large Scale*: As explained earlier, the IoT system expanded to reach about 30 billion devices at the end of 2018. Cisco expected that this number will increase to reach about 50 billion devices at the end of 2020. This large number of connected devices creates a large-scale network to share their collected information and cooperate together to enhance existed services and generate novel applications capable of handling daily life problems of IoT users [53].

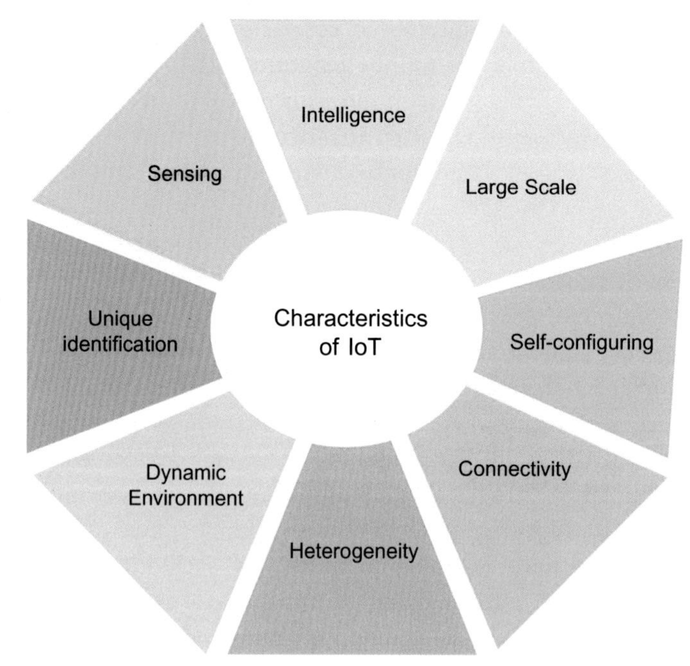

Fig. 10 Common characteristics of the IoT system.

- *Intelligence*: The notion of integrating sensors, computers and communication networks to collect and observe information has found for decades. The modern developments aim to make IoT objects acting intelligently and make autonomous decisions. Most IoT devices act regarding their predetermined actions, but with the convergence of sophisticated hardware and software algorithms, IoT objects become able to respond intelligently and correctly according to different situations and contexts [33].
- *Sensing*: Sensors are one of the main elements of the IoT system, which are used to sense, perceive, and gather information about the surrounding environment. The collected information can be resulted from their recording or after their interaction with the environment. Sensing techniques deliver various abilities to consider human susceptibility about the surroundings. Also, the sensing feature is important aspect toward context awareness which allows devices to adjust themselves to various situations and contexts depending on their operating circumstances [44].
- *Unique Identification*: Each IoT device involves an RFID tag which provides a unique identity for each device. These identities given to IoT devices are used by the manufactures to upgrade devices' software.

The IoT system with billions of connected devices needs a naming architecture that provides unique identities to IoT devices to establish communication paths between different types of devices [54].

- *Dynamic Environment*: The IoT is a dynamic system in nature which makes various things can adapt to environmental changes and act intelligently based on the context. Also, IoT devices gather data with considering dynamic changes in the environment. The status of these devices changes dynamically based on surrounding conditions such as connected or disconnected, sleeping or running [55].
- *Heterogeneity*: There are several manufacturers who want to produce many devices to leverage their connectivity over the Internet. However, they face a problem when it comes to managing the heterogeneity of their devices. The IoT system involves several devices with different hardware platforms, networks, communication protocols and operating systems. Although the heterogeneity of these devices causes many problems, they still able to communicate with each other using different communication networks [56].
- *Connectivity*: The IoT system involves multiple devices that need to be connected to share their information. The IoT has the ability to link and interconnect different objects in the environment to offer new market opportunities for generating new applications and services to help humans in different domains [41].
- *Self-configuring*: The large-scale feature of IoT devices creates a severe problem for various service providers and manufacturers to maintain and update their devices. With the self-configuring features for the IoT, devices can work with each other to deliver a specific operation. Also, these devices could configure themselves and search for the newest software update in association with the device manufacturer with negligible efforts [57].

3.6 IoT Applications

The IoT system can interconnect almost all physical and virtual objects in our environment that yield new services and applications. These applications can be adopted in different domains to increase our quality of life. This section provides a discussion of common IoT applications.

3.6.1 Home automation

Smart home is one of the most popular applications of the IoT system. Thanks to sensor and actuation technologies along with WSNs, people can connect a variety of smart appliances inside their homes to resolve their interests. In a smart home, there are several sensors to enable smart and automated services which operate with minimal human efforts. Also, sensors are used to maintain security and safety [58].

Although the benefits supported by smart home are countless, it introduces some issues regarding security and privacy since all actions and events occurred in the home are being recorded. If an attacker has succeeded to breach the system, it may make the system to act maliciously. So, smart home should be protected in a way that allows smart devices to notify the owner regarding any abnormal action. Also, reliability is another challenge since no administrator is existed to observe the system behavior [59].

3.6.2 Healthcare

The IoT has proven it can provide several benefits for the healthcare domain by creating new application and services that help patients and keep the field innovative. There are multiple wearable devices developed to monitor and track patient's health conditions. These devices allow older patients to live independently without fear. Also, these devices can be utilized to constantly observe and store patients' health conditions and send warning messages in abnormal situations. If the situation is minor, the device itself can recommend a treatment for the patient. While if it is a major situation, the device can send urgent messages to the hospital or ambulances to be immediately dispatched [51].

3.6.3 Smart agriculture

With the existence of multiple sensors within the IoT environment, farmers can use collected data to produce a better return on the investment. The soil parameters such as humidity, salt level and temperature can be collected and measured using the available sensors to increase agriculture production. Furthermore, with the existence of several wireless technologies such as geographical information system and remote sensing, there are many chances to collect relevant information about the soil quickly and efficiently which can help to substitute human effort with automatic machinery to increase agricultural production [60].

There is a significant growth in the adoption of IoT devices in the agriculture domain. It is predicted that the number of IoT devices in agriculture will reach about 75 million by the end of 2020 [61]. There are several advantages for integrating IoT solutions in agriculture. For example, sensors can be used to monitor soil quality, crop's growth progress and weather conditions besides staff performance and equipment efficiency. Also, the IoT system can help to automate various operations across the crop life cycle and accomplish better management over the production method and ensure advanced standards of crop quality [62].

3.6.4 Supply chain and logistics

The IoT system attempts to facilitate real-world operations in business and information systems. Using sensor technologies such as RFID and Near Field Communication (NFC), products can be tracked from the manufacturer to the distribution location. RFID tags attached to the products are used to uniquely identifies each product and collect relevant information automatically to convey it in real-time along with location information. These tags are used to transmit messages showing exactly what products, sizes and style variations as well as temperature and humidity of products. In addition, automated data capture gives real-time visibility of stock and avoids manual counting and human errors. In simple words, the IoT is set to revolutionize the supply chain with both operational efficiencies and revenue opportunities [63].

3.6.5 Smart city

The notion of smart cities refers to the adoption of IoT devices such as sensors, meters, lights, etc., to monitor and collect information about the surrounding city. This information is used to improve public services and city infrastructure. IoT solutions are involved in many areas of smart cities such as smart street lighting, trash management, smart parking and traffic management [64].

For smart traffic, collected sensor information about traffic can be sent to citizens' phones to monitor traffic in real-time and allow drivers to choose the best road to save driving efforts and time. Also, drivers can be warned in the case of accidents to redirect away from congestion. For trash management, IoT sensors are deployed across trash bins to send messages to specific authorities to report bins need to be emptied. Also, these sensors can be used to optimize trash trucks to reduce emerge usage [65].

3.6.6 Smart grid

Using energy efficiently and ultimately saving more money can be achieved through the use of IoT sensors to collect relevant information about energy consumption in the home, for example, suggesting better ways to save energy. Also, IoT sensors information can be used to deliver consumers all relevant information about various energy suppliers in an automated way for choosing the best for consumers.

The concept of smart grids adds intelligence at the power flow cycle from supplier to consumer. This type of intelligence can be used to help consumers to be aware of power consumption and dynamic pricing [66]. Also, one of the main applications of the smart grid is a smart meter which collects, records and analyzes power consumption at different times of the day. This information can be used by consumers to adjust their power consumption and change their lifestyles to reduce costs.

3.6.7 Connected car

Smart car or what it called connected car started to be deployed into our community. This type of cars can access the Internet and share their data with other devices. The number of cars equipped with this facility is increasing every day, which will allow the appearance of several applications for connected cars in the near future [67]. The connected car provides several advantages over the normal one. It can reduce car accidents and decrease car drivers' errors by allowing the driver to operate the car remotely. These driverless cars also can save time and reduce driving stress. Several car manufacturers such as BMW, Ford and Volvo have confirmed that there will be fully autonomous cars by the end of 2021 [62].

3.6.8 Wearables

Wearables have a huge interest in markets all over the world. Many companies started to produce these devices with huge quantities to satisfy increased demands such as Google and Samsung. According to Statista, the number of connected wearable devices is expected to reach 830 million at the end of 2020 [46]. Wearable devices are equipped with sensors and have the ability to connect to the Internet for data sharing. These sensors collect data about the user which is later processed to extract meaningful information. Most common wearable devices are in fitness, health and entertainment [68].

3.7 Challenges of IoT

Although IoT solutions provide countless benefits, they raise many challenges that need to be resolved. Most common issues of the IoT system including Big Data, networking, scalability, heterogeneity, interoperability and security and privacy are discussed in this section. They also summarized in Fig. 11.

Fig. 11 Challenges of IoT.

3.7.1 Big data

Big Data is a quite novel expression that indicates the massive quantity of data whether structured or unstructured, which is hard to process with classical database methods and software techniques. It characterized by what is called 5V's, volume, variety, variability, value and velocity. Big Data has a huge interest from multiple organizations as a new industry domain such as online social networks (Twitter, Facebook, and Instagram) since there is a huge amount of data collected through social networks. For instance, in 2010, Twitter produced in average about 10 terabytes of data per day [69].

With billions of devices and objects, the IoT is one of the major sources of big data. Although Cloud computing can be used to store data permanently, processing this huge quantity of data is an extensive problem especially the performance of various IoT applications is based on the data management service. Also, this huge amount of data raises security and privacy issues since ensuring the data integrity will be a very difficult task to achieve [70].

3.7.2 Networking

The key driver of the IoT system is to connect all objects/devices to share their information. These devices are different in shape and structure which make it use different communication networking protocols [71]. Implementing a networking protocol for the IoT should be built with taking consideration of system performance and usability. This is because the network protocol has a major impact on the behavior of the network. So, choosing the appropriate networking protocol is an issue that needs to be addressed. Furthermore, selecting the appropriate network topology for the protocol is another challenge [59].

3.7.3 Heterogeneity

The IoT is one of the popular examples that describes the heterogeneity issue since it involves billions of different devices in their nature. The main target of the IoT system is to build a common method to abstract the heterogeneity of these devices and accomplishing the best exploitation of their functionality [72].

Since the IoT system is growing significantly, the search for applications that can adapt itself with varying hardware and software of IoT devices will continue to achieve the maximum efficiency for the IoT system. Service providers have to take into their consideration the widespread diversity of network connectivity options, protocols, and communication methods when implementing a service for the IoT system [53].

3.7.4 Interoperability

The interoperability refers to the capability of the system components to cooperate with each other in an efficient manner regardless of their technical specifications. Although the interconnection of IoT heterogeneous devices allows sharing their information which results in creating novel services, it comes at a price. As the acceptance of the IoT system increased and the number of connected objects and networks expands, the interoperability becomes a fundamental priority to interconnect various things efficiently [73].

Despite the presence of several proposed results to address the interoperability problem like open source frameworks, data-over-sound technology and creating common IoT services layer, the interoperability is still a big challenge that needs to be addressed [74].

3.7.5 Scalability

Scalability is one of the significant challenges of the IoT system that requires to be handled to contain the massive increase of connected devices. Scalability signifies the system capability to deal with the potential growth of the system in an efficient manner without affecting the system performance. Being scalable is a mandatory operation for the IoT system to satisfy the varying requirements since people interest varies with time and environmental situations [75]. Therefore, the scalability issue needs more research to test potentials of the IoT system when increasing number of connected devices.

3.7.6 Security and privacy

The enormous increase of IoT devices in our environment leads to increasing the chances to find security vulnerabilities within IoT devices which are poorly secured without any built-in security measures. Exploiting these vulnerabilities results in stealing user information and may put their lives in danger. Also, as IoT sensors are distributed in our surroundings which allow it to collect our sensitive information, marketing behavior, habits and other information which violates our privacy [76].

Therefore, handling security challenges in the IoT system should be a fundamental priority to increase adoption of IoT applications among consumers. Also, IoT users need to be fully confident about the security of their IoT devices and related applications, as they become more integrated into people daily lives activities [77].

4. Conclusion

Recently, blockchain technology has received widespread attention. It has the potential to be involved in almost every industry even if it still in the first stage of approval and still multiple challenges need to be addressed. Blockchain refers to a tamper-proof distributed ledger which enables transactions to be executed in a decentralized environment. It has the capability to solve the main problems of the traditional centralized model. On the other hand, the IoT has emerged as a new evolution of the Internet. It enables different objects and devices in the environment to be connected over the Internet. With the help of sensors and actuators, it collects meaningful information from these objects/devices to improve human productivity and efficiency. The integration of the IoT with blockchain should be the next stage of developments. To be ready for this integration, this chapter presented a discussion of the technical aspects of blockchain and IoT. It started by providing a review of the blockchain technology. Applications and challenges of the blockchain are also discussed. This is followed by providing an overview of the IoT system including its common architecture and essential characteristics. In addition, various applications and challenges of the IoT system were presented.

References

[1] A. Stanciu, Blockchain based distributed control system for Edge Computing, in: 21st International Conference on Control Systems and Computer Science Blockchain, 2017, pp. 667–671.
[2] L. Lamport, R. Shostak, M. Pease, The byzantine generals problem, ACM Trans. Program. Lang. Syst. 4 (3) (1982) 382–401.
[3] E. Khudnev, Blockchain: Foundation technology to change the world, Int. J. Intell. Syst. Appl. (2017)
[4] H.F. Atlam, A. Alenezi, M.O. Alassafi, G.B. Wills, Blockchain with Internet of Things: benefits, challenges, and future directions, Int. J. Intell. Syst. Appl. 41 (10) (2018) 40–48.
[5] N. Kshetri, Blockchain's roles in strengthening cybersecurity and protecting privacy, Telecomm. Policy 41 (10) (2017) 1027–1038.
[6] S. Nakamoto, Bitcoin: A Peer-to-Peer Electronic Cash System, Available:http://www.bitcoin.org/bitcoin.pdf, 2009.
[7] K. Xu, Y. Qu, K. Yang, A tutorial on the internet of things: from a heterogeneous network integration perspective, IEEE Netw. 30 (2) (2016) 102–108.
[8] H.F. Atlam, A. Alenezi, R.J. Walters, G.B. Wills, An overview of risk estimation techniques in risk-based access control for the Internet of Things, in: Proceedings of the 2nd International Conference on Internet of Things, 2017, Big Data and Security (IoTBDS), 2017, pp. 254–260.
[9] Coinbase, What Is the Bitcoin Blockchain?, [Online]. Available: https://support.coinbase.com/customer/portal/articles/1819222-what-is-the-blockchain, 2017. Accessed 20 October 2018.
[10] Oxford, Blockchain | Definition of Blockchain in English by Oxford Dictionaries, [Online]. Available:https://en.oxforddictionaries.com/definition/blockchain, 2018. Accessed 20 October 2018.
[11] F. Stroud, Blockchain, [Online]. Available:https://www.webopedia.com/TERM/B/blockchain.html. Accessed 5 October 2018.
[12] K. Sultan, Conceptualizing blockchain: characteristics & applications, in: U. Ruhi, R. - Lakhani (Eds.), 11th IADIS International Conference Information Systems, 2018, pp. 49–57.
[13] M. Pilkington, Blockchain technology: principles and applications, in: F. Xavier Olleros, M. Zhegu, E. Elgar (Eds.), Handbook of Research on Digital Transformations, 2016.
[14] S.A. Back, M. Corallo, L. Dashjr, M. Friedenbach, G. Maxwell, A. Miller, A. Poelstra, J. Timón, Enabling Blockchain Innovations With Pegged Sidechains, 2014, Applied Energy. https://blockstream.com/sidechains.pdf.
[15] J.J. Sikorski, J. Haughton, M. Kraft, Blockchain technology in the chemical industry: machine-to-machine electricity market, Appl. Energy 195 (2017) 234–246.
[16] M. Ahmad, K. Salah, IoT security: review, blockchain solutions, and open challenges, Futur. Gener. Comput. Syst. 82 (2018) 395–411.
[17] A. Narayanan, J. Bonneau, E. Felten, A. Miller, S. Goldfeder, Bitcoin and Cryptocurrency Technologies, 2016.
[18] G.W. Peters, E. Panayi, Understanding modern banking ledgers through blockchain technologies: future of transaction processing and smart contracts on the internet of money, in: P. Tasca, T. Aste, L. Pelizzon, N. Perony (Eds.), Banking Beyond Banks and Money, New Economic Windows, Springer, Cham, 2016, pp. 239–278.
[19] K. Christidis, M. Devetsikiotis, Blockchains and smart contracts for the internet of things, IEEE Access 4 (2016) 2292–2303.
[20] M. Samaniego, R. Deters, Blockchain as a service for IoT, in: IEEE International Conference on Internet of Things and IEEE Green Computing and Communications IEEE Cyber, Physical and Social Computing and IEEE Smart Data, 2016 2016, pp. 433–436.

[21] Z. Zheng, S. Xie, H. Dai, X. Chen, H. Wang, An overview of blockchain technology: architecture, consensus, and future trends, in: 2017 IEEE 6th International Congress on Big Data, 2017, pp. 557–564.
[22] P. Boucher, S. Nascimento, M. Kritikos, How Blockchain Technology Could Change Our Lives, European Parliamentary Research Service, 2017.
[23] A.M. Antonopoulos, Mastering Bitcoin: Unlocking Digital Cryptocurrencies, first ed., O'Reilly Media, Inc., Sebastopol, CA, 2014.
[24] M. Crosby, Nachiappan, P. Pattanayak, S. Verma, V. Kalyanaraman, Blockchain Technology Beyond Bitcoin, 2015.
[25] C. Holotescu, Understanding blockchain technology and how to get involved, in: The 14th International Scientific Conference eLearning and Software for Education Bucharest, 2018, pp. 1–8 (April).
[26] X. Yue, H. Wang, D. Jin, M. Li, W. Jiang, Healthcare data gateways: found healthcare intelligence on blockchain with novel privacy risk control, J. Med. Syst. (2016) 218–225.
[27] H.F. Atlam, R.J. Walters, G.B. Wills, Fog computing and the Internet of Things: a review, Big Data Cogn. Comput. 2 (10) (2018) 1–18.
[28] K. Christidis, M. Devetsikiotis, Blockchains and smart contracts for the Internet of Things, in: IEEE Access, vol. 4, 2016, pp. 2292–2303.
[29] J. Bruce, The Mini-Blockchain Scheme, [Online]. Available:http://cryptonite.info/files/mbc-scheme-rev3.pdf, 2014.
[30] J. Janish, Bitcoin mining acceleration and performance quantification, in: 2014 IEEE 27th Canadian Conference on Electrical and Computer Engineering (CCECE), 2014, pp. 1–6.
[31] C. Decker, R. Wattenhofer, Bitcoin transaction malleability and MtGox, in: Computer Security DESORICS, Lecture Notes in Computer Science, vol. 8713, Springer International Publishing, 2014, pp. 313–326.
[32] J. Yli-huumo, D. Ko, S. Choi, S. Park, K. Smolander, Where is current research on blockchain technology?—a systematic review, PLoS One (2016) 15–27.
[33] H.F. Atlam, R.J. Walters, G.B. Wills, Internet of Things: state-of-the-art, challenges, applications, and open issues, Int. J. Intell. Comput. Res. 9 (3) (2018) 928–938.
[34] M.U. Farooq, M. Waseem, A review on Internet of Things (iot), internet of everything (ioe) and internet of nano things (IoNT), Int. J. Comput. Appl. 113 (1) (2015) 1–7(0975 8887).
[35] M. Roberto, B. Abyi, R. Domenico, Towards a Definition of the Internet of Things (IoT), IEEE Internet Things, 2015.
[36] H.F. Atlam, R.J. Walters, G.B. Wills, Intelligence of things: opportunities & challenges, in: IEEE 2018 Cloudification of the Internet of Things (CIoT), 2018.
[37] K. Ashton, That 'internet of things' thing, RFiD J. (2009).
[38] K.K. Patel, S.M. Patel, Internet of things-IOT: definition, characteristics, architecture, enabling technologies, application & future challenges, Int. J. Eng. Sci. Comput. 6 (5) (2016) 6122–6131.
[39] H.F. Atlam, A. Alenezi, A. Alharthi, R. Walters, G. Wills, Integration of cloud computing with Internet of Things: challenges and open issues, in: 2017 IEEE International Conference on Internet of Things (iThings) and IEEE Green Computing and Communications (GreenCom) and IEEE Cyber, Physical and Social Computing (CPSCom) and IEEE Smart Data (SmartData), June, 2017, pp. 670–675.
[40] ITU, Overview of the Internet of things, in: Ser. Y Glob. Inf. Infrastructure, Internet Protoc. Asp. Next-Generation Networks—Fram. Funct. Archit. Model, 2012.
[41] RFC 7452, Architectural Considerations in Smart Object Networking, Computer Networks, 2015.
[42] L. Atzori, A. Iera, G. Morabito, The Internet of Things: a survey, Comput. Netw. 54 (15) (2010) 2787–2805.

[43] H. Ma, Internet of Things: objectives and scientific challenges, J. Comput. Sci. Technol. 26 (2011) 919–924.
[44] J. Gubbi, R. Buyya, S. Marusic, M. Palaniswami, Internet of Things (IoT): a vision, architectural elements, and future directions, Futur. Gener. Comput. Syst. 29 (7) (2013) 1645–1660.
[45] P. Guillemin, P. Friess, Internet of Things strategic research roadmap, Eur. Comm. Inf. Soc. Media, Luxemb. (2009).
[46] Statista, Internet of Things (IoT) Connected Devices Installed Base Worldwide From 2015 to 2025 (in Billions), [Online]. Available:https://www.statista.com/statistics/471264/iot-number-of-connected-devices-worldwide/, 2018. Accessed 15 October 2018.
[47] H.F. Atlam, A. Alenezi, R.J. Walters, G.B. Wills, J. Daniel, Developing an adaptive Risk-based access control model for the Internet of Things, in: 2017 IEEE International Conference on Internet of Things (iThings) and IEEE Green Computing and Communications (GreenCom) and IEEE Cyber, Physical and Social Computing (CPSCom) and IEEE Smart Data (SmartData), June, 2017, pp. 655–661.
[48] H.F. Atlam, R.J. Walters, G.B. Wills, Internet of nano things: security issues and applications, in: 2018 2nd International Conference on Cloud and Big Data Computing, 2018, pp. 71–77.
[49] W. Stallings, The Internet of Things: network and security architecture, Internet Protoc. J. 18 (4) (2015) 381–385.
[50] S.H. Shah, I. Yaqoob, A survey: Internet of Things (IOT) technologies, applications and challenges, in: 2016 IEEE Smart Energy Grid Engineering, vol. i, 2016, pp. 381–385.
[51] Cisco, The Internet of Things Reference Model, White Paper, 2014.
[52] M. Muntjir, M. Rahul, H.A. Alhumyani, An analysis of Internet of Things (IoT): novel architectures, modern applications, security aspects and future scope with latest case studies, Int. J. Eng. Res. Technol. 6 (6) (2017) 422–447.
[53] H.F. Atlam, G. Attiya, N. El-Fishawy, Integration of color and texture features in CBIR system, Int. J. Comput. Appl. 164 (April) (2017) 23–28.
[54] P.P. Ray, A survey on Internet of Things architectures, J. King Saud Univ. Comput. Inf. Sci. (2016) 1–29.
[55] H.F. Atlam, G. Attiya, N. El-Fishawy, Comparative study on CBIR based on color feature, Int. J. Comput. Appl. 78 (16) (2013) 975–8887.
[56] M. Adda, J. Abdelaziz, H. Mcheick, R. Saad, Toward an access control model for IOTCollab, in: The 6th International Conference on Ambient Systems, Networks and Technologies, vol. 52, 2015, pp. 428–435.
[57] I. Chatzigiannakis, et al., True self-configuration for the loT, in: 9th IEEE International Conference on Collaborative Computing: Networking, Applications and Worksharing, 2012, pp. 545–551.
[58] I.I. Pătru, M. Carabaş, M. Bărbulescu, L. Gheorghe, Smart home IoT system, in: Netw. Educ. Res. RoEduNet Int. Conf. 15th Ed. RoEduNet 2016—Proc., 2016, pp. 365–370.
[59] B.L. Risteska Stojkoska, K.V. Trivodaliev, A review of Internet of Things for smart home: challenges and solutions, J. Clean. Prod. 140, pp (2017) 1454–1464.
[60] K.L. Krishna, O. Silver, W.F. Malende, K. Anuradha, Internet of Things application for implementation of smart agriculture system, in: Int. Conf. I-SMAC (IoT Soc. Mobile, Anal. Cloud), vol. 25(15), 2017, pp. 54–59.
[61] M.A. Akkaş, R. Sokullu, An IoT-based greenhouse monitoring system with Micaz motes, in: International Workshop on IoT, M2M and Healthcare (IMH 2017), vol. 113, 2017, pp. 603–608.
[62] L. Da Xu, W. He, S. Li, Internet of Things in industries: a survey, IEEE Trans. Ind. Inf. 10 (4) (2014) 2233–2243.

[63] Z. Guo, Z. Zhang, W. Li, Establishment of intelligent identification management platform in railway logistics system by means of the Internet of Things, Procedia Eng. 29 (2012) 726–730.
[64] a. Zanella, N. Bui, a. Castellani, L. Vangelista, M. Zorzi, Internet of Things for smart cities, IEEE Internet Things J. 1 (1) (2014) 22–32.
[65] R. Khatoun, S. Zeadally, Cybersecurity and privacy solutions in smart cities, IEEE Commun. Mag. 55 (3) (2017) 51–59.
[66] H.F. Atlam, M.O. Alassafi, A. Alenezi, R.J. Walters, G.B. Wills, XACML for building access control policies in Internet of Things, in: Proceedings of the 3rd International Conference on Internet of Things, Big Data and Security (IoTBDS 2018), 2018.
[67] M. Kalmeshwar, N. Prasad, Internet of Things: architecture, issues and applications, Int. J. Eng. Res. Appl. 07 (06) (2017) 85–88.
[68] S. Cirani, M. Picone, Wearable Computing for the Internet of Things, IEEE Computer Society, 2015, pp. 35–41.
[69] J. Lin, C. Chen, C. Lin, Integrating QoS awareness with virtualization in cloud computing systems for delay-sensitive applications, Futur. Gener. Comput. Syst. (2013) 478–487.
[70] C. Liu, C. Yang, X. Zhang, J. Chen, External integrity verification for outsourced big data in cloud and IoT: a big picture, Futur. Gener. Comput. Syst. 49 (2015) 58–67.
[71] M. Chen, J. Wan, F. Li, Machine-to-machine communications: architectures, standards and applications, KSII Trans. Internet Inf. Syst. 6 (2) (2012) 480–497.
[72] A. Alenezi, N.H.N. Zulkipli, H.F. Atlam, R.J. Walters, G.B. Wills, The impact of cloud forensic readiness on security, in: Proceedings of the 7th International Conference on Cloud Computing and Services Science (CLOSER 2017), 2017, pp. 511–517.
[73] H.F. Atlam, A. Alenezi, R.K. Hussein, G.B. Wills, Validation of an adaptive risk-based access control model for the Internet of Things, Int. J. Comput. Netw. Inf. Secur. (2018) 26–35.
[74] D. Bubley, Data over sound technology: device-to-device communications & pairing without wireless radio networks, Int. J. Comp. Intell. Res. (2017).
[75] A. Gupta, R. Christie, P.R. Manjula, Scalability in Internet of Things: features, techniques and research challenges, Int. J. Comput. Intell. Res. 13 (7) (2017) 1617–1627.
[76] R.K. Hussein, A. Alenezi, H.F. Atlam, M.Q. Mohammed, R.J. Walters, G.B. Wills, Toward confirming a framework for securing the virtual machine image in cloud computing, Adv. Sci. Technol. Eng. Syst. 2 (4) (2017) 44–50.
[77] M.A. Iqbal, O.G. Olaleye, M.A. Bayoumi, A review on Internet of Things (Iot): security and privacy requirements and the solution approaches, Glob. J. Comput. Sci. Technol. E Network, Web Secur. 16 (7) (2016).

About the authors

Hany F. Atlam is a PhD candidate at the University of Southampton, UK and assistant lecturer in Faculty of Electronic Engineering, Menoufia University, Egypt. He was born in Menoufia, Egypt in 1988. He has completed his Bachelor of Engineering and Computer Science in Faculty of Electronic Engineering, Menoufia University, Egypt in 2011, then completed his master's degree in computer science from the same university in 2014. He joined the University of Southampton as a PhD student since January 2016. He has several experiences in networking as he holds international Cisco certifications and Cisco Instructor certifications. Hany is a member of Institute for Systems and Technologies of Information, Control and Communication (INSTICC), and Institute of Electrical and Electronics Engineers (IEEE). Hany's research interests include and not limited to: Internet of Things Security, Cloud Security, Cloud and Internet of Things Forensics, Blockchain, and Big Data.

Gary B. Wills is an Associate Professor in Computer Science at the University of Southampton. He graduated from the University of Southampton with an Honours degree in Electromechanical Engineering, and then a PhD in Industrial Hypermedia system. He is a Chartered Engineer, a member of the Institute of Engineering Technology and a Principal Fellow of the Higher Educational Academy. He is also a visiting associate professor at the University of Cape Town and a research professor at RLabs. Gary's research projects focus on Secure System Engineering and applications for industry, medicine, and education. Gary published more than 200 publications in international journals and conferences.

CHAPTER TWO

Integrated platforms for blockchain enablement

Md Sadek Ferdous[a,b], Kamanashis Biswas[c], Mohammad Jabed Morshed Chowdhury[d], Niaz Chowdhury[e], Vallipuram Muthukkumarasamy[f]

[a]Shahjalal University of Science and Technology, Sylhet, Bangladesh
[b]Imperial College London, London, United Kingdom
[c]Australian Catholic University, North Sydney, NSW, Australia
[d]Swinburne University of Technology, Melbourne, VIC, Australia
[e]Open University, Milton Keynes, United Kingdom
[f]Griffith University, Gold Coast, Australia

Contents

Advances in Computers, Volume 115
ISSN 0065-2458
https://doi.org/10.1016/bs.adcom.2019.01.001

Abstract

The Internet of Things (IoT) is experiencing an exponential growth in a wide variety of use-cases in multiple application domains, such as healthcare, agriculture, smart cities, smart homes, supply chain, and so on. To harness its full potential, it must be based upon a resilient network architecture with strong support for security, privacy, and trust. Most of these issues still remain to be addressed carefully for the IoT systems. Blockchain technology has recently emerged as a breakthrough technology with the potential to deliver some valuable properties such as resiliency, support for integrity, anonymity, decentralization, and autonomous control. A number of blockchain platforms are proposed that may be suitable for different use-cases including IoT applications. In such, the possibility to integrate the IoT and blockchain technology is seen as a potential solution to address some crucial issues. However, to achieve this, there must be a clear understanding of the requirements of different IoT applications and the suitability of a blockchain platform for a particular application satisfying its underlying requirements. This chapter aims to achieve this goal by describing an evaluation framework which can be utilized to select a suitable blockchain platform for a given IoT application.

1. Introduction

Internet of Things (IoT) has gained huge popularity in recent time and has become a part of our daily life. In IoT, physical objects such as home appliances, vehicles, supply chain items, and infrastructure components can sense the environment around them and adaptively interact with each other in real time. Smart objects in IoT systems are usually heterogeneous in nature and work under different trust or administrative domains [1]. Establishing trust and maintaining security in IoT domain is often regarded as one of the most challenging tasks. IoT devices depend on a variety of underlying network infrastructure which is vulnerable to attacks as evident in several recent cyber attacks [2,3]. In addition, security and privacy of the data in IoT networks is also a significant concern.

Recently, Blockchain (BC) technology has gained popularity in different domains because of its fascinating properties such as resiliency, support for integrity, anonymity, decentralization, and autonomous control. Although crypto-currency is the most widely used application of blockchain technology, the number of other practical applications beyond token-values is emerging. Over the past few years, the application domain of blockchain has expanded steadily, ranging from identity management, governance, IoT networks, financial services to healthcare. Among these applications, the convergence of blockchain and IoT networks holds enormous potentials in the area of IoT device identification, authentication, sensor data storage and seamless secure data transfer. The possibility of this convergence has driven the enthusiasm among the researchers, academia and industry practitioners to disrupt several IoT applications as well as to address the above mentioned issues prevailing in IoT systems.

Toward this aim, several blockchain platforms for IoT systems have been proposed, such as IOTA [4], Waltonchain [5], and OriginTrail [6], with the focus on different IoT application domains. Similarly, some researchers are exploring the applicability of blockchain platforms for different IoT applications such as supply chain, healthcare, smart city, smart home, financial services, automated smart contracts, and quality control and regulation. It is understandable that this diverse set of IoT applications have different requirements. For example, requirements in smart cities are different to that of the wearable fitness tracking in healthcare or goods tracking system in supply chain management. It is often challenging to identify a suitable blockchain platform for a particular IoT application satisfying all its requirements. We believe that little work has been done investigating this aspect. To be more specific, there is a gap in this domain for an evaluation framework which could be utilized to select an appropriate blockchain platform for a given IoT application considering its specific requirements. In this chapter, we address this important issue. Indeed, our motivations are twofolds: (i) to formulate a comprehensive set of requirements for different categories of IoT applications and (ii) to develop an evaluation framework to verify the suitability of a given blockchain platform for a particular type of application according to its identified requirements.

With these intentions, this chapter provides a comprehensive coverage as to how the IoT can be supported using blockchain technology focusing on the platforms and their suitability to fit into specific IoT applications. It begins with a brief discussion on the background (Section 2) of the relevant technologies and their relationship toward integration. The chapter then concentrates on various IoT applications (Section 3) and the requirement analysis of those applications (Section 4). It is then followed by a comprehensive survey on the available blockchain platforms (Section 5). Then the chapter presents an evaluation justifying how a given blockchain platform suits specific IoT applications using an intuitive figure (Section 6) which is then followed by a conclusion (Section 7).

2. Background

In this section, we briefly describe the characteristics of IoT (Section 2.1), blockchain (Section 2.2), and the convergence of IoT and blockchain (Section 2.3).

2.1 Internet of things

The Internet of Things (IoT) is the network of connected objects that are discoverable using standard communication protocols. IoT encompasses everything having connectivity and ability to communicate. The "things" can be anything from sensors to electronic devices, to appliances, to vehicles. The concept is motivated by the idea that objects of our world will talk to each other. Each object might have the communication ability, some "sensing" and "actuating" capabilities.

Depending on the working principle, IoT technology can be one of three types: internet-oriented that acts as a middleware, things-oriented that provides sensing ability and semantic-oriented that enables accessing knowledge. A combination of these three types or just a standalone IoT can be used to build smart applications aiming at solving critical problems in our daily life [7].

Numerous application domains will come in contact with this technology in the near future where the types of applications will vary on the basis of network availability, coverage, scale, heterogeneity, repeatability, user involvement and impact. Gubbi et al. categorized these potential applications into four domains: Personal and Home, Enterprise, Utilities, and Mobile [8].

2.2 Blockchain technology

Bitcoin [9], proposed in 2009, has emerged as the world's first widely used crypto-currency and paved the way for a technological revolution. It is underpinned by a clever combination of existing crypto mechanisms, which is now called blockchain technology or distributed ledger technology, providing its solid technical foundation. In recent years, blockchain has received widespread attention among the industry, the government and academia. It is regarded as one of the fundamental technologies to revolutionize the landscapes of several application domains, by removing the need for a central trusted entity. At the centre of this technology is the blockchain itself which is a database consisting of consecutive blocks of transactions chained together following a strict set of rules. This database is then distributed and stored by the nodes in a peer-to-peer network where each new block of transactions is created and appended at a predefined interval in a decentralized fashion by means of a consensus algorithm. The consensus algorithm guarantees several data integrity related properties (discussed below) in the blockchain. The term blockchain is often synonymised with another term "*Distributed Ledger.*" However, we differentiate between these two terms in the sense that distributed ledger is a more generic term. A blockchain is just an example of a distributed ledger whereas there could be other types of distributed ledger.

Evolving from the Bitcoin blockchain, a new breed of blockchain platforms has emerged which facilitates the deployment and execution of computer programs, known as smart-contracts, on top of the respective blockchain. Such smart-contracts enable the creation of so-called decentralized applications (DApps), which are autonomous programs operating without relying on any human intervention. Being part of the blockchain, smart-contracts and their executions become immutable and irreversible, a sought-after property having a wide-range of applications in different domains.

Some of the major characteristics of blockchain platforms are:

- Distributed consensus on the blockchain state.
- Immutability and irreversibility of the blockchain state.
- Data provenance of transactions guaranteed by cryptographic mechanisms.
- Accountability and transparency of blockchain data and actions.

Equipped with these characteristics, blockchain platforms offer significant advantages over traditional systems for many application domains. Among them, in this chapter, we explore only IoT-focused applications.

Depending on the application domains, different blockchain deployment strategies can be pursued. Based on these strategies, there are two predominant blockchain types: Public and Private. A public blockchain, also sometimes called as the unpermissioned blockchain, allows anyone to join and create and validate blocks as well as to modify the blockchain state by storing and updating data by means of transactions among participating entities. On the other hand, a private blockchain, also may be called as permissioned blockchain, will only allow authorized entities to join and participate in blockchain activities with the aim to ensure some form of accountability for the transaction of data, which might be desirable in some use-cases.

2.3 Toward blockchain-based IoT

During the last decade, the IoT has significantly increased its reach and a large number of devices are interconnected through the Internet, enabling them to send and receive data. It has been predicted that by 2019, 20% of all IoT deployments will be blockchain-enabled according to the IDC report [10]. However, there are many technical hurdles that need to be overcome for successful blockchain and IoT integration. Today, over 5 billion IoT devices all around the world produce massive amount of geographic and demographic data and exchange them on the Internet. These data can be exploited and misused if an appropriate security measure has not been implemented to protect confidentiality, integrity and authenticity of data. Additionally, vulnerable connected devices such as surveillance cameras could also be used by the attackers to facilitate malicious activities. More precisely, each device in an IoT network is a potential point of failure and can be exploited to launch many cyber attacks such as botnets and Distributed Denial of Service (DDoS) attacks [11]. There is no doubt that the use of blockchain to manage access to data from IoT devices may add an additional layer of security to the IoT networks as well as may overcome the single-point-of-failure problem. However, due to the lack of sufficient computing and communication power, it is a big challenge for IoT devices to directly participate in the blockchain network. Another key issue is that although blockchain is defined as the potential missing link between security and the IoT by many researchers, the technology is only as secure as its cryptographic mechanisms. Any human error in coupling this two emerging technologies will lead to security vulnerabilities.

Despite of the above technological hurdles and implementation factors, blockchain and IoT convergence has opened doors of opportunities for many applications such as supply chains and logistics. The inherent properties of blockchain and the self-execution capabilities of smart contract have fostered the potential large scale adoption of this technology to IoT applications where the interconnected devices are able to interact each other and make decisions without any human intervention. For example, machines running smart contracts will record details of all transactions that take place between themselves in a supply chain without any human oversight. Any erroneous event or actions/decision can be identified at the source and immediate corrective actions could be taken to reduce the impacts. Toward this aim, several researchers have explored the suitability of integrating IoT and blockchain in different applications domains [12–19].

3. IoT application domains

This section discusses four IoT application domains, Healthcare (Section 3.1), Supply chain (Section 3.2), Smart city (Section 3.3), and Smart home (Section 3.4), with their relevant blockchain-focused use-cases.

3.1 Healthcare

Blockchain technology has the potential to transform healthcare, placing the patient at the center of the healthcare ecosystem and increasing the security, privacy, and interoperability of health data. This technology could provide a new model for health information exchanges (HIE) by making electronic medical records (EMR) more efficient, disintermediated, and secure. In addition to traditional information systems, IoT devices are also getting popular in health sectors. Health practitioners might take major decisions about a patient based on the real-time data these devices provide. Integrating different IoT devices and exchanging data among them will be critical for future healthcare systems.

Next, we present a few potential use-cases involving healthcare, IoT and blockchain.

3.1.1 Clinical data exchange and interoperability

At present, many of the stakeholders of healthcare systems are working as isolated entities in silos rather than as parts of an integrated care system in most of the countries. In such systems, patient records are stored and maintained by different providers, making it difficult for the patients to access and share their own records. Also, health data is very private in nature and the patients often have to blindly trust the healthcare provider. The problem of data exchange and interoperability will be somewhat aggravated with the introduction of IoT into healthcare domains as it might be required to provide an interoperability support for such high-level of heterogeneous data sources. There have been several research efforts aiming to leverage the blockchain technology to tackle the aforementioned issues. Examples include MedRec from MIT [14] and others [20–23] for sharing medical data, increasing the interoperability [24–26] as well as for improving the transparency in clinical trials [27] and for other specific healthcare related applications [15,28,29]. Unfortunately, researchers are yet to explore the involvement of IoT (e.g., wearable) and blockchain in this scenario.

3.1.2 Drug supply chain

The supply chain is always a significant challenge for a secure and safe drug delivery in the pharmaceutical industry. The falsified, counterfeit, substandard and gray market medicines are a global problem which threatens public health in both developed and developing countries. According to the World Health Organization (WHO) report, the global trade of counterfeit drugs accounts for up to $30 billion per year and one in ten medications sold in low and middle-income countries is either fake or substandard [30]. Many individual companies are exploring the role of blockchain and IoTs in ensuring traceability, securing and optimizing supply chain activities for the manufacture, distribution and dispensing of pharmaceutical goods, medical commodities, and devices. Mediledger [31], Farmatrust [16], and LifeCrypter [32] are currently developing blockchain based solutions to verify provenance of each and every medicine.

3.2 Supply chain

Due to the complexity, associated costs of managing inventories and the lack of transparency in the existing supply chain systems, blockchain has become a potential solution for providing supports ranging from self-executing supply contracts to automated payment and cold chain management. In Supply Chain Management (SCM) systems, blockchain could be used to reduce fraudulent transactions and optimize the supply chain activities. More precisely, IoT-integrated blockchain can facilitate the enterprises to trace their products across supply chains and thereby detect potential fraud and failures. A number of blockchain-based solutions such as Skuchain [17], Provenance [33], Agri-Digital [34], Walmart [35], and FishCoin [36] have been designed to improve efficiency and transparency in SCM systems. In the following sections, we identify a few uses-cases for blockchain and IoT-enabled supply chains that provide end-to-end visibility and control across the supply chain.

3.2.1 Cold chain monitoring

A cold chain or cool chain is a temperature-controlled supply chain that involves the storage, transportation and distribution of perishable goods such as vaccine, chemical, sea-food, meats and dairy products. These goods should always be kept within the recommended range of temperatures to preserve and extend their shelf life. Current research shows that vaccine exposure to temperatures outside the recommended ranges during storage is 33% in developed countries whereas the number is 37.1% in developing countries. On the other hand, exposure to unacceptable temperatures during shipments is 19.3% and 38% in lower income and higher income countries, respectively [37]. This necessitates the importance of implementing an effective and sustainable system to provide safe and reliable management and delivery of perishable goods across the supply chain. Both industry and academia are actively working on this area and have introduced several blockchain and IoT-based solutions such as SkyCell [38], Vacci-chain [39], and modum [40].

3.2.2 Track the origin

The supply chain activities from the point of manufacturers to the point of consumers provide an opportunity for systematic abuse such as mis-handling of goods or even outright fraud. The increasing demands on *history* and *operation transparency* of products have fostered the integration of IoT and blockchain to the supply chain of many industries. Some of the examples include (i) Everledger—a secure blockchain system for tracing and tracking the authenticated provenance of diamonds [41], (ii) GriffithWine—a blockchain-based wine supply chain traceability system [42], and (iii) ArtChain—an open, IoT enabled, expandable title registry system and art asset trading platform [43].

3.2.3 Automated contract

Another potential use-case is the execution of an automated business contract between two corresponding entities once a certain condition is met. IoT has the potential to be an effective enabler to materialize such automation. When applied, any human intervention required may be safely purged which will improve the overall reliability of the system.

3.3 Smart City

The concept of a smart city has numerous definitions depending on the context and meaning of the word *smart*. Sometimes it refers to being intelligent while occasionally it indicates the ability to generate and exchange real-time data. In general, a smart city incorporates people, IoT devices, technology and data to provide better services and living experiences for its citizens [18]. Real-time data from these devices and immutable historical data on blockchain is jointly going to open up vast opportunities for the researchers who would get an extra edge to look at the smart city use-cases from a new perspective called data-to-decision which ties up sensor data with Artificial Intelligence (AI) in making real-time decisions [44]. Smart Dubai is arguably the first blockchain-power initiatives in building a city that ticks almost all boxes of a smart city. A number of other cities including Singapore, Hong Kong, and several Chinese cities are ready to follow suit [45].

In the following, we discuss how an IoT-integrated blockchain platform can play a vital role in different smart city use-cases.

3.3.1 Transportation

Managing the transportation system has been a great challenge for any modern city. A blockchain-based and IoT-integrated transport system management could play a pivotal role in addressing this challenge. Such a system would enable continuous sensing of passengers and vehicles to facilitate many applications and services based on machine learning algorithms in various areas such as designing timetables for metro trains and public buses, anticipating commuters demand in different parts of the city, assigning drivers shifts and so on. It could also improve existing applications such as real-time traffic updates, adjusting the duration of traffic signal lights, suggesting alternative routes and monitoring vehicle's fitness, road tax and other regulatory requirements, by using off and on-vehicle sensing devices.

3.3.2 Utility services

Power and water management has been one of the most important elements of a smart city. This includes smart grid, smart water supply and their administration at both stakeholder and consumer end. The growing awareness in favor of using renewable energy at households and usage of the zero-emission electric vehicle for commuting introduce potentials for blockchain and IoT to play a crucial role together. For example, a blockchain-based utility sharing system would assist grids to decide how much production is necessary for a certain time of the day or to make decisions as to where to distribute water in the first place in the event of disruption or shortage. The same blockchain can be incorporated in a community-driven electricity sharing system where neighbors would provide electricity, generated using Solar PVs in their rooftops, to their neighbors in need and vice versa.

3.3.3 Citizen engagement

A smart city helps to improves citizens' lifestyle by engaging them in activities and recreations. Public parks, libraries, museums, sports, cinemas, shopping center, etc. are various forms of citizen engagement commonly found in modern cities. The real-time data could be collected using IoT-enabled sensors installed at these public spaces while blockchain would provide the backbone of historical information that aids machine learning or similar methods to make recommendation (e.g., books, exhibitions), form groups (e.g., workout challenges), suggest products (e.g., clothes, movies, foods), and so on. This would ultimately improve the quality of citizen engagement experience and introduce new perspectives.

3.4 Smart home

Smart home, also known as *Home automation*, is one of the most widely anticipated applications of IoT. In essence, a smart home device must be equipped with IoT devices and sensors to function as desired. The underlying motivation of a smart home is that this will enable the home owners to have unparalleled control, remotely from anywhere in the world, over every single electronic device in their homes so that the overall comfort, convenience, and security can be increased in an energy efficient way. The desire and provision to exercise such control from anywhere in the world has fuelled the popularity of smart home concept. Indeed, there are already a number of smart home devices in the consumer market. With the introduction of different types of hardware and novel services, it is safe to assume that the market will continue to grow in future. This assumption can be backed with the prediction that the size of the world wide smart home market will be around 53 billion USD by 2022 [46]. Blockchain has the potential to add trust and traceability to the overall smart home concept and its operation.

Next, we explore a few use-cases to illustrate the usefulness of a smart home IoT applications involving blockchain.

3.4.1 Smart appliances

Smart appliances are in the fore-front of showcasing the utilities of IoT in the smart home setting. Indeed, it is envisioned that every single piece of home appliance will also be an IoT device in future, paving way for creating novel applications in order to provide a greater level of control and comfort for everyone at home. These will enable the residents of a smart home to monitor and control any home appliances remotely. A more futuristic scenario would be to establish M2M (Machine-to-Machine) communications between different smart appliances within the home to initiate autonomous behavior for facilitating daily chores. All these would require a resilient network that can support heterogeneous devices and facilitate autonomic code execution. A smart-contract supported blockchain can be a potential tool to enable this. It has been reported that Wallmart has investigated the possibility of integrating smart appliances using blockchain technology [47]. Researchers are also investigating probable smart home applications utilizing blockchain technology [19].

3.4.2 Safety and security

One of the major goals of a smart home is to ensure the safety and security of its residents. A smart home would have the capability to detect any electrical and fire safety hazards automatically and raise alarms accordingly and contact with the emergency service. A smart home should ensure the security of its residents from external intrusion, even when they are away from home and send alarms to its residents as well as to different law enforcement agencies when any intrusion is detected. These would require continuous monitoring and seamless integration with different authorities. A blockchain integrated IoT platform would be a great enabler to realize this vision.

4. Requirement analysis

In this section, we formulate different types of functional, security, and privacy requirements that are important to realize different IoT applications stated above. The functional requirements are necessary components, and functionalities required to provide services to users whereas different security and privacy requirements are created to address various security and privacy issues of the users and the organizations respectively. It is often useful to identify the key characteristics of these applications in order to formulate different requirements. Next, we present a few of key characteristics for these applications.

- *Multifacet data sources:* IoT devices for different applications will be installed in different environments and will belong to heterogeneous networks, thus representing an ever-scattered multifacet data sources.
- *Continuous data stream:* Such IoT devices will preferably collect data in a continuous and time-sensitive manner.
- *Data rate and volume:* Data can be generated at a significantly huge rate and volume by an array of IoT devices within a single application.
- *Distributed stakeholders:* Stakeholders involved in an IoT application might be distributed around different locations utilizing different systems in their end.

In our next step, we present a set of functional requirements that have been formulated considering the key characteristics discussed above. The functional requirements are presented below.

- F1—Scalability: Scalability refers to the ability to grow in size and functionalities without degrading the performance of the original system. This is crucial for any application discussed in this chapter.
- F2—Multiple Sources: The systems must be able to handle data generated from multiple heterogeneous sources and transmit such data with minimal to no latency.
- F3—Data Sharing: All the stakeholders within each application should be able to exchange and share information, generated by heterogeneous data sources, internally within the organization as well as externally to other entities without any intermediary.
- F4—Interoperability: A system developed for a particular application should be interoperable among a wide range of stakeholders of the application [48]. Interoperability will ensure that data can be shared among respective entities without any issue even though every stakeholder might utilize different types of systems in their end.
- F5—Identity Management: Every IoT devices and other human entities must be properly identified within the systems. Therefore, a proper identity management framework must be embedded into the systems of every use-case within an application.
- F6—Transparency: Systems for these applications should be able to create an auditable chain of custody/activities from the data producer to the data consumer. This guarantees transparency and is crucial in supply chain and healthcare systems. It plays a key role in ensuring customer satisfaction and compliance. Supporting this requirement might enable additional business cases in other applications.
- F7—Traceability: For supply chains, traceability has become increasingly important due to the growth in consumers' interest about the origin of the products or services. A digital supply chain should enable the consumers to trace and track the provenance of a product or service. This requirement might enable additional business cases in other applications.
- F8—External Interface: Systems in a particular application domain should expose an external interface by which it can be connected to the entities of other application domains so as to enable novel business cases. An example of an external interface would be to connect

smart-home applications to the smart infrastructures of a smart city, to other smart homes and so on.

- F9—Payment Mechanism: Some application domains such as supply chain will inherently require a payment system built into its system. Whereas, the support of payment will enable additional business cases in other application domains.
- F10—Performance: Systems for these applications should maintain certain level of performance in terms of response time, available storage or capacity so that they have the capability to process high volume and high frequency data generated by plethora of IoT devices. This is particularly important for smart city and supply chain application. For other applications it might enable additional business cases.
- F11—Reliability: The reliability requirement of an application ensures the availability of the system, up-to-date and accurate information as well as consistent flow of information among all entities in the system.

It is crucial that systems for these IoT applications must satisfy a set of security and privacy requirements because of their involvement with sensitive data in different use-cases. It is mandatory that these systems satisfy the core CIA (Confidentiality, Integrity, and Availability) security requirements regarding the data and services involved. In addition, these systems must satisfy the following security (denoted with *S*) and privacy requirements (denoted with *P*).

- S1—Secure Transmission: Data generated in an application must be transmitted, both internally and externally, securely—that is with appropriate crypto mechanisms.
- S2—Fine-grained Access Control: Among these applications, healthcare and smart home system would require authorizing the right user and providing the appropriate access to the data. In addition, the system should have fine-grained access control mechanism to provide access to only specific required data rather than a general access right in smart home and healthcare systems. It would also be useful for other applications to introduce novel business scenarios.
- S3—Data Provenance and Integrity: The provenance and integrity of data generated from a specific source must be guaranteed.
- S4—Fault Detection and Patching: There must be mechanisms to identify and trace every faulty IoT device in the system. Once detected, an

alarm must be raised and if possible, the corresponding device should be remotely patched. This requirement is crucial for all applications except Healthcare.

- P1—Privacy Protection: Systems must protect the privacy of users and organizations with appropriate mechanisms.

4.1 Summary

It is evident from our above analysis that the most of the formulated requirements are applicable to the majority of the application domains. However, some requirements are most suitable to particular applications than others. Next, we summarize the requirements for different IoT applications in Table 1. As per the table, we differentiate between explicit and implicit requirements which denote the mandatory and optional requirements respectively for a particular application.

Similarly we summarize the security and privacy requirements for different applications in Table 2. In the table, the symbol "✓✓" is used to denote that a specific requirement is crucial for an application whereas a single "✓" is used to denote that the requirement is desirable but not mandatory for the corresponding application.

Table 1 Summary of functional requirements for different applications.

Application	Explicit requirement	Implicit requirement
Healthcare	F1–F6, F11	F7–F10
Supply chain	F1–F7, F9–F11	F8
Smart city	F1–F5, F10, F11	F6–F9
Smart home	F1–F5, F11	F6–F10

Table 2 Summary of security and privacy requirements for different applications.

Application	S1	S2	S3	S4	P1
Healthcare	✓✓	✓✓	✓✓	✓	✓✓
Supply chain	✓✓	✓	✓✓	✓✓	✓✓
Smart city	✓✓	✓	✓✓	✓✓	✓✓
Smart home	✓✓	✓✓	✓✓	✓✓	✓✓

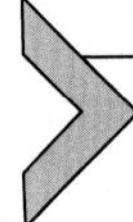

5. Survey on existing blockchain platforms for IoT systems

In this section, we explore different blockchain platforms that have recently emerged to support different IoT applications. The selected blockchain platforms are: Waltonchain (briefly discussed in Section 5.1), OriginTrail (Section 5.2), Slock.it (Section 5.3), Moeco (Section 5.4), IOTA (Section 5.5), IBM Watson (Section 5.6) and NetObjex (Section 5.7). Then, in Section 5.8, we summarize these platforms using a set of key properties.

It is to be noted that, among these platforms, IOTA theoretically is not a blockchain platform, rather it represents a distributed ledger platform. However, because of its relevance to the scope of this chapter, we have decided to include it in our analysis. A brief description of each platform is presented next.

5.1 Waltonchain

Waltonchain is a new blockchain platform for the IoT industry [5]. The platform is named thus in order to commemorate and recognize the contribution of Charles Walton, the inventor of RFID (Radio Frequency Identification) technology and to advance his vision for the ubiquitous deployment of the RFID technology in the form of IoT. With this motivation, Waltonchain would like to disrupt the current IoT industries by integrating the transparency, accountability, and provenance properties of the blockchain with the RFID-enabled IoT hardware. Indeed, the core platform consists of RFID hardware (both RFID tags and reader), the Waltonchain public blockchain platform and the software platform that interfaces the hardware with the blockchain. The ultimate goal is to create a novel service delivery model called Value IoT (VIoT) which will be suitable for a wide range of IoT applications such as supply chain tracking, product authentication and identification and so on.

Waltonchain utilizes two types of blockchain: a parent blockchain, called *Waltonchain* and different *child* blockchains. Waltonchain is a public blockchain allowing anyone to participate in the consensus mechanism. A child chain, on the other hand, can be public or private depending on a use-case. There can be as many child chains as required, each for enforcing business logic for a specific industry or use-case. A new child chain is created using a special transaction which is recorded in the parent blockchain.

The Waltonchain platform introduces a hybrid consensus algorithm called WPoC (Waltonchain Proof of Contribution). WPoC is a combination of three different consensus algorithms: Proof of Work (PoW), Proof of Stake (PoS), and Proof of Labor (PoL) [5]. PoW is similar to what is used in Bitcoin [9]. PoS is a stake-based consensus algorithm [49]. PoL, on the other hand, has been described as a brand new consensus algorithm for cross chain data transmission and token exchange [5]. Unfortunately, there is not much detail information is available regarding this algorithm.

5.2 OriginTrail

OriginTrail is a decentralized, public blockchain supported data sharing platform for multiorganizational environments [6]. This platform integrates blockchain technology with digital supply chains to enable supply chain data immutability as well as integrity. The idea is to provide a standard blockchain-based solution with an incentivized protocol to ensure product standards and safety of the consumers.

OriginTrail implements the Electronic Product Code Information Service (EPCIS) framework to facilitate the layered, extensible and modular design across the entire structure. The OriginTrail ecosystem can be viewed as a four-layers system. On top of the blockchain layer, there are two system layers namely the network and data layers which implement an off-chain decentralized peer to peer network known as OriginTrail Decentralized Network (ODN). On top of the network layer, there is a decentralized application layer which interfaces between the users and the system to provide data input facilities.

The current version of OriginTrail implements Proof of Work (PoW) which runs on top of Ethereum blockchain. It is envisioned that the final version would accommodate different blockchains with different consensus algorithms.

5.3 Slock.it

Slock.it is an IoT platform on top of Ethereum blockchain [50]. Its vision is to establish a truly decentralized sharing economy which will enable a direct interaction between a producer/owner and a consumer of IoT objects. The principle of sharing economy is to allow people share their unused or less-used physical or virtual resources, such as rooms or flats, cars, electricity or even time, for financial incentives. The traditional approach requires a lot of human intervention with a big issue of trust and transparency. The existing applications of sharing economy such as Uber and Airbnb are not decentralized. They rely on their monopolistic centralized providers which charge a considerable fee, yet the security, trust and transparency issues are prevalent in such applications.

Slock.it aims to address these issues by providing a platform consisting of IoT objects, Slock software platform and smart-contract supported Ethereum blockchain [51] introducing fully automated Machine-to-Machine, Machine-to-Human and Human-to-Human interactions. Each IoT object will interact with each other via a smart contract (or a set of smart contracts) deployed in the Ethereum blockchain. A person can interact with each IoT object using any preferred device such as his or her mobile phone. Also, Slock does not have its own blockchain. Instead, it utilizes Ethereum as its underlying blockchain platform. Hence, it relies on Ethereum's current consensus mechanism, rewarding process and other properties.

5.4 Moeco

Moeco is a blockchain platform envisioned by its creator as the "DNS of things." This platform integrates several network standards and offers connectivity to billions of devices globally through participating gateways [52]. Moeco uses crowdsourcing approach to integrate existing networks and private gateway owners in its infrastructure. Anyone having the communication connectivity can become a gateway service provider. For example, someone having a wireless router at home or a smartphone with the internet connectivity can join as a gateway provider and start serving vendors, who are business providers owning sensor devices, to facilitate their sensor devices looking for connectivity. Thus, Moeco removes the need for traditional fixed setup of using cellular or other form of networks. It works as if the "ebay" of connectivity. Vendors use connectivity on the fly from the available gateways. Such an infrastructure is advantageous for both parties as vendors can avoid huge establishment cost while gateway owners earn money by making available their unspent bandwidth.

This system utilizes two blockchains namely transport blockchain and invoice blockchain. The transport blockchain is responsible for delivering data packages. The data package transportation and payment validation take place in the Moeco network based on the Exonum framework which is another blockchain developed to represent, store, exchange and secure any digital asset [53], and uses a consensus algorithm similar to PBFT [54] while the payment is arranged using ERC 20 Ethereum Moeco tokens in the Ethereum network. Because it is an overlay over Ethereum, features of the parent blockchain including the consensus mechanism get extended to Moeco.

5.5 IOTA

IOTA is a special type of cryptocurrency and a distributed ledger designed for the Internet of Things (IoT) [4]. Scalability, free and fast transactions and ability to validate unlimited number of transactions simultaneously make IOTA suitable for a use-case like IoT. Unlike using the proof-of-work and building blocks, IOTA uses the *Tangle*, a consensus-building data structure made of Directed Acyclic Graph (DAG).

The Tangle is a consensus-building system which removes the need for miners and data blocks on the platform. Each network member (machine) willing to execute a transaction must actively participate in the network consensus by approving two past transactions. This way each transaction links to the two transactions it verifies, and over time, it will be linked to future transactions that verify it. For each verification, the verifier performs a small work linking the transactions into the overall Tangle. This approach solves the scalability problem as the network no longer relies on building blocks for a blockchain. Since every new machine on the network contributes its computing power to the network when it submits a transaction, the cost of using the network is as good as the electricity consumed by the machine for verifying the required two transactions. The latest addition in IOTA is Qubic which is a protocol specifying IOTA's solution for quorum-based computations, outsourced computations and smart contracts [55]. This protocol is expected to empower machines operating in IOTA with the ability to develop and run autonomous contracts involving two communicating IoT devices, thus enabling a Machine-to-Machine (M2M) communication.

5.6 IBM watson

IoT Watson is an integrated technology which combines Watson IoT platform and blockchain [56]. Watson use Hyperledger Fabric [57] framework to provide blockchain services. It can capture data in real-time by using IoT devices. It also provides data analytic and visualization services to the user.

The Hyperledger project is a collaborative effort between IBM and Linux Foundation to create an enterprise-grade, open-source distributed ledger framework and code base. The aim of this project is to provide an open standard blockchain platform so that any enterprise can build their own solution. There are several active on-going projects under the the umbrella of Hyperledger project, namely, Burrow, Fabric, Sawtooth, Iroha and Indy. Fabric is the most relevant platform from this group. It is a permissioned blockchain infrastructure with modular architecture which enable to configure different types of consensus algorithms. It also supports execution of smart contact (called "chaincode" in Fabric) and membership services provided by a Certificate Authority, managing X.509 certificates which are used to authenticate member identity and roles.

There are three types of nodes in Hyperledger Fabric. These are: Orderer node, Peer node, and Client node. Client nodes act on behalf of end-users and create and thereby invoke transactions. Peer nodes maintain the ledger and receive ordered update messages for committing new transaction to the ledger. Endorsers are special type of peers where their task is to endorse a transaction by checking whether they fulfill necessary and sufficient conditions. Orderer nodes provide communication channel to clients and peers over which messages containing transaction can be broadcast. Being a permissioned blockchain, it can support a considerable number of transactions. However, the number will be dependant on the use-case and how it is deployed. Fabric uses modular architectural design to support different consensus mechanisms to be plugged-in. Currently, Fabric supports only two consensus mechanisms, SOLO and Kafka, with more consensus mechanisms to be added in future.

5.7 NetObjex platform

NetObjex is a decentralized digital asset management platform that uses IoT and blockchain to provide services for four major market segments namely, Supply chain and logistics, Manufacturing industry, Smart city and Automotive industry [58]. The platform uses IoTs for data acquisition and dissemination and supports a wide range of communication protocols such as Cellular, mid-range protocols (LoRA, Sigfox, NB-IOT), Wifi, Ethernet, BLE, specialized protocols (DSRC). Additionally, it enables enterprises to share information securely through blockchain and to enforce business rules through smart contracts. To ensure legitimate access to sensitive information, NetObjex stores them in cryptographically secure ledgers by the use of blockchain technology.

The NetObjex platform provides a flexible, plug and play environment for the users to develop and deploy their own smart products. It integrates a number of big data repositories, distributed ledger technologies, and edge devices with its core blockchain middleware element to facilitate interoperability and cross-communications among different components.

The NetObjex platform implements a standardized mechanism for smart devices to communicate globally with each other. To interact between digital assets owned by different organizations within a single ecosystem, the platform introduces a technology fabric through its *IoToken* mechanism. It also provides supports for IoToken native crypto-currency for inter-device transactions.

5.8 Summary

This section summarizes the analysis of the selected blockchain platforms against a set of properties. Our initial goal is to compare and contrast the identified blockchain platforms using the chosen properties. These properties represent the key characteristics of a blockchain platform and will be essential for developing blockchain-supported IoT applications. Ultimately, we would like to leverage this analysis to develop an evaluation framework for blockchain-enables IoT applications. At first, we briefly discuss the properties.

- *Public/private* property is used to indicate if the platform supports only a public or private blockchain, or both.
- *Transaction speed* presents Transactions per Second (TPS) supported by the platform.
- *Block creation time* stipulates the average block creation time in the platform.
- *Block size* specifies the designated block size in the platform.
- *Fee* indicates if there is any fee associated for using the platform.
- *Consensus* outlines the supported consensus algorithm in the platform.
- *Network size* stipulates the size of the network for the platform.
- *Smart contract* indicates if the platform supports smart contract facility.
- *Secure channel* designates if the communication in the platform can be facilitated under a secure channel.
- *Verified identity* is used to show if the platform supports any verified identity service.

Utilizing these properties, we present the summary in Table 3. We have used the "✓" to indicate a certain property is satisfied and the "x" to indicate the property is not supported by the respective platform. The "—" is used to signify that the property is not applicable for the platform. Moreover, the term "+Eth" is used to indicate that the respective platform inherits the

Table 3 Comparison of blockchain platforms using relevant properties.

Properties	IOTA	Walton chain	Origin trail	Slock.It	Moeco	IBM watson	NetObjex platform
Public	✓	✓	✓	✓	✓	x	✓
Private	x	✓	x	x	x	✓	✓
Transaction speed	500–800 TPS	100 TPS [59]	+Eth	+Eth	+Eth	160–3500	Platform dependent
Fee	x	✓	✓	✓	✓	x	Platform dependent
Block creation time	—	30s	✓	+Eth	+Eth	Variable	Platform dependent
Consensus	Tangle	WPoC	PoW	+Eth	+Eth	PBFT. However, other algorithms can be plugged in	Platform dependent
Network size	Large	Large	+Eth	+Eth	+Eth	Small to medium	Platform dependent
Block size	—	225	+Eth	+Eth	+Eth	Can be configured using BatchTimeout and Batch-Size	Platform dependent
Smart contract	✓	✓	+Eth	+Eth	+Eth	✓	Platform dependent
Secure channel	x	x	x	x	x	✓	x
Verified identity	—	✓ (For private network)	✓	x	x	✓	✓

values of the properties from Ethereum. Finally, numerical values or textual explanations, where appropriate, have been provided for other properties for corresponding platforms.

6. Evaluation and discussion

Since there are a number of blockchain platforms designed to provide different functionalities, it is important to evaluate their applicability with respect to the identified requirements of the selected IoT applications. This core set of requirements are then analyzed to evaluate the suitability of different blockchain platforms for different IoT application scenarios. Our analysis has resulted in an evaluation framework which is presented in Fig. 1.

This figure is illustrated based on a few observations. The first observation is that all blockchain platforms embody the principle of accurate dissemination of information, in the form of transactions, among large nodes. In that sense, all such platforms support the notion of Scalability (requirement F1) and Reliability (F11) requirements. Hence, we have grouped them together within a same category in Fig. 1. Similarly, the second observation is that all blockchain platforms satisfy the notion of Data Sharing (F3) and other related requirements such as Multiple Sources (F2), Interoperability (F4) and External Interface (F8). Hence, these are grouped under the same category in Fig. 1. Similarly, Transparency (F6) and Provenance have been grouped together under the same category in the figure. We have excluded Sloc and Moeco from this evaluation framework because of their lack of generality for our selected IoT applications.

Now, we analyze the evaluation framework which serves two main purposes:

- to identify which requirements are satisfied by which platforms and
- to choose (a) suitable platform(s) satisfying different requirements.

At first, we describe how the evaluation framework (Fig. 1) serves the first purpose. It can be seen that five platforms, IOTA, OriginTrail, Waltonchain, IBM Watson and NetObjex, support scalability/reliability which is a key concern for IoT enabled blockchain applications. There are many applications that need to be implemented in a closed and private environment to provide privacy protection and Waltonchain, IBM Watson and NetObjex are the best choices for such applications. In addition to user identity, device identity is also very important for many application domains such as smart city and smart home. Not only this, these identities need to be properly verified. Only IBM Watson and NetObjex have the capability to support

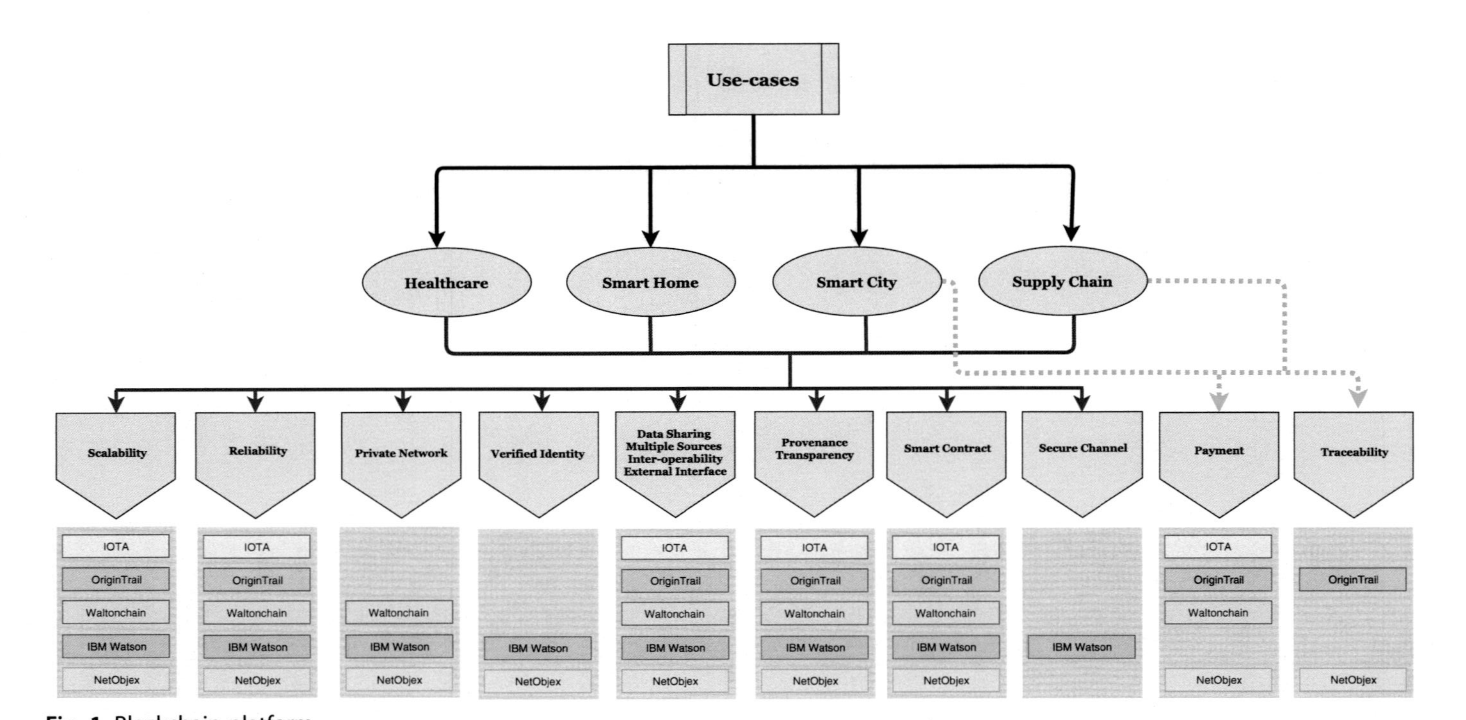

Fig. 1 Blockchain platform.

verified identity each IoT device and person in a network. It is not surprising that every platform supports data sharing since blockchain has already been established as a secure data sharing platform for its decentralized characteristic. With respect to the secure channel requirement, only IBM Watson allows users to explicitly define the HTTPS secure channel.

Payment and traceability are two key requirements for supply chain and smart city applications. For example, smart parking involves the payment system whereas traceability is important for effective management and distribution of resources. IOTA, OriginTrail, Waltonchain and NetObjex platform provide the payment services by using their native crypto-currencies. However, only OriginTrail and NetObjex platform are designed to trace the origin of a product since these two platforms are specially designed for supply chain applications.

Next, the suitability of Fig. 1 for the second purpose is presented. The evaluation framework has been visualized in such a way that the color or the position of each platform can be utilized to evaluate a platform against a set of selection criteria of requirements for a specific application. For example, if a certain application requires to satisfy the scalability, support for private network, verified identity and traceability, NetObjex is the only platform that can support all these.

Careful readers might notice the absence of an important requirement Performance (requirement F10) from the evaluation framework in Fig. 1. The performance is different than other requirements in the sense that it is normally quantitative in nature. In different contexts, it can be measured in different ways. In the scope of this chapter, we measure it using the transaction speed (TPS) property as TPS represents the most widely-used performance parameter in the blockchain community. With this characteristics, it becomes a bit less-intuitive to combine the Performance requirement with other requirements which are mainly qualitative in nature.

However, we propose to use the Performance requirement in a tie-break situation among several platforms. For example, if an application needs to support Scalability, Data Sharing capability with payment support, there are a few options to choose from: IOTA, OriginTrail, Waltonchain, and NetObjex. In such cases, we propose to use the Performance requirement (based on TPS) to select the best one from these. This mechanism also provides additional flexibilities. We can even consider different quantitative consensus characteristics as part of the Performance requirement to impose other quantitative selection criteria to select the best platform for an application.

7. Conclusion

IoT and blockchain are two emerging technologies that are expected to have immense impact in the society around the world. Each of these technologies has their own sets of applications and some significant shortcomings. Combining these two technologies, however, might address many of these shortcomings. Not only that, this combination opens up doors of opportunities for novel applications with additional advantages. This chapter aims to explore this avenue. In particular, this principal motivation of this chapter has been to create an evaluation framework which can be utilized to evaluate different blockchain platforms for their suitability in different IoT applications.

Towards this aim, this chapter has explored four different IoT applications: Healthcare, Supply chain, Smart city, and Smart home. For each of these applications, different use-cases have been analyzed. Based on this, several functional, security and privacy requirements have been identified. Next, seven IoT-focused blockchain platforms have been examined to identify their inherent properties. Finally, combining the requirements of the IoT applications and properties of the selected blockchain platforms, an evaluation framework has been created which is presented as a figure (Fig. 1). The graphical representation provides an intuitive visualization to identify the suitable blockchain platform(s) for a particular application under certain requirements.

There are still a lot of challenges that need to be tackled before these two emerging technologies can be successfully merged. One major challenge will be to identify a suitable blockchain platform for a particular use-case (e.g., cold chain monitoring) within an application domain (e.g., supply chain). The current evaluation framework in this chapter has this limitation that it can only identify suitable platforms for a generic IoT application domain. We aim to address this in our future work. Even with this limitation, we believe our effort presented in this chapter will represent a step forward towards addressing this challenge effectively for researchers and practitioners in this domain.

References

[1] L. Liu, X. Liu, X. Li, Cloud-based service composition architecture for internet of things, in: Internet of Things, Springer, 2012, pp. 559–564.

[2] Mirai (malware), https://en.wikipedia.org/wiki/Mirai_(malware), [Online: accessed 1 December 2018].

[3] Trending, IoT Malware Attacks of 2018, https://securingtomorrow.mcafee.com/consumer/mobile-and-iot-security/top-trending-iot-malware-attacks-of-2018/, [Online: accessed 1 December 2018].
[4] IOTA IOTA White Paper, https://iota.org/IOTA_Whitepaper.pdf, [Online: accessed 1 July 2018].
[5] Waltonchain Waltonchain White Paper, V2.0, https://waltonchain.org/templets/default/doc/Waltonchain-whitepaper_EN_20180525.pdf, [Online: accessed 1 November 2018].
[6] OriginalTrail OriginalTrail White Paper, https://origintrail.io/storage/documents/OriginTrail-White-Paper.pdf, [Online: accessed 1 July 2018].
[7] L. Atzori, A. Iera, G. Morabito, The internet of things: a survey, Comput. Netw. 54 (15) (2010) 2787–2805.
[8] J. Gubbi, R. Buyya, S. Marusic, M. Palaniswamia, Internet of things (iot): a vision, architectural elements, and future directions, Futur. Gener. Comput. Syst. 29 (7) (2013) 1645–1660.
[9] S. Nakamoto, Bitcoin: A Peer-to-Peer Electronic Cash System, 2008.
[10] C. MacGillivray, V. Turner, L. Lamy, K. Prouty, R. Segal, A. Siviero, M. Torchia, D. Vesset, R. Westervelt, R. Yesner, IDC FutureScape: Worldwide Internet of Things 2017 Predictions, IDC, 2016.
[11] A. Panarello, N. Tapas, G. Merlino, F. Longo, A. Puliafito, Blockchain and iot integration: a systematic survey, Sensors 18 (8) (2018), https://doi.org/10.3390/s18082575. URL. http://www.mdpi.com/1424-8220/18/8/2575.
[12] M. Chowdhury, M.S. Ferdous, K. Biswas, Blockchain platforms for IOT use-cases, in: 2nd Symposium on Distributed Ledger Technology, 2018.
[13] M.S. Ferdous, M.J.M. Chowdhury, K. Biswas, N. Chowdhury, Immutable autobiography of smart cars, in: 3rd Symposium on Distributed Ledger Technology, 2018.
[14] A. Azaria, A. Ekblaw, T. Vieira, A. Lippman, Medrec: using blockchain for medical data access and permission management, in: Open and Big Data (OBD), International Conference on, IEEE, 2016, pp. 25–30.
[15] S. Angraal, H.M. Krumholz, W.L. Schulz, Blockchain technology: applications in health care, Circ. Cardiovasc. Qual. Outcomes 10 (9) (2017) e003800.
[16] FarmaTrust, https://www.farmatrust.com/, [Online: accessed 1 November 2018].
[17] Skuchain Platform, http://www.skuchain.com/, [Online: accessed on 30 November 2018].
[18] A. Cocchia, Smart and Digital City: A Systematic Literature Review, Springer, 2014, pp. 13–43. Chapter 2.
[19] A. Dorri, S.S. Kanhere, R. Jurdak, P. Gauravaram, Blockchain for IOT security and privacy: the case study of a smart home, in: Pervasive Computing and Communications Workshops (PerCom Workshops), 2017 IEEE International Conference on, IEEE, 2017, pp. 618–623.
[20] P.T.S. Liu, Medical record system using blockchain, big data and tokenization, in: International Conference on Information and Communications Security, Springer, 2016, pp. 254–261.
[21] Q. Xia, E.B. Sifah, A. Smahi, S. Amofa, X. Zhang, BBDS: blockchain-based data sharing for electronic medical records in cloud environments, Information 8 (2) (2017) 44.
[22] A. Dubovitskaya, Z. Xu, S. Ryu, M. Schumacher, F. Wang, Secure and trustable electronic medical records sharing using blockchain, in: AMIA Annual Symposium Proceedings, vol. 2017, American Medical Informatics Association, 2017, p. 650.
[23] Y. Ge, J.J. Carr, Method and Apparatus for Personally Controlled Sharing of Medical Image and Other Health Data, US Patent App. 12/827,717. Jan. 27, 2011.

[24] C. Brodersen, B. Kalis, C. Leong, E. Mitchell, E. Pupo, A. Truscott, L. Accenture, Blockchain: Securing a New Health Interoperability Experience, (2016) ed: Accenture LLP.
[25] K. Peterson, R. Deeduvanu, P. Kanjamala, K. Boles, A blockchain-based approach to health information exchange networks, in: Proceedings of the NIST Workshop Blockchain Healthcare, vol. 1, 2016, pp. 1–10.
[26] P. Zhang, J. White, D. C. Schmidt, G. Lenz, Applying Software Patterns to Address Interoperability in Blockchain-Based Healthcare Apps, arXiv preprint (2017) arXiv:1706.03700.
[27] T. Nugent, D. Upton, M. Cimpoesu, Improving data transparency in clinical trials using blockchain smart contracts, F1000Res. 5 (2016) 2541.
[28] T.-T. Kuo, H.-E. Kim, L. Ohno-Machado, Blockchain distributed ledger technologies for biomedical and health care applications, J. Am. Med. Inform. Assoc. 24 (6) (2017) 1211–1220.
[29] A. Dubovitskaya, Z. Xu, S. Ryu, M. Schumacher, F. Wang, How blockchain could empower ehealth: an application for radiation oncology, in: VLDB Workshop on Data Management and Analytics for Medicine and Healthcare, Springer, 2017, pp. 3–6.
[30] Tens of Thousands Dying From 30 Billion Fake Drugs Trade, https://www.reuters.com/article/us-pharmaceuticals-fakes/tens-of-thousands-dying-from-30-billion-fake-drugs-trade-who-says-idUSKBN1DS1XJ [Online: accessed 28 November 2017].
[31] MediLedger Project, https://www.mediledger.com/, [Online: accessed 1 November 2018].
[32] LifeCrypter, https://medium.com/@philippsandner/blockchain-technology-in-the-pharmaceutical-industry-3a3229251afd, [Online: accessed 1 November 2018].
[33] Provenance Platform, https://www.provenance.org/, [Online: accessed on 30 November 2018].
[34] Agridigital Platform, https://www.agridigital.io/ [Online: accessed on 30 November 2018].
[35] M. Smith, In Wake of Romaine *E. coli* Scare, Walmart Deploys Blockchain to Track Leafy Greens, https://news.walmart.com/2018/09/24/in-wake-of-romaine-e-coli-scare-walmart-deploys-blockchain-to-track-leafy-greens, [Online: accessed on 30 November 2018].
[36] Fishcoin: A Blockchain Based Data Ecosystem for the Global Seafood Industry, https://fishcoin.co/files/fishcoin.pdf, [Online: accessed on 30 November 2018] (2018).
[37] C.M. Hanson, A.M. George, A. Sawadogo, B. Schreiber, Is freezing in the vaccine cold chain an ongoing issue? A literature review, Vaccine 35 (17) (2017) 2127–2133. Building Next Generation Immunization Supply Chains. https://doi.org/10.1016/j.vaccine.2016.09.070. http://www.sciencedirect.com/science/article/pii/S0264410X16309471.
[38] Safest Pharma Containers, https://skycell.ch/index.html, [Online: accessed on 30 November 2018].
[39] K. Biswas, V. Muthukkumarasamy, W.L. Tan, Vacci-chain: a safe and smarter vaccine storage and monitoring system, in: Symposium on Distributed Ledger Technology—SDLT'2017, 2017.
[40] Modum Whitepaper, https://modum.io/sites/default/files/documents/2018-10/17090520white20paper20v.201.1.pdf, [Online: accessed on 30 November, 2018].
[41] Everledger Platform, https://www.everledger.io/, [Online: accessed on 30 November 2018].
[42] K. Biswas, V. Muthukkumarasamy, W.L. Tan, Blockchain based wine supply chain traceability system, in: Future Technologies Conference (FTC), 2017.
[43] Artchain GLobal, https://www.artchain.world/, [Online: accessed on 30 November 2018].

[44] H.G. Miller, P. Mork, From data to decisions: a value chain for big data, IT Prof. 15 (1) (2013) 57–59.
[45] M.S. Khan, M. Woo, K. Nam, P.K. Chathoth, Smart city and smart tourism: a case of Dubai, Sustainability 9 (12) (2017) 2279.
[46] Smart Home—Statistics & Facts, https://www.statista.com/topics/2430/smart-homes/, [Online: accessed 18 October 2018].
[47] Walmart Files Blockchain Patent for Smart Appliance Management, https://www.ccn.com/walmart-files-blockchain-patent-for-smart-appliances/, 2018. [Online: accessed on November 30, 2018].
[48] T.K. Hui, R.S. Sherratt, D.D. Sánchez, Major requirements for building smart homes in smart cities based on internet of things technologies, Futur. Gener. Comput. Syst. 76 (2017) 358–369.
[49] QuantumMechanic, Proof of Stake Instead of Proof of Work, https://bitcointalk.org/index.php?topic=27787.0, 2011. [Online: accessed on 5 November, 2018].
[50] Slock.it platform, https://slock.it/, [Online: accessed on November 5, 2018].
[51] Slock.it, Decentralizing the Emerging Sharing Economy, https://blog.slock.it/slock-it-decentralizing-the-emerging-sharing-economy-cf19ce09b957, 2015. [Online: accessed on November 5, 2018].
[52] Moeco, Moeco Whitepaper, Technical Report v 0.9, Moeco.io, 2018.
[53] Exonum, Exonum Whitepaper, Technical Report, Exonum, 2018.
[54] M. Castro, B. Liskov, Practical byzantine fault tolerance, in: Proceedings of the Third Symposium on Operating Systems Design and Implementation, New Orleans, USA, 1999.
[55] Qubic IOTA, https://qubic.iota.org/, [Online: accessed on 30 November 2018].
[56] IBM Watson Platform, https://www.ibm.com/watson/, [Online: accessed on 30 November 2018].
[57] Hyperledger Fabric, https://www.hyperledger.org/projects/fabric, [Online: accessed on 30 November 2018].
[58] NetObjex Platform, https://www.netobjex.com/, [Online: accessed on 30 November 2018].
[59] Waltonchain March AMA Part 1, https://medium.com/@Waltonchain_EN/waltonchain-march-ama-part-1-a4dc391ce231, 2018. [Online: accessed on November 10, 2018].

About the authors

Md Sadek Ferdous is an Assistant Professor, at Shahjalal University of Science and Technology, Sylhet, Bangladesh. He is also a Research Associate in the Centre for Global Finance and Technology at Imperial College Business School investigating how the blockchain technology can be leveraged for self-sovereign identity. He completed his PhD on Identity Management from the University of Glasgow. He holds a double masters from Norwegian University of Science and Technology (NTNU) and the

University of Tartu, Estonia in security and mobile computing. He has several years of experience of working as a postdoctoral researcher in different universities in different European and UK-funded research projects. He was part of the team at the University of Southampton, UK which formulated a novel concept called Federation-as-a-Service (FaaS) which utilizes blockchain technologies for the governance and the increased security of federated private clouds among public government agencies. He has co-authored more than 40 research papers published in top conferences such as ICDCS, TrustCom and HPCS and top journal such as Elsevier Journal of Pervasive and Mobile Computing. His current research interests include blockchain technology, identity management, trust management and security and privacy issues in cloud computing and social networks.

Kamanashis Biswas received PhD in ICT from Griffith University, Australia and Masters in Computer Science (Specialization in Security Engineering) from Blekinge Institute of Technology (BTH), Sweden. Prior to his PhD, he worked as a faculty member in Department of Computer Science and Engineering at Daffodil International University, Bangladesh for about four and half years. He is currently working as a Lecturer in Information Technology at Peter Faber Business School, Australian Catholic University. His research interests include blockchain technology, design and development of lightweight cryptographic schemes, energy efficient secure routing algorithms, intrusion detection systems (IDS) and clustering schemes in wireless sensor networks. He has published more than 20 research papers in various conferences, symposiums and journals including IEEE and Springer. He is a member of IEEE and reviewer for many Journals and International conferences.

Mohammad Jabed Morshed Chowdhury is currently a PhD Candidate at Swinburne University of Technology, Melbourne, Australia. He has earned his double Masters in Information Security and Mobile Computing from Norwegian University of Science and Technology, Norway and University of Tartu, Estonia under the European Union's Erasmus Mundus Scholarship Program. He has published his research in top venues including TrustComm, HICSS, and REFSQ. He is currently working with Blockchain in his PhD research and published research work related to Blockchain. Prior to PhD research, he has worked as the senior lecturer at Daffodil International University, Bangladesh.

Niaz Chowdhury is a postdoctoral Research Associate at the Knowledge Media Institute (KMI), The Open University in the United Kingdom. His primary area of research includes the Internet of Things (IoT), Blockchain, Machine Learning, Data Science and Privacy. He has diverse, yet well-connected research experiences gathered from three nations in the British/Irish Isles: Ireland, Scotland and England. Prior to his current position at KMI, he completed another postdoc in the Department of Computing and Communication in the same university where he worked in the smart city project MK-Smart. Dr Chowdhury obtained his PhD from the School of Computing Science of the University of Glasgow in Scotland, as a recipient of the Scottish ORS Scholarship in conjunction with the Glasgow University College of Science and Engineering Scholarship. He was also a research scholar at the School of Computer Science in Trinity College Dublin where he received Govt. of Ireland IRCSET Embark Initiative Scholarship. His earlier background is in Computer Science and Engineering at East West University, Bangladesh where he pursued a bachelor and a masters degree with the Gold Medal distinction.

Vallipuram Muthukkumarasamy (Muthu) obtained PhD from Cambridge University, England. He is currently attached to School of Information and Communications Technology, Griffith University, Australia as Associate Professor. Muthu has pioneered the Network Security research and teaching at Griffith University and has been leading the Network Security Research Group at the Institute for Integrated and Intelligent Systems (IIIS). His current research areas include blockchain technology, cyber security, wireless sensor networks, trust management, key establishment protocols and medical sensor networks. Muthu has been successful in attracting national and international funding for his research activities. He has a passion for innovation and successful collaboration with NICTA, Gold Coast University Hospital, and IBM Security Research lab. Recent project on application of blockchain technology involves industry and inter-disciplinary elements. Muthu has published over 100 high quality peer-reviewed papers in leading International Journals and Conferences, including IEEE Transactions on Information Forensics and Security, IEEE Sensors, Elsevier Journal of Digital Signal Processing, and International Journal of Information Assurance and Security. Muthu's excellence in teaching has been recognized by a number of awards. His leadership, as the Deputy Head L&T, enabled the School to become the No. 1 in Australia for Student Satisfaction.

CHAPTER THREE

Intersections between IoT and distributed ledger

Hany F. Atlam[a,b], Gary B. Wills[a]

[a]Electronic and Computer Science Department, University of Southampton, Southampton, United Kingdom
[b]Computer Science and Engineering Department, Faculty of Electronic Engineering, Menoufia University, Shebeen El-Kom, Egypt

Contents

Abstract

The Internet of Things (IoT) is growing exponentially. It allows not only humans but also all various devices and objects in the environment to be connected over the Internet to share their data to create new applications and services which result in a more convenient and connected lifestyle. However, the current centralized IoT architecture faces

Advances in Computers, Volume 115
ISSN 0065-2458
https://doi.org/10.1016/bs.adcom.2018.12.001

several issues. For instance, all computing operations of all nodes in the network are carried out using a single server. This creates a single point of failure in which if the server goes down, the entire system will be unavailable. Also, the IoT centralized architecture is an easy target of various types of security and privacy attacks, since all IoT data collected from different devices is under the full authority of a single server. Therefore, adopting one of the Distributed Ledger Technologies (DLTs) for the IoT may be the right decision. One of the popular types of DLTs is the blockchain. It provides an immutable ledger with the capability of maintaining the integrity of transactions by decentralizing the ledger among participating nodes in the blockchain network which eliminates the need for a central authority. Integrating the IoT system with the blockchain technology can provide several benefits which can resolve the issues associated with the IoT centralized architecture. Therefore, this chapter provides a discussion of the intersection between IoT and DLTs. It started by providing an overview of the DLT by highlighting its main components, benefits and challenges. The centralized IoT system is also discussed with highlighting its essential limitations. Then, the integration of blockchain with IoT is presented by highlighting the integration benefits. Various application and challenges of integrating blockchain with IoT are also discussed.

1. Introduction

The Internet of Things (IoT) represents a revolutionary technology that enables almost everything everywhere to be connected over the Internet. The IoT enables various devices and objects around us in the environment to be addressable, recognizable and locatable via cheap sensor devices. These devices can be connected and communicate with each other over the Internet using either wired or wireless communication networks [1]. These devices involve not only normal electronic devices or technological development products like vehicles, phones, etc., but also other objects such as food, animals, clothes, trees, etc. The key purpose of the IoT system is to allow various objects to be connected in anyplace, anytime by anyone preferably using any path/network and any service [2].

Although the IoT system provides countless benefits in various domains, it faces several issues with the current centralized model in which all IoT devices and objects are identified, authenticated and controlled by a centralized server. This model faces many obstacles. For instance, it carries out all processing operations and controls all nodes in the network, which creates a single point of failure in which if the server goes down, the entire system will be unavailable [3]. Also, security is another issue for the centralized model since all sensitive information stored in one location and under the reasonability of a single server which makes it an easy target for various types of attacks. Moreover, protecting the data privacy seems to be questionable, since the real-time data of IoT devices are collected and stored in a remote server outside the user control and with the authority of the centralized server only. In addition, the centralized architecture faces a scalability issue as it fits only for small businesses, but it will be an impractical solution for large organizations having many branches in a different location all over the world [4].

On the other hand, Distributed Ledger Technology (DLT) has gained a great attention in recent years as an innovative approach that provides a transparent and verifiable record of transactions. DLT combines a group of untrusted nodes in a distributed and decentralized environment. It has a massive potential to change how governments, organizations and institutions work. It can bring myriad advantages to various government activities such as tax collection, benefits associated with social security, passport issuance, licenses and voting. It can also provide several advantages to other applications such as music, finance, cyber security, public services, healthcare, etc. DLT provides an immutable ledger that cannot be changed or altered and eliminates the need for a centralized trusted third party [5].

With several limitations in the centralized IoT architecture, moving the IoT system into one of the distributed ledger technologies may be the right decision. One of the popular types of DLTs is the blockchain. It is defined as a distributed and decentralized ledger of transactions to manage a continuously growing group of records. To store a transaction in the ledger, the majority of participating nodes in the blockchain network should agree and record their consent. A set of transactions are grouped together and allocate a block in the ledger, which is chained of blocks [3]. To link the blocks together, each block encompasses a timestamp and hash function to the previous block. The hash function validates the integrity and nonrepudiation of the data inside the block. Moreover, to keep all participating nodes of the blockchain network updated, each user holds a copy of the original ledger and all nodes are synchronized and updated with newly change.

Integrating the IoT with blockchain will have many advantages. For instance, adopting the decentralized architecture for the IoT system can solve many issues especially security and single point of failure, since the blockchain provides a decentralized and distributed environment where there is no need for a central authority to manage the execution of operations and control communication between various nodes in the network. This, in turn, provides a trusted environment where participating nodes are the only entities to accept or discard a transaction based on their consent [6].

Moreover, the blockchain provides better security for various IoT applications since it provides an immutable and tamper-proof ledger to protect data against malicious attacks in which any data update or modification will not be added to ledger unless the majority of participating nodes verify it [7].

This chapter provides a discussion of the intersection between IoT and DLT. It started by presenting the main differences between centralized, decentralized and distributed systems. This followed by providing an overview of the DLT by highlighting its main components, advantages and challenges. The centralized IoT system by examining its essential limitations is also presented. Then, the integration of blockchain with IoT is presented by examining the benefits of the integration process. Various application and challenges of integrating blockchain with IoT are also discussed.

The remaining of this chapter is structured as follows; Section 2 presents the main differences between centralized, decentralized and distributed systems; Section 3 provides an overview of the DLT including its components, advantages and challenges; Section 4 discusses the centralized IoT system and its limitations; Section 5 discusses the intersection of blockchain with IoT with highlighting benefits, new applications and challenges resulted after integrating blockchain with IoT; Section 6 is the conclusion.

2. Centralized, decentralized and distributed systems

This section provides an overview of centralized, decentralized and distributed systems by presenting the main differences in the network structure, advantages and disadvantages of each approach.

2.1 Centralized system

The centralized system is built depending on a central server to manage a set of nodes. The nodes are simply nodes which can perform operations from neighboring nodes across the network. The central server handles all requests coming from various nodes and assigns tasks to various nodes in the network. Typically, the communication between various nodes and the central server is like Transmission Control Protocol (TCP) connection in which when a connection has been created, messages are sent between the central server and the connected node. These messages can be such as registering the node and getting node address [8].

The centralized system has several advantages. For example, the entire responsibility of the network is placed under the full control of the central server so, it is easier to manage, maintain and control. In addition, the use of a central server saves the costs of having multiple hardware equipment. In other words, building a centralized system only need a central computer with the required hardware and software elements while other nodes could be just terminals. These terminals served only as an input/output method to connect various users to the central server. Also, if one terminal goes down, the user can simply use another terminal to access the stored files on the central server. Moreover, the centralized system provides better physical security since all data are stored in a central place, which makes it easier to protect against physical damage and reduces duplication and ensures data integrity since data are controlled by the central server which provides a uniform service for all nodes in the network.

On the other hand, the centralized system has multiple shortcomings. For example, since the central server carries out all processing operations and controls all nodes connected to it, if the server crash, the entire system will be unavailable. This problem is called a single point of failure so if the single point, central server, is failed, all the system will go down. Moreover, since all computing achieved across the central server, the hardware specification should be good enough to serve all nodes in the network, if not, the nodes may wait for a long time to get their job done. Also, the central system is unfeasible in large organizations which have many branches in different sites all over the world [9].

2.2 Decentralized system

A decentralized system is designed based on peer-to-peer communications between different nodes in the network without the need for a centralized server to manage operation execution and make decisions on behalf of other nodes. So, each node makes their own independent decision based on its targets which may collide with other nodes targets. Nodes can connect and communicate with each other to share information and provide various services to other nodes [10].

There are two basic structures for a decentralized system, pure decentralized structure and organizational structure. In the pure structure, all nodes are responsible for taking their own decisions without having any authority to manage communication and operations between communicating nodes. While the organizational structure has a supernode for a small network to manage communication and take decisions on behalf of this subnetwork. These supernodes are connected with similar nodes in the network to share information and make decisions [11].

The decentralized system has multiple advantages. For instance, it reduces the possibility of controlling and operating the entire system with only one central server, instead, it allows each node to take part in decision-making operations. In addition, the decentralized system carries decision-making operations closer to the scene of action which results in quick decisions that can save a lot of money. Also, the decentralized system is scalable and facilitate the expansion of organizations which results in opening a new business in different geographic locations. Moreover, in contrast to the centralized model, the failing of one node will not affect the entire system, so there is no single point of failure.

The decentralized system is not straightforward. It suffers some difficulties and challenges. For example, it increases costs as each node needs enough processing power and memory to execute operations and make decisions independently. Also, it increases the coordination problem between various subnetworks [3].

2.3 Distributed system

A distributed system is a set of autonomous nodes interconnected together to form a single and integrated coherent network with a huge processing and storage power to achieve a certain goal. The peers are communicated by passing messages to one another. The combination of storage and processioning capabilities allows the distributed system to perform complex and large tasks faster than other systems by dividing tasks into multiple subtasks and distributed it among network nodes which process and execute their subtasks and return the result to the main node to collect and constitute the final result [12].

The distributed system provides countless benefits. For instance, it increases the overall performance of the system by distributing the computational load across different nodes which provides less load at each node, which in turn increases the performance of the entire system. Also, it provides a reliable network of nodes where if a node goes down, the entire system will not be affected which eliminates the problem of single point of failure associated with the centralized system. In addition, since distributed systems operate with a diversity of different nodes, it provides a scalable system which can adjust their processing resources in light of assigned tasks and operations.

On the other hand, the distributed system has some shortcomings. It has severe security and privacy issues since the system is based on sharing tasks

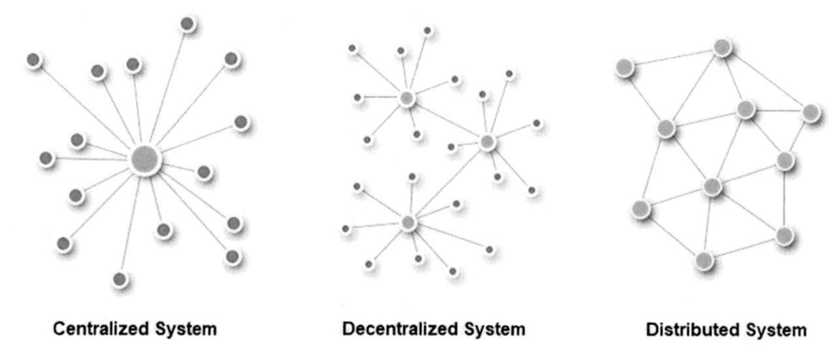

Fig. 1 Structure of centralized, decentralized and distributed systems.

Table 1 Comparison between centralized, decentralized, distributed systems.

Feature	Centralized system	Decentralized system	Distributed system
Scalability	Low	Moderate	Infinite
Security	High	Moderate	Low
Maintenance	Easy to maintain	Moderate	Difficult to maintain
Stability	Highly unstable	Possible recovery	Very stable
Complexity	Less complex	Moderate	Complex
Point of failure	Single point of failure	Finite number of failures	Infinite number of failures

and data between multiple nodes. If one of these nodes is malicious, it can cause serious issues. Also, the system communicates using messages which can be lost easily. Also, large data need a high bandwidth for data transmission which results in more costs to change existing network connections [13].

Fig. 1 shows the structure of centralized, decentralized and distributed systems. Also, Table 1 provides a comparison between centralized, decentralized, distributed systems in terms of scalability, security, maintenance, stability, complexity and point of failure.

3. Distributed ledger technology

Distributed ledgers are a multipurpose technology that is built to share data among different nodes in different locations all over the world. This technology provides several benefits to various applications. This section provides an overview of the DLT by discussing different definitions, main elements and advantages of DLT. Then, presenting challenges of adopting this technology, distinguishing differences between blockchain and DLT and lastly discussing main technologies of DLT.

3.1 Definitions of DLT

DLT provides a universal data structure by combining a group of untrusted nodes in a distributed and environment. It provides an immutable ledger that cannot be changed or altered and eliminates the need for a centralized trusted third party. So, a centralized server is not required to manage operations and ensures trust between communicating parties instead, a distributed ledger is responsible for maintaining the trust by tracking the ownership of different nodes in the network. DLT has a huge potential to change how governments, organizations and institutions work. It can bring countless benefits to government activities such as tax collection, benefits associated with social security, passport issuance, licenses and voting. It can also provide several advantages to other applications such as music, finance, cyber security, public services, healthcare, etc. [14].

A distributed ledger is a form of database shared across multiple locations including organizations and countries. This ledger is shared and synchronized between different nodes of the network. It keeps all transactions of the participating nodes in the network [15]. There are other definitions for the DLT. For example, Investopedia [16] described the DLT as "*The technological infrastructure and protocols that allows simultaneous access, validation and record updating in an immutable manner across a network spread across multiple entities or locations.*" Also, Financial Conduct Authority [17] defined the DLT as "*A set of technological solutions that enables a single, sequenced, standardized and cryptographically-secured record of activity to be safely distributed to, and acted upon by, a network of varied participants.*" Also, Bank for International Settlements (BIS) [18] defined the DLT as "*the processes and related technologies that enable nodes in a network (or arrangement) to securely propose, validate and record state changes (or updates) to a synchronized ledger that is distributed across the network's nodes.*"

3.2 Components of DLT

Several governments started to adopt DLTs to provide various types of public services. This technology allows nodes to update their records in a shared database without the need for a central authority to validate the operation or even enforce their own standards. Also, this technology eliminates the issue of single point of failure through the decentralization feature which provides an extensive increase in security regarding storing transactions and ensuring their integrity [19].

There are four components to implement a DLT, these components, as summarized in Fig. 2, include:

- Shared ledger: It is a shared database for storing all transactions that belong to nodes participating in the network. Since the ledger can be deployed at different locations, it has to be updated and synchronized with other copies of the ledger in the network in a very short time without noticeable latency.
- Cryptography: Transactions between two communicating nodes are recorded, maintained and secured cryptographically. Every node participating in the network can create a transaction in a secure way without the need for a central authority. Cryptography plays a vital part in the DLT through authenticating approved nodes, validating records and facilitating consensus on the ledger update. In other words, there is a cryptographic digital signature for each participating node to authenticate himself before adding or changing a transaction [18].
- Consensus Mechanism: This is the process used by all participating nodes in the network to validate the contents of the ledger.

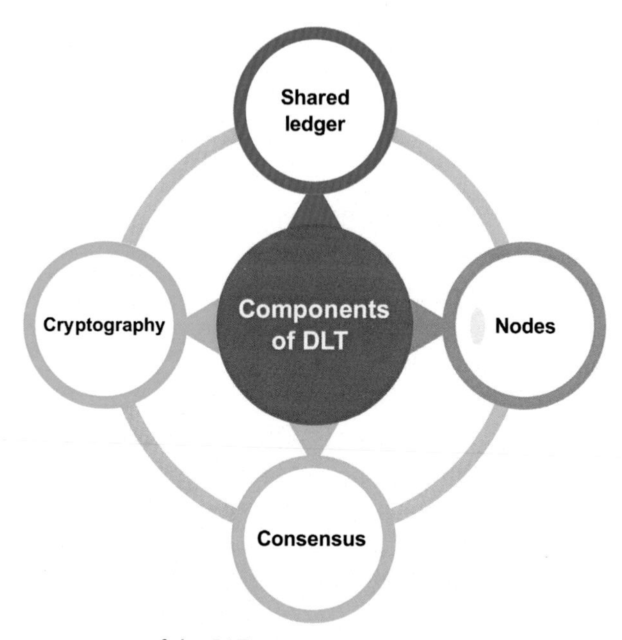

Fig. 2 Main components of the DLT.

Consensus generally includes two phases: validation and agreement on ledger update. There are multiple consensus mechanisms. However, the most common mechanisms are Proof of Stake (PoS) and Proof of Work (PoW). The key difference between various consensus mechanisms is the way they delegate and reward the verification of a transaction [20].

- Nodes: They represent the participating users in the network. Nodes have different roles in the network including system administrator, asset issuer, proposer, validator and auditor. The system administrator role is used to control the access to the system and provides certain management services. While asset issuer role is permissioned to issue new assets. Proposer role is used to propose updates to the ledger, whereas validator role confirms the validity of a proposed change in the ledger. The lowest role is auditor which allows the user to only view the ledger without the ability to make changes or updates.

3.3 Advantages of DLT

DLT enables the participating nodes to store transactions in a shared database that can be accessed in a secure manner. These transactions are distributed between nodes in the network to access and use it without relying on a trusted central system. DLT as a new technology can add enormous advantages to different applications over the centralized ledger and other kinds of shared ledgers. This section provides potential benefits of the DLT, which are as follows:

- Availability: DLT provides a high level of availability as it can run on a continuous basis theoretically. The distributed and shared nature of the system facilitates the recovery of both data and processes in the case of an attack which can reduce the need for expensive recovery plans. However, the availability of the DLT remains untested, especially when large volumes are involved [14].
- Automation and Programmability: DTL supports automation in programming, so when a certain condition is verified, the programming actions are executed automatically. This feature relies on smart contracts which build digital contracts as a software code by implementing contracts terms as programming conditions and actions. Actions are automatically executed as soon as conditions were verified. Although smart contracts can be built on the centralized system, the actions cannot be executed only if they approved by the central system, which can take a long time [21].

- Immutability and verifiability: One of the critical advantages of the DLT is the ability to guarantee the integrity of transactions by creating immutable and verifiable ledgers. In the traditional centralized architecture, a trusted third party needs to exist for ensuring information integrity. Whereas in the DLT, the data cannot be changed until the majority of participating nodes in the network approve it [22].
- Decentralization: DTL is a decentralized technology in nature. It has a shared ledger giving all participating nodes the ability to hold an original copy of the ledger without the need to be controlled or managed through a single central authority. This gives the opportunity for all nodes to participate and transact equally. This also converted to a lower cost, better scalability and faster time for creating and validating a transaction.
- Transparency: DLT offers a high level of transparency by sharing transaction details between all participants nodes involved in those transactions. Also, there is no need for a central authority which improves business friendliness and guarantees a trusted workflow.
- Efficiency: DLT reduces the efforts needed to do reconciliation and handle disputes manually. The existing systems with separate ledgers can lead to inconsistent master and transaction data resulting in faulty and duplicated data. Also, identifying and correcting these data will take a significant loss of time. This not only slows down the process but also forms a source of contract uncertainty. While the DLT is expected to bring significant efficiencies to this process through distributed and immutability features [23].
- Security and Resilience: DLT delivers a good level of security since it uses a public key infrastructure that protects against malicious actions. The participating nodes of the network place their trust in the integrity and security features of the consensus mechanism. In addition, the use of distributed and decentralized ledger eliminates the single point of failure which affects the system resilience [24].
- Cost Reduction: DTL is based on a shared ledger which shares its contents with the participating nodes in the network in which each participating user holds a copy of the original ledger without the need for a central authority. This reduces costs associated with distributing and maintaining the ledger. In other words, the use of a distributed ledger eliminates maintenance costs of individual ledgers and reduces the need for costly business continuity plans [25]. Also, it reduces costs spent to ensure data integrity as the DLT is an immutable system in nature. According to Natarajan et al. [21], DLT could save about $15–20 billion per year for the financial industry only.

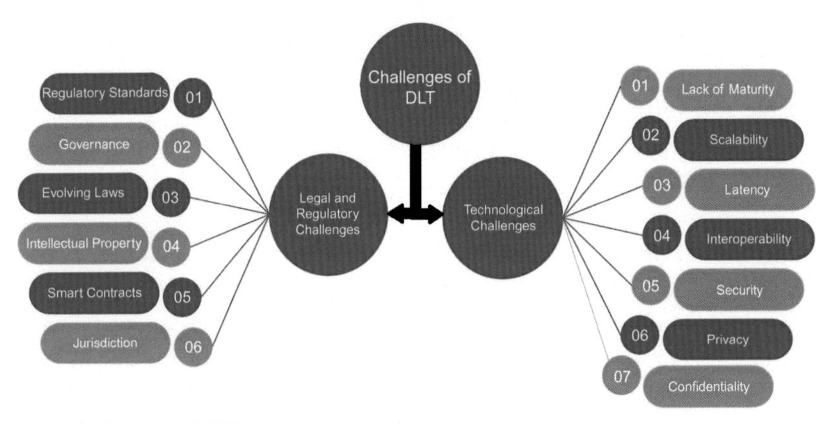

Fig. 3 Challenges of DLT.

3.4 Challenges of DLT

DLT is still in the first stage of approval, and there are many technological and legal issues that need to be addressed. This section provides a discussion of the most common challenges of DLT, as summarized in Fig. 3.

3.4.1 Technological challenges

The first type of DLT challenges is the technological issues which include lack of maturity, scalability, latency, interoperability, security and governance.

- Lack of Maturity: DLT is still in the early stage of evolution. There are serious concerns about the robustness and resilience of the DLT especially for large amount of transactions, availability of standardized hardware and software applications besides availability of skilled professionals. There is a lack of understanding among businesses, consumers and authorities about how the technology operates the potential use cases for DLT and the likely short- and medium-term market development potential. Although huge organizations like Microsoft and IBM started creating DLT products and services which can provide the necessary trust and confidence in the DLT and its huge expansion in various applications, there is still a big gap of research that needs to be handled to get the full benefits of such a revolutionary technology [26].
- Scalability: Current versions of distributed ledgers face concerns regarding the scalability of the DLT in terms of transaction volume and speed of verifications. Existing ledgers have limited transaction speed and block size. Although these issues can be resolved over time, the main issue that needs to be resolved is the capability of the system to handle the global

scale. The failure to handle this issue could result in expenses of a more centralized and less transparent platform, which can eliminate many of DLT benefits [21].

- Latency: With increasing number of transactions, the ledger is growing rapidly in size, which leads to slow transaction time. Also, the increase of participating nodes in the network adds more latency as internode latency logarithmically increases as each new user is added. Therefore, the DLT is incompetent to scale to process more transactions [27].
- Interoperability: There are different DLT systems that need to be interoperable with other ledgers. Also, the integration of the DLT with existing infrastructures will require not only large expenses but also extensive coordination and collaboration. Providing a solution that enables different systems to work efficiently with each other should be the main target to address the interoperability issue associated with the DLT [28].
- Security: The DLT eliminates the need for a centralized server by allowing all participating nodes in the network to hold an original copy of the ledger. However, this presents issues especially in the case of cyber attacks. Distributing access and management rights across multiple nodes may introduce a security threat. Also, the encryption level and network security may differ broadly, so if one user is breached, the entire network will be in danger. Hence, network security has to be very strong [29].
- Privacy: The DLT is built based on sharing information about various transactions between all participating nodes in the network. If the information in a transaction involves private or sensitive information such as medical data or an account number, it will be visible for all participating nodes in the network. Moreover, since nodes in the ledgers can be from different geographic locations, the transfer of data between these different locations will heavily depend on rules and requirement of each location, which create a legal issue that needs to be resolved.
- Confidentiality: It is a similar issue like privacy where information shared among participating nodes in the ledger become public and everyone can view this information. This applies to both public and private ledgers. So, if a company uses the distributed ledger to keep their confidential information, it will create a risk of a confidentiality attack or loss of trade secret protection. Therefore, there is a need to discover novel approaches to prevent confidential and sensitive information from being kept in the distributed ledger to protect it from any future confidentiality breach [30].

3.4.2 Legal and regulatory challenges

The second type of DLT challenges is the legal and regulatory issues which include regulatory standards, governance, evolving laws, intellectual property, smart contracts and jurisdiction.

- Regulatory Standards: Regulatory standards for several applications are necessary but are still in the early progress stages. A legal standard for the DLT is required to ensure the authenticity of data stored in the distributed ledger. Also, a standardized regulation is obligatory for data protection and authenticating the identity of legal nodes within the network. Although many regulators across the world are actively searching the technology, more regulatory standards for the DLT are yet to appear [19].
- Governance: In the DLT environment where no central entity is involved, there are several issues about ensuring active governance of the overall infrastructure. In the centralized infrastructure, regulators have used effective governance arrangements. But for the DLT whether permissionless or permissioned, it is unclear whom will have an issue and how to apply governance arrangements. The existence of administrator in permissioned DLT can be subject to specific governance arrangements but depending on the nature of the DLT.
- Evolving Laws: It is obvious that laws pause technology innovation, this is definitely the case with the DLT. Regulations regarding information sharing need to be changed to protect companies as well as their investors and their customers. The DLT provides an auditable and transparent environment for several applications and enables new products and services to grow significantly, but there is a lack of laws and regulations for such a technology.
- Intellectual Property: With the appearance of a new technology such as DLT, there are a 1000 patent applications by companies that utilize the benefits of this new technology. Although the core technology is open source, companies have built patented applications in which they require to protect their intellectual property rights, so there will be multiple patent infringement lawsuits as patent holders seek to enforce their exclusive rights to their patents. Therefore, adopters of the DLT need to be very careful as their application or implementation could potentially be violating an existing patent [31].
- Smart Contracts: There are several legal and law issues that appear with the emergent of new technologies, for example, a smart contract, which is based on a software code to implement the terms and conditions of a

contract. Smart contracts may contest the nature of traditional legal principles of contract law such as contract formation and termination. This will add more difficulty for courts to work with the new technology. Also, as smart contracts are software codes, their use may introduce enforceability questions if trying to investigate them within the traditional contract definition. Moreover, as smart contracts are built with the decentralized feature, resolving any future disputes arise over a contract will be another issue in the absence of a central authority [32].

- Jurisdiction: According to the Cambridge dictionary [33], Jurisdiction is defined as the authority of a court or official organization to make decisions and judgments. As DLT ledgers can connect different participating nodes from multiple jurisdictions around the world, it creates challenges from a jurisdictional perspective. The principles of a contract are different across jurisdictions, so defining the suitable governing law to resolve any future dispute will be very difficult.

3.5 Distributed ledger vs blockchain

Many people are confused about how to distinguish between DLT and blockchain. People often believe that they are the same thing, which is not the case. The reason behind this confusion is the popularity of the blockchain without mentioning the parent technology, DLT. The comparison between DLT and blockchain is simply like comparing an apple to a fruit. In simple terms, blockchain is one of the technologies that depends on the distributed ledger and other related features. There is an extensive variety of distributed ledger models, with different degrees of centralization and different types of access control for various business requirements.

There are a set of differences between distributed ledgers and blockchain. For instance, blockchain is a sequence of blocks chained together using hash functions and timestamp, while distributed ledgers do not need such a chain. Also, distributed ledgers do not require consensus mechanism and deliver better scalability. In summary, all blockchains are distributed ledgers, but not all distributed ledgers are blockchains.

3.6 Distributed ledger technologies

DLT refers to an innovative mechanism of storing and sharing data between multiple data storages. This technology enables transactions and data to be stored, shared, and synchronized across a distributed network of nodes. Each node acts as both a client and a server at the same time, holding the same

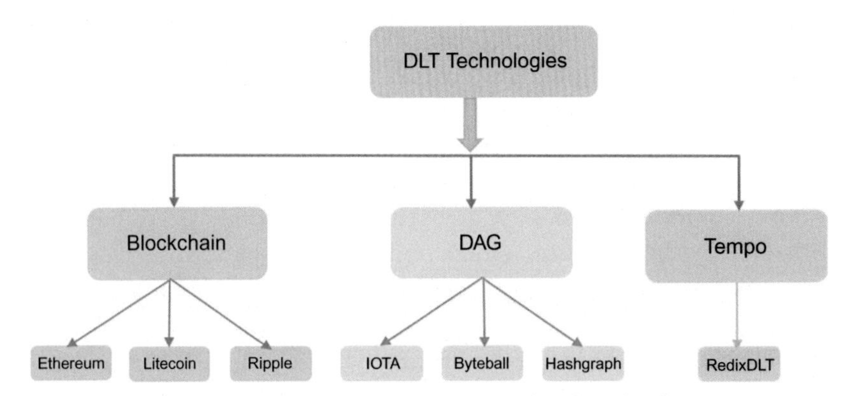

Fig. 4 DLT technologies and common applications of each technology.

copy of the distributed ledger. This technology enhances efficiency and eliminates reconciliation costs [21].

DLTs can be classified according to their applied data structure. Among various structures of the DLT, there are three main structures that have been widely adopted: blockchain, Directed Acyclic Graph (DAG) and Tempo, as shown in Fig. 4 [34].

3.6.1 Blockchain

The first and most popular type of the DLT is blockchain. Blockchain is a distributed and decentralized ledger of transactions used to manage a constantly increasing set of records. To store a transaction in the ledger, the majority of participating nodes in the blockchain network should agree and record their consent. A set of transactions are grouped together and allocate a block in the ledger, which is chained of blocks. To link the blocks together, each block encompasses a timestamp and hash function to the previous block. The hash function validates the integrity and nonrepudiation of the data inside the block. Moreover, to keep all participating nodes of the blockchain network updated, each user holds a copy of the original ledger and all nodes are synchronized and updated with newly change [3].

Blockchain delivers a high level of transparency by sharing transaction details between all participants nodes involved in those transactions. In a blockchain environment, no need for a third party which improve business friendliness, guarantees a trusted workflow and blockchain eliminates the single point of failure which affects the entire system. Moreover, blockchain provides better security since it uses public key infrastructure that protects against malicious actions. The participating nodes of the blockchain network place their trust in the integrity and security features of the consensus mechanism [24].

3.6.2 Directed acyclic graph (DAG)

The second type of the DLT is DAG. It is a directed graph data structure that uses a topological ordering. The sequence can only go from earlier to later. DAG is often applied to problems related to data processing, scheduling, finding the best route in navigation and data compression [14].

DAG is simply involving multiple nodes connected to each other with edges. An edge is a connection between nodes with a specific direction. DAG is a noncircular structure, so it is not possible to face the exact node twice when moving from node to node by edges.

Blockchain adds blocks sequentially to constitute what is called chain of blocks, while DAG uses blocks' acyclic graph which parallelizes the validation process which results in higher throughput. In addition, DAG transactions are connected from one to another in which each transaction confirms the next one, while blockchain requires proof of work from minors for each transaction [35]. Fig. 5 shows the structure of blockchain and DAG.

DAG has several advantages over blockchain. First, DAG is a zero-transaction fee since there is no need for minors to compete to create or approve a transaction. Second, DAG provides a higher level of scalability, as it becomes faster and more secure with the growth of network boundaries. Third and lastly, DAG is a partition tolerant, which allows a portion of the network to split off the main network for a period of time and continue to run without the Internet connectivity. These portions can be reconnected to the main network when the Internet connection was established.

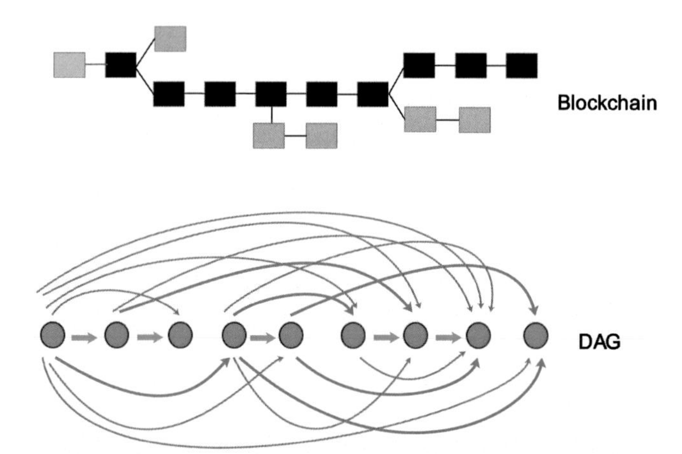

Fig. 5 Blockchain consists of blocks chained together using hash function while DAG transactions are connected from one to another in which each transaction confirms the next transaction.

This feature is very beneficial for areas around the world that have limited Internet connectivity or an unreliable electricity supply [36].

Although DAG provides countless benefits, it is not straightforward and associated with some flows. For example, DAG requires loads of traffic to operate efficiently which will be a big problem for small business networks. Also, the DAG network becomes more vulnerable to attacks with decreasing network traffic. Moreover, although one of the main benefits of DAG over blockchain is scalability, it does not be approved at a large scale since it's a quite new technology [14].

There are several applications that use DAG such as IOTA, Byteball, and Raiblocks, Hashgraph and Hylochain. IOTA is the most popular application among DAG applications. It is a cryptocurrency built for IoT. It is so-called as it is built to facilitate transactions of IoT. IOTA solves challenges associated with blockchain regarding transaction fees and scalability. It gets rid of blocks and chaining process of blocks of the blockchain in which a user can send a transaction to the IOTA ledger only if the user verifies the previous two transactions [37].

3.6.3 Tempo

Tempo ledger is an essential part of Redix [38], which is a DLT platform that works efficiently with the IoT. Tempo uses partitions of the ledger to accomplish the appropriate ordering of actions that occur in the whole network. The Tempo ledger comprises of three main elements; a networked cluster of nodes, a global ledger database which is distributed across the nodes and an algorithm for generating a cryptographically secure record of temporally ordered events [39].

The Tempo ledger involves events which are represented by objects, which are called atoms. It is a distributed database that keeps all atoms in the network. It is built to be horizontally scalable, supports semistructured data, and can update entries.

Table 2 provides a comparison between three main technologies of the DLT in terms of data structure, verification time, applications and projects for each technology, scalability, transaction fees, mining process, energy consumption and popularity.

4. Centralized IoT system

The IoT system allows almost all devices and objects in the environment to be connected and communicates with each other using either wired or wireless technologies. Since each object generates data about

Table 2 Comparison between blockchain, DAG and tempo.

Item	Blockchain	DAG	Tempo
Data structure	Distributed ledger with blocks chained together using the Hash function	Directed graph data structure that uses a topological ordering	Tempol Ledger consisting of a distributed database and consensus algorithm
Verification time	Several minutes	Minutes	<5 Seconds
Applications	Bitcoin, Litecoin, Ripple, Ethereum, etc.	IOTA, Byteball, and Raiblocks, Hashgraph and Hylochain	RedixDLT
Scalability	Less scalable	Scalable but untusted	Scalable but untusted
Transaction fees	High	Low	Low
Mining process	Mining is required	No mining is required	No mining is required
Energy consumption	High energy consumption for mining process	Less energy consumption since no mining is involved	Less energy consumption since no mining is involved
Popularity	Very well known, first launched 2008	Not well known yet, launched 2017	Not well known yet, launched 2018

their surroundings, these data can be integrated with other data from other devices to provide a meaningful information for various services and applications. Although the IoT notion is simple, it provides countless benefits that create new applications and services to facilitate our way of living [40].

This section provides a discussion of the centralized IoT system and its structure by highlighting limitations of the centralized model in the context of IoT.

4.1 Structure of centralized IoT

Currently, the majority of IoT solutions are based on the centralized client-server approach in which all IoT devices and objects are connected and authenticated through cloud servers. Every connection between different

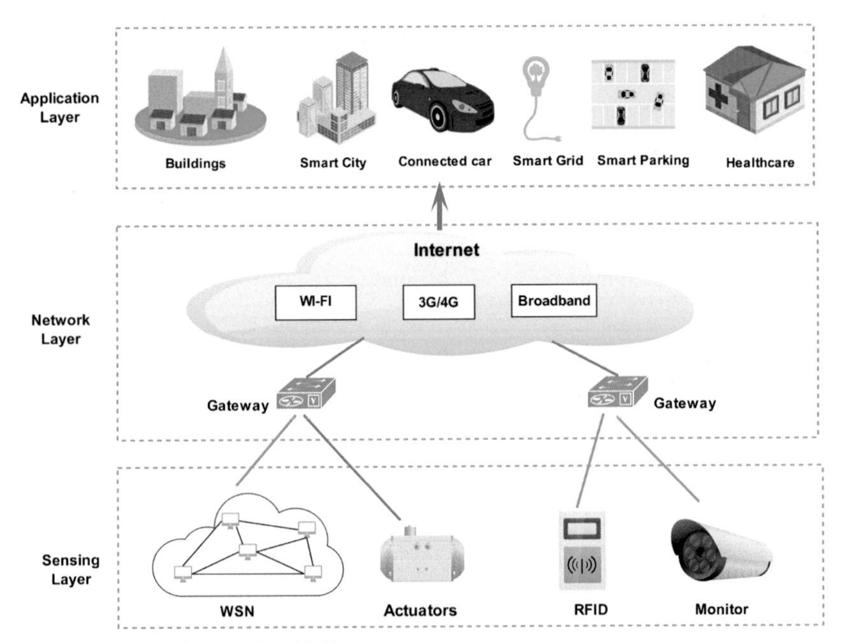

Fig. 6 Structure of centralized IoT system.

devices must be achieved through centralized servers. According to Fernández-Caramés and Fraga-Lamas [4] and Lu et al. [41], the centralized IoT architecture consists of three main layers: sensing, network and application layer, as shown in Fig. 6.

The basic layer of the IoT architecture is the sensing layers which involves different types of sensors, RFIDs, actuators and Wireless Sensors Networks (WSNs). This layer collects all relevant data about the surrounding environment to provide a meaningful information about our physical world. All devices and objects in the sensing layer are not connected to each other directly, but with a centralized gateway instead [42]. The network layer is used to connect all IoT objects and devices to the Internet. It contains gateways that act as interlayer communication points between the sensing layer and network layer. Several communication technologies and protocols are used in the layer, such as 3G/4G, ZigBee, Wi-Fi, Bluetooth and Broadband to transport data between the sensing layer and the application layer. The application layer involves various IoT applications that can benefit from data collected by sensors such as smart city, healthcare, connected cars, smart parking, smart grid and other.

4.2 Limitations of centralized IoT model

The centralized architecture of the IoT system provides a good start for connecting a wide range of various objects and devices all over the world under the responsibility of a centralized server which manages and control all communication between devices and provides the required identification and authentication for different devices and objects. However, it is unable to support large-scale IoT networks which need to be extended in the near future especially with the huge increase of adopting IoT solutions [43].

The number of IoT devices is increasing every day and there are several indications that this increase will continue especially with integrating IoT system with new technologies that can deliver more improvement for IoT services and applications. Cisco has reported that the number of IoT devices is about to reach 50 billion in 2020 [44]. Therefore, with these expectations, the constraints associated with the centralized architecture of the IoT system need to be addressed to continue the adoption of IoT solutions in the future.

There are several limitations associated with the IoT centralized architecture. These limitations include:

- Scalability: It is a major issue for the centralized system since it based on managing and controlling all processes using a central authority. This system structure can scale well but only for small networks. Deploying a centralized system for large business organizations with many branches in different locations will be impractical. It will be hard to transport decisions to different locations based on the management hierarchy. In the IoT context, since there is a massive increase in the number of IoT devices, there are many doubts about the capability of the centralized architecture of the IoT to scale and operate efficiently with the increasing demands [45].
- Cost: Since all computing operations are executed through the central server, the hardware and software capabilities should be good enough to serve all nodes in the network. There is a huge amount of communications between nodes and the centralized server which need to be handled which require high processing power to serve multiple nodes at the same time. Also, it requires maintaining large data storages that able to store data of different devices in the network. In the IoT context, there are high costs related to the deployment and maintenance of centralized servers which increased with increasing number of IoT devices in the network [4].

- Privacy: The centralized system is vulnerable to data manipulation. Collecting real-time data of different devices and store it in one place with the authority of the centralized server can violate the data privacy. The collected data may contain sensitive information about nodes such as their financial accounts, passwords, etc. Since these data are stored in one place, it can be easily breached. On the other hand, there are several examples of privacy violation by service providers. For example, some service providers sell information about their customers to marketing companies that can use this information to analyze nodes' behavior. Also, if an energy company found that their smart meter data analysis will be the evidence that might result in high costs or lawsuits. They will edit or even delete these data [46]. Therefore, privacy is another issue in the centralized IoT system that needs to be handled.
- Security: Security is a nightmare for any system. It is a major issue in the centralized system since all data stored in one location and all operations are executed through a central server which makes it an easy target for various types of attacks especially to Denial of Service (DoS) and Distributed Denial of Service (DDoS) attacks. The enormous increase of IoT devices in our environment leads to increasing the chances to exploit security vulnerabilities within IoT devices which are poorly secured. Therefore, both IoT devices (source of data) and centralized cloud server (data storage location) are an easy target for security attacks [47].
- Single Point of Failure: Since the centralized server carries out all processing operations and controls all nodes connected to it in the network. This creates a single point of failure in which if the server goes down, the entire system will be unavailable. Avoiding this issue can be done by adding redundant switches, network connections and servers as a backup to provide an alternative path when the original server goes down. However, this solution creates problems with synchronization between the original server and backup as well as it requires high expenses to install a backup server.
- Access and Diversity: Nodes can access the network for different needs. However, centralized systems require their nodes to access the information on the network consistently using identical processes. This kind of networks may not provide the flexibility needed by various nodes with diverse needs. In addition, a centralized system uses a single operating system for the whole network. While this can have advantages for some nodes, it limits diversity within the network and can prevent some nodes from accessing the network. Since the IoT system is dynamic and

heterogeneous in nature with diverse devices and objects, so ensuring access of various heterogeneous devices should be a fundamental priority for the centralized IoT architecture [48].

- Inflexibility: Since the centralized server carries out all processing operations and controls all nodes connected to the network, there are huge workloads coming from different nodes in the network. Although the centralized server schedules the workload to avoid peak-load concerns when people across an organization need to use it simultaneously, the tight schedule and delay associated with this process limit the flexibility of the user while doing their own work.

5. Intersection of blockchain with IoT

This section provides a discussion of integrating IoT with blockchain by highlighting benefits of the integration process. Applications and challenges of integrating IoT with blockchain will also be discussed.

5.1 Integration of blockchain with IoT

The IoT system facilities the development of various applications and services by allowing different devices and objects to share their data over the Internet which enhance people quality of life through digitization of various services. Over the last few years, Cloud computing technologies have provided the IoT system with the required functionality to analyze and process information to convert it into real-time actions and knowledge.

The extraordinary progress of the IoT has unlocked novel opportunities in different domains; however, one of the major concerns that stand as a barrier for the promised distribution of IoT devices is the lack of trust and confidence. This is because the existing IoT architecture relies on a centralized system in which a third-party or service provider manages and controls all data collected from IoT devices without clear boundaries about how the collected data is being used. The centralized server acts as a black box and the network participants do not have a pure vision of where and how their data is being utilized [49].

On the other hand, blockchain provides autonomous, distributed, decentralized and trustless environment. In contrast to the centralized architecture which presents several issues regarding single point of failure and scalability, the blockchain uses a decentralized and distributed ledger to utilize the processing capabilities of all the participating nodes in the blockchain network which provide more efficiency. Also, as there is no need for a third party or a central authority, this improve business friendliness and guarantees a trusted workflow. In addition, blockchain delivers a high level of transparency by sharing transaction details among all participants nodes involved in those transactions.

There are several similarities and differences between IoT and blockchain. Table 3 provides a summary of comparison between IoT and blockchain.

Integrating the IoT with blockchain can bring several benefits to both technologies. For instance, adopting the decentralized architecture for the IoT system can resolve many issues particularly security. The peer-to-peer communication model can be used to process billions of transactions between IoT devices which can critically decrease the costs regarding installing and maintaining large centralized data centers and distribute computation and storage among billions of IoT devices. The decentralization feature will also eliminate the whole network from being unavailable if one node goes down [19]. Fig. 7 shows the centralized architecture of the IoT system and the decentralized IoT system after integrating IoT with blockchain.

Blockchain has the capability to provide an easy infrastructure for two nodes to directly convey a piece of property such as money or data between one another with a secured and reliable time-stamped contractual handshake. By the use of smart contracts, the agreement between communicating

Table 3 Comparison between IoT and blockchain.

Items	Blockchain	IoT
System structure	Decentralized	Centralized
Resources	Resource consuming	Resource restricted
Privacy	Ensures the privacy of the participating nodes	Lack of privacy
Latency	Block mining is time-consuming	Demands low latency
Scalability	Scale poorly with a large network	IoT considered to contains a large number of devices
Bandwidth	High bandwidth consumption	IoT devices have limited bandwidth and resources
Security	Has better security	Security is one of the big challenges of IoT

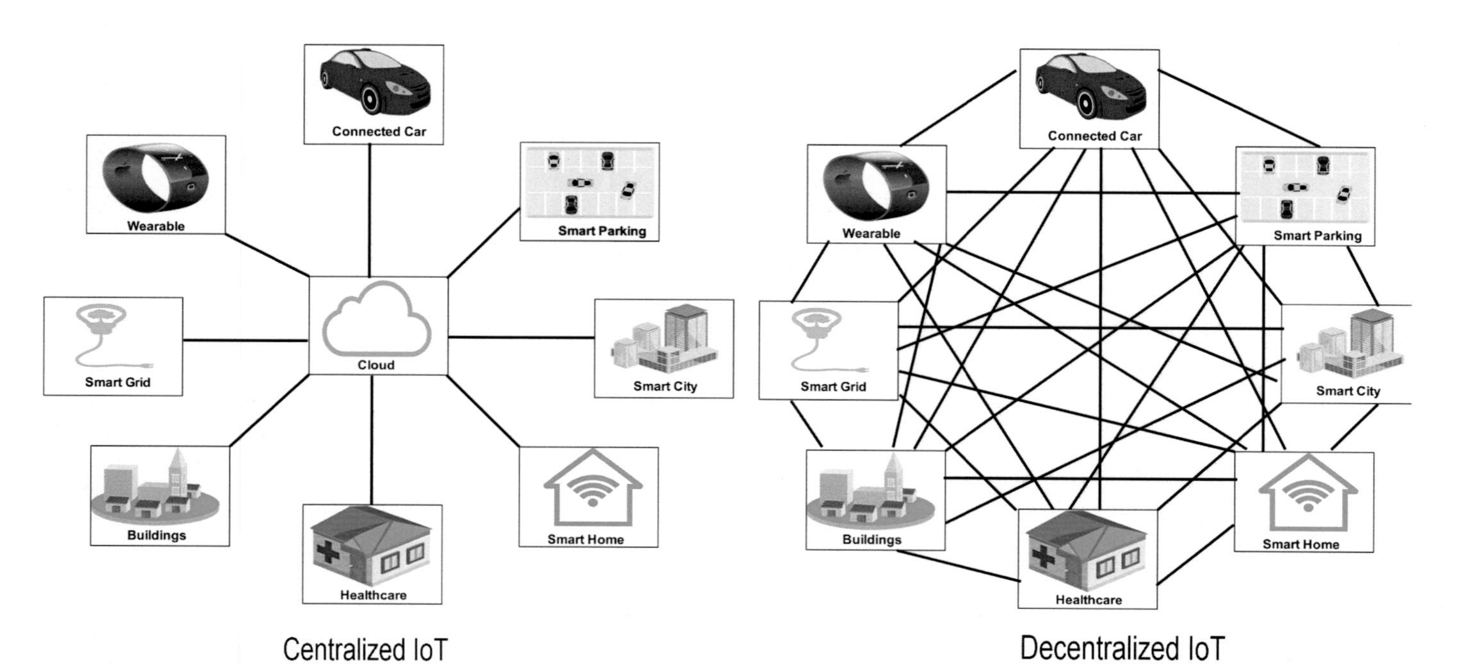

Fig. 7 Centralized IoT where a central authority manages and controls all operations of the communicating nodes and after integrating blockchain with IoT and build decentralized IoT where no central authority or a single point of failure.

Table 4 How blockchain can address the challenges of IoT.

IoT challenge	How blockchain can address the challenge
Security	Blockchain provides an immutable and secure environment for various types of IoT devices. It also ensures data integrity since any change should be verified by the majority of participating nodes in the blockchain network [45]
Point of failure	Blockchain uses decentralized and distributed communication between participating nodes in the network which eliminates the issue of single point of failure.
Third party authority	Blockchain provides a decentralized and distributed environment for the IoT devices so there no need for a centralized server or service provider to build the required trust between communicating nodes in the IoT system
Address space	In contrast to IPv4 with 32-bit and IPV6 with 128-bit address space, blockchain has 160-bit address space which allows blockchain to generate and allocate addresses for about 1.46×10^{48} IoT devices offline [51]
Susceptibility to manipulation	Since blockchain provides a decentralized and immutable environment which allow to detect and prevent any malicious action. The update is only approved after the consent of most participating nodes in the blockchain network
Ownership and identity	Blockchain can provide a trustworthy, authorized identity registration, ownership tracking and monitoring. It has been used in monitoring and tracking products, goods and assets successfully [51]
Data Integrity	Blockchain provides an immutable and tampered-proof ledger that cannot be updated unless the majority of participating nodes provide their consent and verify the update
Authentication and access control	The blockchain smart contracts have the capability of providing decentralized authentication rules and logic that can enable an efficient authentication for IoT devices
Flexibility	With various commercial and open source choices for blockchain, it's possible for IoT organizations to use it to realize several targets without spending a huge amount of money on research and development
Costs and capacity constraints	Since there is no need for a centralized server in the blockchain, IoT devices can communicate securely, exchange data with each other and carry out actions automatically through smart contracts [6]

parties can be formulated and stored as an immutable record of history which enables autonomous functions with no need for a centralized authority [50]. As a result, the blockchain will unlock a sequence of IoT situations that were hard, or even impossible to perform without it.

The Integration of blockchain with IoT provides a good solution for many issues of the IoT system. Table 4 provides a summary of IoT challenges and how integrating blockchain with IoT can solve these issues.

Integrating blockchain with IoT will create a system with more benefits and fewer issues. According to Tapscott and Tapscott [52], the integration of blockchain with IoT will create a system with the following features.

- Responsive, working in different situations and adapt to changing conditions.
- Resilient without a single point of failure.
- Robust with an ability to contain billions of nodes and transactions without affecting the performance of the network.
- Reductive with optimized costs and increased efficiency.
- High availability in real-time and provide a smooth data flow.
- Revenue-generating, providing opportunities to new business models.
- Radically openness, endlessly evolving and capability of updating the network with new inputs.
- Reliable, ensuring data integrity and trustworthiness of nodes.

Adopting blockchain with IoT is not a theoretical assumption. It practically occurs when IBM with a partnership with Samsung has designed a platform called ADEPT (Autonomous Decentralized Peer-To-Peer Telemetry) which utilize the design of bitcoin to construct a distributed network of devices [53]. In addition, there are many other practical projects that investigate the potential of integrating blockchain with IoT. For example, Chain of Things [54] which provides an integrated blockchain and IoT hardware solution to solve IoT challenges regarding identity, security, and interoperability. Also, Slock.it [55] provides the transparency and auditability features to the IoT objects by integrating blockchain with IoT. It resolves the problem of connecting a device to the blockchain and improves the essential features for nonblockchain designers working on IoT systems by providing an interoperable and decentralized platform. Moreover, Waltonchain [56] provides a trustworthy and traceable business network with complete data sharing and absolute information transparency. It is designed by integrating RFID and blockchain technologies.

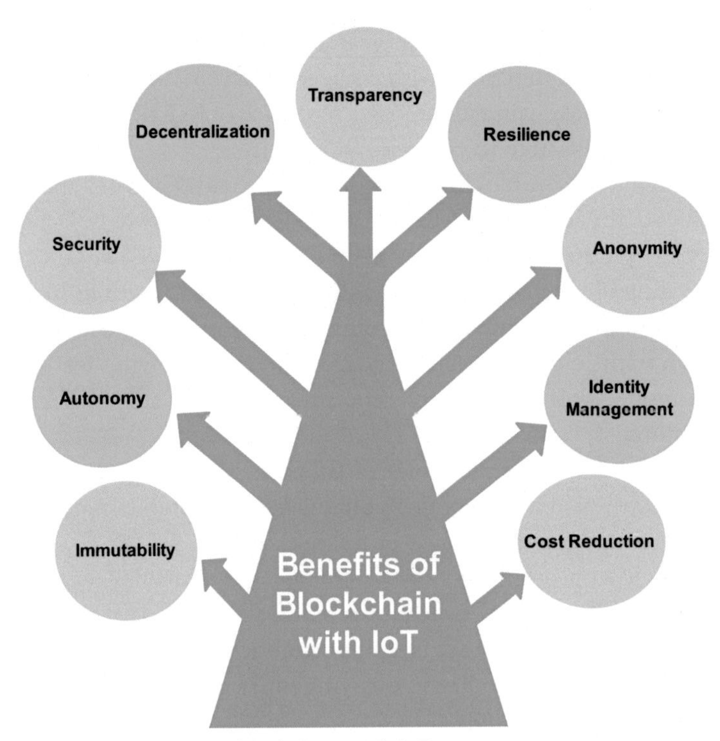

Fig. 8 Benefits of integrating blockchain with IoT.

5.2 Benefits of integrating blockchain with IoT

There are many benefits of adopting blockchain with IoT such as publicity, immutability, decentralization, resiliency, etc. This section provides a summarized discussion of advantages of integrating blockchain with IoT, as indicated in Fig. 8.

- Transparency: There is a high level of transparency in the blockchain network by allowing all the participating nodes to view transactions and blocks in the distributed ledger without any modification since the content of a transaction is protected by the participant's private key. Since the IoT is a dynamic and heterogeneous system in nature, the transparency feature will allow various devices to share their data with other devices in the network and at the same time data integrity is guaranteed.
- Decentralization: Blockchain provides a decentralized and distributed environment where there is no need for a central authority to manage the execution of operations and control communication between

various nodes in the network. There is no centralized authority to verify a transaction or put certain rules to accept a transaction. This, in turn, provides a trusted environment where participating nodes in the network are the only entities that will decide to accept or discard a transaction based on the agreement of the majority of participating nodes [53].

- Security: It is one of the main challenges standing as a barrier for the successful implementation of IoT devices. But with integrating IoT with blockchain, this issue will be significantly reduced, as blockchain provides better security since it uses public key infrastructure that protects against malicious actions. The participating nodes of the blockchain network place their trust in the integrity and security features of the consensus mechanism. In addition, blockchain eliminates the single point of failure which affects the entire system [57].
- Autonomy: The blockchain technology allows the development of new generation of devices with the capability to provide smart autonomous assets and hardware as a service. For example, autonomous cars which need the blockchain to provide a secure environment and data tamper-proof to increase safety and security since cyber-attack is one of the essential issues of adopting autonomous cars. Adding the autonomy feature to the IoT system with the huge development in machine learning techniques will enable the appearance of novel autonomous IoT applications [6].
- Identity Management: The IoT system with a growing number of devices and objects is suffering from identity theft that allows malicious entities to steal legal owner's identities and exploit their sensitive information. Integrating blockchain with IoT can solve this issue since the blockchain technology has the potential to control and manage the user identity and credentials of nodes easily and in a more secure manner. In addition, blockchain can provide trusted distributed authentication and authorization of devices for IoT applications [58].
- Immutability: Blockchain provides a tamper-proof capability in which if a participating user request to add a transaction, the transaction is only added to the block if the majority of participating nodes in the blockchain network verify it. An automatic checking is reliably done for each user to generate a fast and protected ledger that is significantly tamper-proof the transactions and blocks. The traditional security solutions applied on the centralized server in the IoT system do not protect data integrity, but with the immutability feature of the blockchain, data integrity will be guaranteed [59].

- Resilience: Blockchain provides a resilient environment where transactions are stored in different locations. In addition, blockchain allows each participating user in the network to hold an original copy of the distributed ledger that encompasses all transactions which have ever made in the network. Therefore, blockchain is better to handle and tolerate an attack, since if a user is compromised, it will not be affected as blockchain will be maintained by every other user in the network. Holding a copy of data for each IoT data will enhance the information sharing; however, it presents new processing and storage challenges [60].
- Anonymity: Although the blockchain is a public and distributed ledger among various nodes in the network, it can provide an anonymous identity for nodes to keep their privacy. For example, the buyer and seller can use an anonymous and unique address number to process a transaction to keep their identities uncovered. Although this feature has caused a lot of problems for blockchain especially it can provide a cover for criminals and malicious nodes in money laundering and other suspicious activities, it can provide several advantages for many other applications like electoral voting systems [61].
- Cost Reduction: Since blockchain is a decentralized technology, it has the potential to reduce costs associated with installing and maintaining large centralized servers and other networking equipment. Adopting peer-to-peer communication to process billions of transactions between IoT devices will distribute computing and storage requirements across billions of devices that form IoT networks [29].

5.3 Applications of blockchain with IoT

The decentralized, autonomous and trustless attributes associated with the blockchain technology provide several benefits for various IoT applications. This section provides a discussion of common IoT applications that can be strengthened through the blockchain technology.

5.3.1 Smart city

A smart city is built to provide various services to citizens using new information and communication technologies to provide a better lifestyle and enhance the quality of life. This achieved by having a network of sensors and smart objects that sense the surrounding environment and collect relevant data to create new services and applications for citizens.

With newly built intelligent and interconnected services at cities, it becomes a desirable target for attackers. Since the existing centralized architecture stores all data in one location, centralized server, in which if the attacker has breached the server, all data will be under his own malicious behavior. So, the malicious attack on smart cities' services can cause serious issues that can result in real damage to the citizens' lives. Therefore, efficient security measures should be applied to mitigate against these attacks [62].

The integration of blockchain with IoT specifically in the smart city context can provide a solution to this problem. The decentralized feature of the blockchain can be used to split data into multiple chunks and distribute them across several smart devices in the smart city network and make the owner is the only one who can rebuild the original data. Also, blockchain can be used to provide certification of the data produced by IoT devices [63].

5.3.2 Smart home

A smart home is generally a home with integrated home automation system to improve quality of life by providing a safe, secure and comfortable place to live in. Also, it enables the owner to control home's appliances without being physically at home. IoT devices at a smart home collect a huge amount of data that can be utilized to enhance the owner's experience. The collected data can be subjected to powerful algorithms of machine learning to enable IoT devices to learn from the collected data and make autonomous decisions based on their surrounding conditions without requiring manual input from the owner. Therefore, with machine learning, IoT devices can constantly enhance themselves and provide highly efficient services [43].

Although smart home provides magnificent services which really improve people life, there is still a security nightmare in which an attacker can access smart devices and tamper their data or functions that can cause real problems that can literally cause the owner to lose his life. Since smart devices come with poor built-in security measures besides smart homeowners tend to work with default or weak passwords which can be easily breached. To solve this issue, the blockchain provides a magical solution by making IoT devices immutable and tamper-proof. This, in turn, protects data collected by IoT devices and control data to be accessed by other third parties. A common case of using blockchain for smart home is Comcast which uses a permissioned ledger to enable the homeowner to grant or deny access permissions remotely through a mobile application [64].

5.3.3 Supply chain

Supply chain is one of the IoT applications that involves integrated planning and performing several operations. This includes material flow, information flow as well as financial capital flow. The management of flow process of goods, products, services, and information involving the storage and movement of raw materials as well as full-fledged finished goods from one point to another is called as supply chain management [65].

While this process seems easy in theory, but practically there is a huge amount of paperwork involved for each transaction between a supplier and retailer. This creates a big problem especially with growing the business size. Blockchain along with smart contracts can solve issues of the supply chain and reduce costs, time and efforts. It delivers the required scalability for supply chain by providing a large database that can be accessed from multiple locations around the world. Also, the blockchain technology can be adopted in a private manner to maintain data integrity across multiple participants. It also provides higher standards of security and the ability to be customized according to the data feed [5].

5.3.4 Healthcare

With the existence of connected devices and objects, real-time monitoring can save lives especially in emergency situations such as diabetes, asthma attacks, heart failure, etc. With a smartphone application, the patient real-time data can be collected and transferred to the doctor on a regular basis to monitor the patient remotely and independently. Also, these smart devices can be programmed to take autonomous decisions. For example, in emergency situations, it can call the ambulance or the doctor of the patient, while in less emergency situations, it can provide a suitable prescription for the patient [66].

The existing healthcare system faces several issues especially in security and privacy of healthcare data collected and transmitted in real-time. Adopting the blockchain technology in the healthcare sector can add numerous improvements. For instance, blockchain can provide an immutable ledger of medical records that cannot be altered once it created and signed. This, in turn, can increase the integrity of medical records. Another benefit of the blockchain is the consent management. Since the existing healthcare system has different privacy and consent regulations at every stage, the blockchain can be utilized to store the patient consent for purposes of data sharing, so any third-party can check the blockchain to decide whether to share the patient information or not [67].

5.3.5 Smart grid

Smart grid is one of the critical applications of the IoT system. It is a part of the IoT framework that is used to monitor and track lighting, parking spaces, traffic signals, road warnings, traffic congestion, and early detection of things such as earthquakes and extreme weather. By definition, a smart grid is a network of transmission lines that involves transformers, smart meters, sensors, distribution automation and software that are deployed to businesses and homes across the city. Providing a bidirectional communication between connected devices and hardware to sense and quickly reflect the user requirements should be one of the main targets of the IoT smart grid [1].

Since smart grids are based on automation and remote access, they become a target for security issues in which attackers can access the devices and cause serious damage for either devices or homes. The adoption of the blockchain can resolve this issue by providing a shared and encrypted ledger which is immutable to changes made by malicious nodes or attackers. It can also be utilized to verify identities and authorize the access by storing and recording transactions in the immutable ledger and make data exchanges between distributed gadgets smooth and cost-efficient [68].

5.4 Challenges of blockchain with IoT

Although the convergence of blockchain with IoT brings countless benefits, it also faces several challenges that need to be addressed to increase the adoption of the blockchain technology in various IoT applications. This section provides common challenges standing in the way of integrating blockchain with IoT, as summarized in Fig. 9.

5.4.1 Security

As the IoT system contains billions of heterogeneous devices and objects connected over the Internet with poor built-in security measures. This makes it a desirable target for most security attacks. To improve the security of the IoT system, most security experts and researchers have confirmed that integrating blockchain with IoT can resolve this issue; however, one of the key challenges of providing a secure IoT system is the reliability of the data generated by IoT devices. Indeed, blockchain provides an immutable and tamper-proof ledger that ensures data integrity; however, if the data are corrupted before storing it in the blockchain, it will stay corrupted. The data can be corrupted either by a malicious intruder or the devices themselves and their sensors fail to work correctly from the start. Therefore, IoT devices need to be tested before their integration with the blockchain and apply appropriate techniques to detect device failures immediately [69].

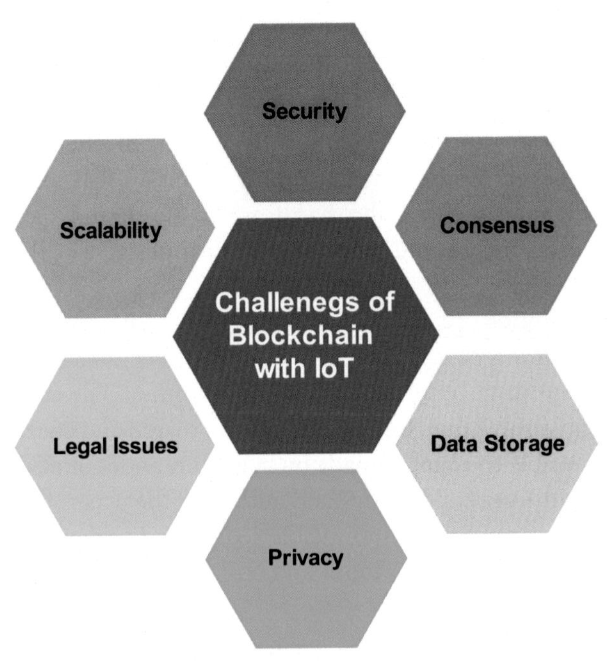

Fig. 9 Challenges of Integrating blockchain with IoT.

5.4.2 Scalability

One of the main challenges of integrating blockchain with IoT is the capability of the blockchain to scale to meet the security requirements of a dynamic network containing billions of devices. Although the integration between blockchain and IoT provides myriad benefits such as authentication, data integrity, fault tolerance and security, it comes at the price of scalability. The existing blockchain platforms such as Bitcoin and Ethereum can process only seven and 20 transactions per second, respectively. This processing speed cannot handle the processing requirements of the IoT system with billions of transactions. Also, the limited bandwidth associated with the blockchain technology cannot enable real-time transaction processing. Although there were several proposed blockchain scaling solutions over the last years, the scalability is still the major issue facing the convergence of blockchain with IoT [53].

5.4.3 Privacy

Several IoT applications collect and work with sensitive and classified data, for example, e-health connected devices that track and monitor the patient healthcare data. These data should be anonymously stored and protected against any data modification. The integration of blockchain with IoT has promised to provide a secure and immutable environment; however, the data privacy faces several issues especially in transparent and public blockchains where all participating nodes in the blockchain network can view values of all transactions, so the blockchain cannot guarantee transactions' privacy [70]. Also, as indicated by Barcelo [71], the user's transaction can be linked to disclose user's personal information. Therefore, the privacy issue of the blockchain needs more research.

5.4.4 Data storage

In the IoT system, there are billions of devices that create Terabytes of data in real-time which needs to be stored for processing to extract meaningful information. Indeed, the integration of blockchain with IoT eliminates the need for a centralized authority to store and process the IoT collected data [72]. However, blockchain is not designed to store large amounts of data like those produced in the IoT system. The storage capacity is considered one of the big problems of the blockchain technology, but with integrating blockchain with IoT, this issue will become much greater. Also, as distributed ledger keeps all transactions that have ever made in the network, the size of the ledger will increase with increasing number of participating nodes in the network. The storage issue is still one of the biggest challenges for the integration of blockchain with IoT; therefore, more research and new approaches are needed to resolve this issue [4].

5.4.5 Legal issues

The IoT is a dynamic system in nature which have expanded over the boundaries of countries to connect and communicate data between various devices all over the world. Like all new technologies, the IoT is affected by the country's laws or regulations regarding data privacy. However, most existing laws do not cope with technological developments and became obsolete and need to be reviewed. Creating novel laws and standards can facilitate the certification of security features of devices to build a more secure and trusted IoT network. In this sense, laws that deal with information privacy and information handling are still a big challenge to be tackled in the IoT and will be an even bigger challenge after integrating blockchain with IoT. The blockchain can collect various nodes from different countries without having any legal or compliance code to follow, which is a serious issue that needs to be handled by issuing the appropriate laws which can cope with technological progress of both IoT and blockchain [49].

5.4.6 Consensus

The IoT system involves billions of devices and objects with limited resources such as computing power and storage. These resource-constrained devices cannot participate in the consensus mechanisms such as PoW and PoS associated with the blockchain technology that require high processing power for the mining process. Since the mining process is a complex operation that needs a huge amount of energy, integrating sophisticated Artificial Intelligence (AI) techniques with blockchain can be very efficient in optimizing energy consumption. So, therefore, more research is needed to investigate how to adopt consensus mechanisms in the IoT system [6].

6. Conclusion

The IoT has proven it can provide several benefits in various domains. It has evolved to include the perception of realizing a global infrastructure of interconnected networks of physical and virtual objects. These objects are interconnected using either wired or wireless networks to share information between various IoT devices to create novel applications and services. However, the current IoT centralized architecture faces many issues regarding security and scalability. One of the recommended solutions to resolve these challenges is the blockchain technology. It uses a decentralized and distributed ledger to utilize the processing capabilities of all the participating nodes in the blockchain network, which reduce latency and eliminate the single point of failure. Integrating the IoT system with blockchain can bring countless benefits. For instance, the decentralization feature of blockchain can process billions of transactions between IoT devices. This chapter provided a discussion of the intersection between IoT and DLTs. It started by providing an overview of DLT by highlighting its main components, benefits and challenges. The centralized IoT system is also discussed with highlighting its essential limitations. Then, the integration of blockchain with IoT is presented by highlighting the benefits of the integration process. Various application and challenges of blockchain with IoT are also discussed.

References

[1] H.F. Atlam, R.J. Walters, G.B. Wills, Internet of things: state-of-the-art, challenges, applications, and open issues, Int. J. Intell. Comput. Res. 9 (3) (2018) 928–938.

[2] K.K. Patel, S.M. Patel, Internet of things-IOT: definition, characteristics, architecture, enabling technologies, application & future challenges, Int. J. Eng. Sci. Comput. 6 (5) (2016) 6122–6131.

[3] H.F. Atlam, A. Alenezi, M.O. Alassafi, G.B. Wills, Blockchain with Internet of Things: benefits, challenges, and future directions, Int. J. Intell. Sys. Appl. 10 (2018) 40–48.
[4] T.M. Fernández-Caramés, P. Fraga-Lamas, A review on the use of blockchain for the internet of things, IEEE Access 6 (2018) 32979–33001.
[5] H. Wu, Z. Li, B. King, Z. Ben Miled, J. Wassick, J. Tazelaar, A distributed ledger for supply chain physical distribution visibility, Information 8 (4) (2017) 137.
[6] A. Reyna, C. Martín, J. Chen, E. Soler, M. Díaz, On blockchain and its integration with IoT. Challenges and opportunities, Futur. Gener. Comput. Syst. 88 (2018) 173–190.
[7] E. Karafiloski, A. Mishev, Blockchain solutions for big data challenges: a literature review, in: 17th International Conference on Smart Technologies, 2017, pp. 763–768. July.
[8] M.Å. Hugoson, Centralized versus decentralized information systems: a historical flashback, IFIP Adv. Inf. Commun. Technol. 303 (2008) 106–115.
[9] M. Tommasi, F. Weinschelbaum, Centralization vs. decentralization: a principal-agent analysis, J. Public Econ. Theory 9 (2007) 369–389.
[10] K. Kawano, M. Orimo, K. Mori, Autonomous decentralized system test technique, in: Conference Proceedings of the Thirteenth Annual International Computer Software & Applications, 1989, pp. 52–57.
[11] H.F. Atlam, G. Attiya, N. El-Fishawy, Comparative study on CBIR based on color feature, Int. J. Comput. Appl. 78 (16) (2013) 975–8887.
[12] L. Swierczewski, The Distributed Computing Model Based on the Capabilities of the Internet, arXiv Prepr. (2012). arXiv1210.1593.
[13] J. Rehman, What Are Advantages and Disadvantages of Distributed Operating Systems, IT release, 2018. [Online]. Available: http://www.itrelease.com/2015/09/what-are-advantages-and-disadvantages-of-distributed-operating-systems/. [Accessed: 26-Oct-2018].
[14] F.M. Benˇ, I.P. Žarko, Distributed ledger technology: blockchain compared to directed acyclic graph, in: 2018 IEEE 38th International Conference on Distributed Computing Systems (ICDCS), 2018, pp. 1569–1570.
[15] M. Hancock, E. Vaizey, Distributed Ledger Technology: Beyond Blockchain, UK Government Office for Science, 2016.
[16] Investopedia, Distributed Ledger Technology, 2018. [Online]. Available: https://www.investopedia.com/terms/d/distributed-ledger-technology-dlt.asp. [Accessed: 25-Oct-2018].
[17] Financial Conduct Authority, Discussion Paper on Distributed Ledger Technology, Discussion Paper DP17/3, 2017, www.fca.org.uk/publication/discussion/dp17-03.pdf, [Accessed 30-Oct -2018].
[18] B. Cœure, Distributed Ledger Technology in Payment, Clearing and Settlement: An Analytical Framework, 2017. [Online], Available: https://www.bis.org/cpmi/publ/d157.pdf. [Accessed 30-Oct-2018].
[19] P. Ferraro, C. King, R. Shorten, Distributed ledger technology for smart cities, the sharing economy, and social compliance, IEEE Access 6 (2018) 62728–62746.
[20] J.J. Sikorski, J. Haughton, M. Kraft, Blockchain technology in the chemical industry: machine-to-machine electricity market, Appl. Energy 195 (2017) 234–246.
[21] H. Natarajan, S.K. Krause, H.L. Gradstein, Distributed Ledger Technology (DLT) and Blockchain, World Bank Group, Washington, D.C., 2017.
[22] Z. Zheng, S. Xie, H. Dai, X. Chen, H. Wang, An overview of blockchain technology: architecture, consensus, and future trends, in: 2017 IEEE 6th International Congress on Big Data, 2017, pp. 557–564.
[23] H.F. Atlam, A. Alenezi, R.J. Walters, G.B. Wills, An overview of risk estimation techniques in risk-based access control for the internet of things, in: IoTBDS 2017—Proceedings of the 2nd International Conference on Internet of Things, Big Data and Security, 2017.

[24] K. Sultan, U. Ruhi, R. Lakhani, Conceptualizing blockchain: characteristics & applications, in: 11th IADIS International Conference Information Systems, 2018, pp. 49–57.
[25] ESMA, The Distributed Ledger Technology Applied to Securities Markets, 2016. [online], Available: https://www.esma.europa.eu/system/files_force/library/dlt_report_-_esma50-1121423017-285.pdf. [Accessed 30-Oct-2018].
[26] A. Deshpande, K. Stewart, L. Lepetit, S. Gunashekar, Distributed Ledger Technologies/Blockchain: Challenges, Opportunities and the Prospects for Standards, British Standards Institution, 2017.
[27] W. Meng, E.W. Tischhauser, Q. Wang, Y. Wang, J. Han, When intrusion detection meets blockchain technology: a review, IEEE Access 6 (2018) 10179–10188.
[28] H.F. Atlam, M.O. Alassafi, A. Alenezi, R.J. Walters, G.B. Wills, XACML for building access control policies in internet of things, in: IoTBDS 2018—Proceedings of the 3rd International Conference on Internet of Things, Big Data and Security, vol. 2018, 2018. March.
[29] K. Christidis, G.S. Member, Blockchains and smart contracts for the internet of things, IEEE Access 4 (2016) 2292–2303.
[30] A. Stanciu, Blockchain based distributed control system for edge computing, in: 21st International Conference on Control Systems and Computer Science Blockchain, 2017, pp. 667–671.
[31] W.-T. Tsai, L. Feng, H. Zhang, Y. You, L. Wang, Y. Zhong, Intellectual-property blockchain-based protection model for microfilms, in: 2017 IEEE Symposium on Service-Oriented System Engineering (SOSE), 2017, pp. 174–178.
[32] M. Alharby, A. van Moorsel, Blockchain based smart contracts: a systematic mapping study, Comput. Sci. Inf. Technol. (2017) 125–140.
[33] Cambridge Dictionary, Jurisdiction | Cambridge English Dictionary, [Online]. Available: https://dictionary.cambridge.org/dictionary/english/jurisdiction. [Accessed: 31-Oct-2018].
[34] E. Benos, R. Garratt, P. Gurrola-Perez, The Economics of Distributed Ledger Technology for Securities Settlement, Bank of England, 2017. Staff Work Paper no 670.
[35] S. Cho, T. Elhourani, S. Ramasubramanian, Independent directed acyclic graphs for resilient multipath routing, IEEE/ACM Trans. Networking 20 (1) (2012) 153–162.
[36] M. Kalisch, B. Peter, Estimating high-dimensional directed acyclic graphs with the PC-algorithm, J. Mach. Learn. Res. 8 (2007) 613–636.
[37] IOTA, The Next Generation of Distributed Ledger Technology, 2018. [Online]. Available: https://www.iota.org/. [Accessed: 04-Nov-2018].
[38] Radix, [Online]. Available: https://www.radixdlt.com. [Accessed: 04-Nov-2018].
[39] D. Hughes, Radix-Tempo, Radix White Pap, 2017.
[40] H.F. Atlam, R.J. Walters, G.B. Wills, Fog computing and the internet of things: a review, Big Data Cogn. Computing 2 (2) (2018) 1–18.
[41] L. Yueming, Y. Li, S. Yin, Design and implementation of IoT centralized management model with linkage policy, in: Third International Conference on Cyberspace Technology (CCT 2015), 2015, pp. 5–9.
[42] H.F. Atlam, R.J. Walters, G.B. Wills, Internet of nano things: security issues and applications, in: 2018 2nd International Conference on Cloud and Big Data Computing, 2018, pp. 71–77.
[43] H.F. Atlam, R.J. Walters, G.B. Wills, Intelligence of things: opportunities & challenges, in: IEEE 2018 Cloudification of the Internet of Things (CIoT), 2018, pp. 1–8.
[44] N. Kshetri, Can Blockchain Strengthen the Internet of Things? IEEE Computer Society, 2017, pp. 68–72.
[45] H. Halpin, M. Piekarska, Introduction to security and privacy on the blockchain, in: 2017 IEEE European Symposium on Security and Privacy Workshops (EuroS&PW), 2017, pp. 1–3.

[46] M. Conoscenti, D. Torino, A. Vetr, D. Torino, J.C. De Martin, Peer to peer for privacy and decentralization in the internet of things, in: 2017 IEEE/ACM 39th IEEE International Conference on Software Engineering Companion Peer, 2017, pp. 288–290.
[47] H.F. Atlam, G.B. Wills, IoT security, privacy, safety and ethics, in: Digital Twin Technologies and Smart Cities, Springer International Publishing AG (in press), 2019.
[48] H.F. Atlam, A. Alenezi, R.J. Walters, G.B. Wills, J. Daniel, Developing an adaptive risk-based access control model for the internet of things, in: 2017 IEEE International Conference on Internet of Things (iThings) and IEEE Green Computing and Communications (GreenCom) and IEEE Cyber, Physical and Social Computing (CPSCom) and IEEE Smart Data (SmartData), 2017, pp. 655–661.
[49] N. Fabiano, Internet of things and blockchain: legal issues and privacy. The challenge for a privacy standard, in: Proceedings—2017 IEEE International Conference on Internet of Things, IEEE Green Computing and Communications, IEEE Cyber, Physical and Social Computing, IEEE Smart Data, iThings-GreenCom-CPSCom-SmartData 2017, vol. 2018, 2018, pp. 727–734.
[50] H.F. Atlam, A. Alenezi, A. Alharthi, R. Walters, G. Wills, Integration of cloud computing with internet of things: challenges and open issues, in: 2017 IEEE International Conference on Internet of Things (iThings) and IEEE Green Computing and Communications (GreenCom) and IEEE Cyber, Physical and Social Computing (CPSCom) and IEEE Smart Data (SmartData), 2017, pp. 670–675. June.
[51] M. Ahmad, K. Salah, IoT security: review, blockchain solutions, and open challenges, Futur. Gener. Comput. Syst. 82 (2018) 395–411.
[52] D. Tapscott, A. Tapscott, Blockchain Revolution: How the Technology Behind Bitcoin Is Changing Money, Business, and the World, Penguin Random House, New York, 2016.
[53] M. Samaniego, R. Deters, Blockchain as a service for IoT, in: 2016 IEEE International Conference on Internet of Things (iThings) and IEEE Green Computing and Communications (GreenCom) and IEEE Cyber, Physical and Social Computing (CPSCom) and IEEE Smart Data (SmartData), 2016, pp. 433–436.
[54] Chain of Things, [Online]. Available: https://www.chainofthings.com/. [Accessed: 09-Nov-2018].
[55] Slock.it, [Online]. Available: https://slock.it/. [Accessed: 09-Nov-2018].
[56] Waltonchain, [Online]. Available: https://waltonchain.org/. [Accessed: 09-Nov-2018].
[57] A. Dorri, S.S. Kanhere, R. Jurdak, Blockchain in Internet of things: challenges and solutions, arXiv1608.05187. 2016.
[58] E.F. Jesus, V.R.L. Chicarino, C.V.N. De Albuquerque, A.A.D.A. Rocha, A survey of how to use blockchain to secure internet of things and the stalker attack, Secur. Commun. Netw. 2018 (2018), 9675050.
[59] Z. Zheng, S. Xie, H.-N. Dai, X. Chen, H. Wang, Blockchain challenges and opportunities: a survey, Int. J. Web Grid Serv. 14 (4) (2016) 1–24.
[60] X. Liang, J. Zhao, S. Shetty, D. Li, Towards data assurance and resilience in IoT using blockchain, in: Proceedings—IEEE Military Communications Conference MILCOM, 2017, pp. 261–266.
[61] E. Heilman, F. Baldimtsi, S. Goldberg, Blindly signed contracts: anonymous on-blockchain and off-blockchain bitcoin transactions, in: International Conference on Financial Cryptography and Data Security, 2016, pp. 43–60.
[62] A. Panarello, N. Tapas, G. Merlino, F. Longo, A. Puliafito, Blockchain and IoT integration: a systematic survey, Sensors 18 (8) (2018) 1–37.
[63] N.Z. Aitzhan, D. Svetinovic, Security and privacy in decentralized energy trading through multi-signatures, blockchain and anonymous messaging streams, IEEE Trans. Dependable Secure Comput. 15 (5) (2018) 840–852.

[64] S. Williamson, How IoT, Blockchain, and AI Can Join Forces to Improve the Smart Home Experience, 2018. [Online]. Available: https://medium.com/swlh/how-iot-blockchain-and-ai-can-join-forces-to-improve-the-smart-home-experience-7cdbdab75214. [Accessed: 10-Nov-2018].
[65] B. Bhandari, Supply chain management, blockchains and smart contracts. SSRN Electron. J. (2018). https://doi.org/10.2139/ssrn.3204297.
[66] H.F. Atlam, A. Alenezi, R.K. Hussein, G.B. Wills, Validation of an adaptive risk-based access control model for the internet of things, Int. J. Comput. Netw. Inf. Secur. 10 (2018) 26–35.
[67] X. Yue, H. Wang, D. Jin, M. Li, W. Jiang, Healthcare data gateways: found healthcare intelligence on blockchain with novel privacy risk control, J. Med. Syst. 40 (2016) 218–225.
[68] M. Mylrea, S.N.G. Gourisetti, Blockchain for smart grid resilience: exchanging distributed energy at speed, scale and security, in: 2017 Resilience Week (RWS), 2017, pp. 18–23.
[69] R. Roman, J. Zhou, J. Lopez, On the features and challenges of security and privacy in distributed internet of things, Comput. Netw. 57 (10) (2013) 2266–2279.
[70] H.F. Atlam, G. Attiya, N. El-Fishawy, Integration of color and texture features in CBIR system, Int. J. Comput. Appl. 164 (2017) 23–28.
[71] J. Barcelo, User Privacy in the Public Bitcoin Blockchain, White Paper. 2014.
[72] A. Alenezi, N.H.N. Zulkipli, H.F. Atlam, R.J. Walters, G.B. Wills, The impact of cloud forensic readiness on security, in: Proceedings of the 7th International Conference on Cloud Computing and Services Science (CLOSER 2017), 2017, pp. 511–517.

About the authors

Hany F. Atlam is a PhD Researcher at the University of Southampton, UK and lecturer at Faculty of Electronic Engineering, Menoufia University, Egypt. He was born in Menoufia, Egypt in 1988. He has completed his Bachelor of Engineering and Computer Science in Faculty of Electronic Engineering, Menoufia University, Egypt in 2011, then completed his master's degree in computer science from the same university in 2014. He joined the University of Southampton as a PhD student since January 2016. He has several experiences in networking as he is a Cisco Certified Network Associate (CCNA) and a Cisco Certified Academy Instructor (CCAI). Hany is a member of Institute for Systems and Technologies of Information, Control and Communication (INSTICC), and Institute of Electrical and Electronics Engineers (IEEE). Hany's research interests include and not limited to: Internet of Things Security, Cloud Security, Cloud and Internet of Things Forensics, Blockchain, and Big Data.

Gary B. Wills is an Associate Professor in Computer Science at the University of Southampton. He graduated from the University of Southampton with an Honors degree in Electromechanical Engineering, and then a PhD in Industrial Hypermedia system. He is a Chartered Engineer, a member of the Institute of Engineering Technology and a Principal Fellow of the Higher Educational Academy. He is also a visiting associate professor at the University of Cape Town and a research professor at RLabs. Gary's research projects focus on Secure System Engineering and applications for industry, medicine and education. Gary published more than 200 publications in international journals and conferences.

Blockchain technology for decentralized autonomous organizations

Madhusudan Singh[a,b], **Shiho Kim**[b]
[a]School of Technology Studies, Endicott College of International Studies, Woosong University, Daejeon, South Korea
[b]Yonsei Institute of Convergence Technology, Yonsei University, Seoul, South Korea

Contents

Advances in Computers, Volume 115
ISSN 0065-2458
https://doi.org/10.1016/bs.adcom.2019.06.001

Abstract

With continuously changing operational and business needs of the organizations, Decentralized Autonomous Organizations (DAO) is the current need of the organizations. Centralized Autonomous Organization (CAO) lack transparency and are managed by few efficient managers whereas Decentralized autonomous Organization's (DAO) is novel scalable, self-organizing coordination on the blockchain, controlled by smart contracts and its essential operations are automated agreeing to rules and principles assigned in code without human involvement. In this chapter we discuss the needs for Decentralized Autonomous Organizations (DAO) and key efforts in this field. We then introduce a prospective solution employing blockchain Ethereum, which incorporates a Turing complete programming language with smart contract computing functionality. A solution is elaborated that permits the formation of organizations where participants preserve straight real-time check of contributed collects and governance policies are formalized, automatized and imposed using software. Basic code for smart contract is composed to make a Decentralized Autonomous Organization (DAO) on the Ethereum blockchain. We also explain the working of DAOs code, centering on fundamental establishment and governance characteristics, which includes organization, formation and voting rights. DAOs are considered to agree to the expectation of the business work in the future. But there is still lack of operational base for DAOs in the blockchain community.

1. Introduction

As we know, the meaning of Organization is defined in dictionary; a group whose members work together for a shared purpose in a continuing way. The concept of an organization is mentioned prior articles.

"*… organization is a particular pattern of structure, people, tasks and techniques.*"

Source: Leavitt [1].

"*… A system which is composed of a set of subsystems…*"

Source: Katz and Kahn [2].

Each organization must follow an absolute rule, which is governed by the government of the incorporated organization. An organization consists of a group of shareholders, who represent the ownership of the organization in a percentage format.

Blockchain Technology is novel and increasingly crucial technical foundation for the internet. The blockchain technology is impacting organizational behavior to develop a new organizational form called "Decentralized Autonomous Organization (DAO)." A DAO is a computer program, which runs on a peer to peer network, incorporating with governance and decision-making rules. DAOs can be programmed to operate autonomously, without human involvement, or the code can provide direct, real-time control of the DAO and the funds controlled by it. The earliest DAOs are software-controlled community organization experiments which seek to re-implement certain aspects of traditional corporate governance, replacing voluntary compliance with a corporation's charter with actual compliance using pre-agreed computer code. A Decentralized Organization (DO) would be a set of humans interacting according to rules specified in the code. The addition of the word Autonomous is used for an organization whose primary functions are automated, whereas human activities and interactions are not so necessary.

1.1 Centralized autonomous organization (CAO)

Centralized Autonomous Organization (CAO) conducts their decisions from the top management down thru successive levels with only an elite group of best managers responsible for general judgment. This hierarchy is termed as the chain of command.

The judgment process is gradual in **Centralized** Autonomous Organization (CAO) which further makes the progress slow, as shown in Fig. 1.

1.2 Problems with CAO

Most of the **Centralized** Autonomous Organization (CAO) is centralized. They lack transparency and is managed by a small number of effective managers. They assure property via a legal system. To overcome these problems, we require a Distributed Autonomous Organization (DAO). With DAO, we can terminate the former structure but still we require an elite group of management personnel that work with each other in a team.

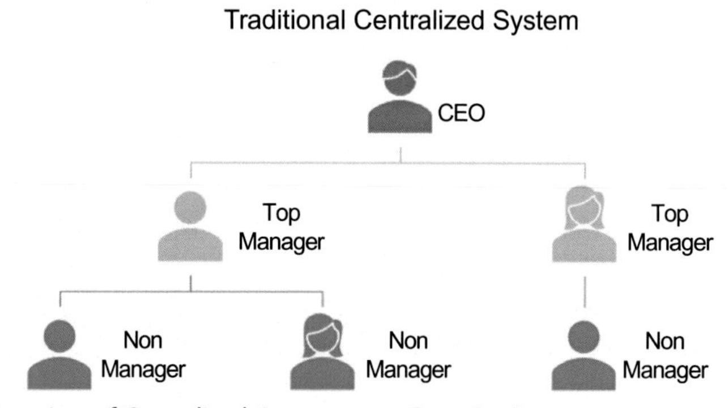

Fig. 1 Overview of Centralized Autonomous Organization.

1.3 Use case/application of CAO

Centralized Autonomous System focus on management administrative body and judgment process in a separate administrative unit with information running from best managers to several business enterprises. DAO constructs appear more alike several small-scale cooperation's of a separate construct, introducing administrative repetitions and more held together chains of control. Realizing the divergence among the two essentially dissimilar design doctrines can assist to produce an efficient foundation for the enterprise.

(a) Comparison of management foundations

In a centralized structure, every manager is authorized to deal with a broader array of workers, divisions and business operations. Management modes can become self-assured in centralized foundations, as managers have very less time to communicate with individual assistants.

In decentralized structure, every manager is authorized to deal with lesser employees and business operations, and several managers contribute the similar work entitles and duties in several fields of the business. Decentralized foundations permit managers to build conclusions on a smaller level, which can be perfect for conditions in which independent teams must adjust to specific workplace considerations, for example, direct sales.

- Choices and Information Flows

  In centralized organizational structures, choices are created at the high level and conveyed at lower level via the levels. The middle and lower level manager's decisions are limited for deciding how to enforce the decisions passed to them.

  Decentralized structures are just opposite. In decentralized organization the managers at the lower level and employees have the power to create important decisions for themselves and their work groups, and information on their choices is accounted to top levels of management.

(b) Applications of centralized structures

Small scale enterprises have less number of personnel's and so they are controlled in a centralized fashion. In the small-scale enterprises, the owner of the enterprise might be the manager of the company alone and he is right away accounted by other employees in the company.

The example of a centralized organizational structure is trucking company. All functional judgements are formed by company managers. The managers send information to individual drivers through starters.

(c) Applications of decentralized structures

The example of a decentralized organizational structure is Franchise organizations. Franchise organizations operate on product production and selling conclusions at the best, and a good deal of freedom is given in functioning their independent stores.

1.3.1 Blockchain technology background

Blockchain has revolutionized the exchange of information and media after the Internet. Blockchain technology is pertained to as a path-breaking innovation and the forerunner of a fresh economic period. Blockchain call forth a new type of recent system called the Blockchain Economic System. The blockchain economic system conventions will be determined by the smart contracts, whenever stimulatory transactions are enforced autonomously. Evidently, the blockchain economy has taken the shape of a novel organizational structure called the decentralized autonomous organizations (DAO). Blockchain is the basis for the Decentralized Autonomous Organizations (DAOs) which provides novel stages of crowd coordination by eradicating the trust and fault problems.

1.4 Use case/application of blockchain

Blockchain technology has the ability to alter our lifestyle in many fields. Blockchain has the tendency to affect our lifestyle via possible blockchain use cases such as Cryptocurrency, Supply Chain Management, Digital Identity, Voting, Healthcare, Fundraising (ICOs, or Security Token Offerings), Notary, Food Safety, Intellectual Property (IP), Decentralized Autonomous Organization (DAO), Distributed Cloud Storage, Sharing Economy, Music, Smart Contracts, Charity, Automobiles, Real Estate, Internet of Things, Cyber Security, and Insurance.

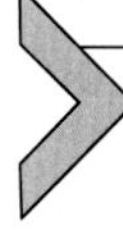

2. Decentralized autonomous organizations (DAO)

2.1 Introduction of DAO

A Decentralized autonomous Organization (DAO) is an organization whose essential operations are automated agreeing to rules and principles assigned in code without human involvement. A DAO is a novel scalable, self-organizing coordination on the blockchain, controlled by smart contracts. DAOs are considered to agree to the expectation of the business work in the future. But there is still lack of operational base for DAOs in the blockchain community (Fig. 2).

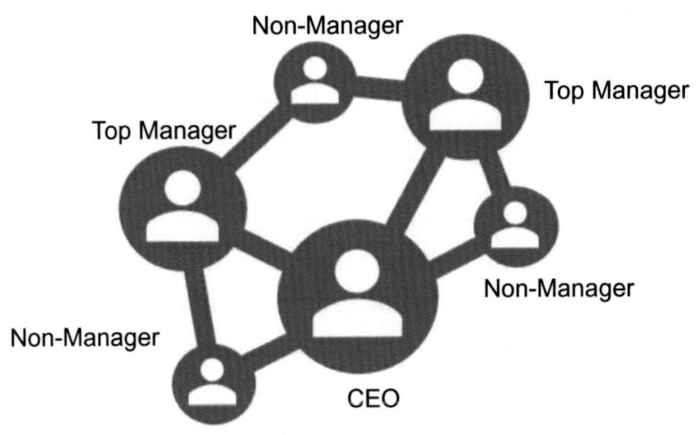

Fig. 2 Decentralized Autonomous Organization Overview.

The DAO project was introduced and developed by Jentzsch, to provision organizational solution [3]. It was implemented on the Ethereum blockchain. He elaborated the description of the operation of the distributed collection in his whitepaper detailing the control of distributed funds by participants in formalized, software-based organization systems, established on the applied organization model produced by their Slock.It coders. It was developed for shareholders to vote on where the company should invest its funds.

"The DAO" was the first substantial test of this model and is the most eminent example of a DAO. The first public DAO was established on Ethereum on April 30, 2016. The motive was to establish a novel track in business administration for the improvement of the participants with the constant decision of unbeatable code. In the initial stage of its commencement, it acquired considerable attention from media on increasing the equivalent of USD 168 million from several investors, creating the world's largest crowdfunding project. But sooner "The DAO" was maliciously misused and compromised to a considerable level due to a flaw in the DAO's code on June 17th.

DAOs used the Ethereum Blockchain using Ether as the cryptocurrency for executing contracts. The computational power used for validating transactions in the Ethereum network is payed using the Ether cryptocurrency.

Advanced solutions in distributed algorithm design can be aggregated with advanced perceptions of reasoning and learning from disseminated market person to form novel structure of self-organizing, robust, and resilient governance [4].

2.2 Need for decentralized autonomous organization

(a) Why is this important?

DAO is a system constructed on Ethereum blockchain. DAO is an example of how platforms change former modes of coordinating to digital, connected world, but rather change novel kinds of coordination and governance. DAO is a system on which several kinds of actions can be constructed from investment funds to common data storage. It can be used to arrange a self-governing ridesharing system, where there are contending actions for corresponding, commercialism, interface, etc., all functioning smoothly unitedly. It alters novel government displays, such as "futarchy," which employs anticipation, commercialize to select among rules.

DAO can also be considered as a reaction to the transmutation in work, similar as platform cooperatives. As work turns more alike an unsafe commitment than a stabilize origin of financial gain, organizational base can assist deal with the novel realism. While platform co-ops resolve the trouble by employing digital platform to change reasonable allocation of amount and ability, DAOs attempt to attain the same via smart contracts, code and blockchain.

Decentralized Autonomous Organizations (DAO) provision a secure and trusted distributed ledger to route financial transactions using the blockchain technology over internet using trusted timestamp in a distributed database to make fraud difficult. There is no requirement of the mutually acceptable third party, which simplifies the process of financial transaction and there is no need to replicate contracts in different records which further reduces the costs of the transaction in the blockchain network [5].

The idea of a "Decentralized Organized Company" was first introduced by Daniel Larimer on September 7, 2013 [6] and was implemented in Bitshares in 2014, EOS in 2018 [7].

Vitalik Buterin specified that after the launch of DAO, it will be coordinated and executed without human managerial intervention if the Turing complete platform supported the smart contracts [8]. Ethereum met the Turing threshold, to render such DAOs [7,8]. Participants operate their entities and their personal records in the open platform Decentralized autonomous organizations [9].

2.3 Algorithms/tools for DAO

(a) Algorithms/tools for DAO

We require considerable and precise classical legitimate tools for development of DAO and smart contracts. It is better to acquire benefit from the custom-made tools which are acknowledged by the administrative authority. There are some tools which abide by the rules and regulations by law. One such example is the EU's eIDAS Regulation, which provisions a basis for e-recognition and tools for executing and applying trust services like digital signature rather than using classical proof of identity. Utilizing such tools can greatly simplify the use of smart contracts and also make it possible to define the legal consequences of any smart contract [10].

Stafford Beer's Viable System Model (VSM) [11] operations and its relations provision a tool to depict the practicality of a Decentralized Autonomous Organization [6].

However, the VSM breaks abruptly in abiding by the specific construction and components of several kinds of organizations. Although the VSM provisions a standard for practicality, however, the standards are only explained functionally [6].

The VSM is a reputable and comparatively ancient model; however, the VSM is hard to employ due to its complex nature [12]. VSM is an effective tool to depict organizations [6]; however, it is hard to operationalize. Therefore, in research the VSM is enforced as a conceptual tool [13] to depict a system, instead of operationalizing it in reality.

The VSM suggestion was suggested after the hard fork as there were not much capable on-chain lawful control tools available during hard fork [14]. There were several alternatives talked about the working of ETC with the community. The selected alternative allows for DGD bearers to Claim back their ETC, without delivering substantial lawful and fiscal troubles between DigixGlobal and DigixDAO [15].

"CarbonVote" is other option as DigixDAO governance model is not accessible for voting. A CarbonVote eliminates the possibility of hacking during voting process and is simple secure and provides convenience [15].

A lightwallet comprises of software which provisions a complete spectrum of tools within a user interface in the browser to communicate with blockchain.

Spectrum can be used by projects for advising proposals where DGD bearers can vote in this light wallet [16].

(b) The DAO stacks

The DAO stack provisions the fundamental tools for the development, functioning and governance of DAOs, both internal and external within a wider communication system. Below figure shows the components of the DAO stack (Fig. 3).

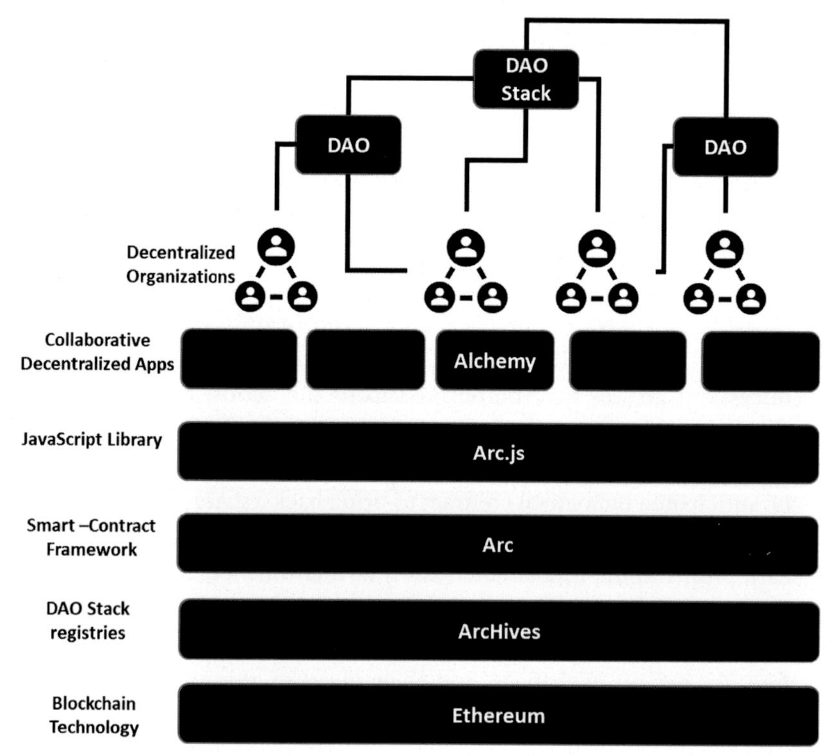

Fig. 3 The DAO stack.

The DAO stack system comprises of discrete interoperable DAOs, and each DAO communicates with other DAO maintaining an open distributed collaboration. All DAOs consists of serial smart contracts which are deployed by Arc. An **Arc** is a solid model of governance for development, configuration, deployment and functioning DAO onto the Ethereum blockchain. DAO relies on the IPFS for data storage and data recovery. There are two ways to interact with these DAOs. Direct way is via executing Blockchain transaction and indirect way is by trusting a front-end to the fundamental blockchain system [17].

Alchemy is developed by DAO stack as a collaborative DApp [18]. It enables a person to create a new DAO for collaboration with other DAO in the DAO stack system. It is based on **Arc.js**, which is a JavaScript library that controls the Arc framework via Web3.js. It is built to give comfort to front-end JavaScript developers for developing collaborative applications above Arc without the requirement to directly communicate with the Ethereum blockchain. The **ArcHives** is a public group of records, organized by the DAO stack community for effective network communication [17,19].

2.4 Challenges in DAO

(a) The DAO and the challenges

The DAO symbolize a revolution, an explosion that declares a novel stage in history in which attacks can be created on markets and fields other than defended by rules in which individuals may coordinate themselves and their assets more severally of their captains. Only this novel world is comparatively unexploited and filled of difficulties. First of all, in the instance of the DAO, its financial support might unknowingly have acquired a charge far more ambitious than committing in the decentralized systems that are their livelihood where produces and grosses may be derived from the action of comparatively inevitable rules, math and smart contract principle. Rather, they will be enabling mostly in off-chain values; normal organizations that will be anticipated by societal contract to bring back a share of net profit they finally yield. A central trouble is that it does not provision for capitalists to earn profits in the fundamental assess of their finances as they come up. To realize, believe capitalists in decentralized plans such as Ethereum that buy its measure item, such ether (ETH). These items are exchanged in a free manner on exchanges, and their measure goes up as the business measures the possible of the task extremely high, even earlier actual item speed and require for not theoretical keeps of measure arrive into act. In case, as the Ethereum web has successful formulated the measure of ETH has gone up as capitalists await the appearance of distributed require for the item produced by schemes such as the DAO. Such items are able to think over the measure of advancement as it encounters and are liquid like in public exchanged capitals.

By comparison, though commence agencies can develop quickly in value they can acquire long time to make gains. The demand with the DAO model is that benefits essentially be built by investigation and shared back by social contract to generate reward for item bearers, and this will probably need years to appear; however, the businesses are following a sensible belief that most capitalists in The DAO are attempting brings back in a much briefer time slot. The risk then is that when commitments are attained and the realism that invested funds have been lock for unpleasantly endless and irregular time period has changed the item bearers will hurried to the exit. The effect will be a steep fall in the measure of choosing items on the markets driving left bearers to act confined in to their commitments or form significant decreases.

Naturally, our concern around how and when brings back will be produced is completely immature unless the initiatives the DAO commits finally give benefits in the initial invest. This will partially depend on the power of the DAO's voting system for producing better commitment conclusions in the beginning form. There are several different notions on whether it will accomplish.

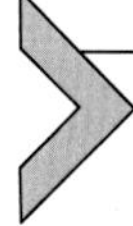

3. Blockchain for DAO

3.1 Smart contracts for DAO

The Decentralized Autonomous Organization (DAO) was an investment fund assembled in 2016 to be supervised completely by smart contracts, with no human managerial intervention. Investors vote on the fund's investment process. A lot of journalist illustrated how DAO would change the investment process [20].

Fig. 4A shows pseudocode for DAO-like contract, instancing permission for fund withdrawal by the investor. In Line 2, the function acquires the customers address, then in Line 3, checks out if customer has adequate funds for withdrawal. If true then via an external function call, the funds are sent to the customer in Line 6. In line 8, the customer's balance is decreased on successful transfer [21].

A Pseudocode for DAO-like exploit

```
1. Function withdraw (uint amount)
2. {
3.  customer = msg.sender
4.   if (balance [customer] >= amount
5.   {
6.      if
   (customer. call . sendMoney(amount))
7.      {
8.        balance[customer] -=amount;
9.      }
10. }
```

B Pseudocode for DAO-like exploit

```
1.    Function sendMoney (uint amount)
2.    {
3.     victim = msg.sender;
4.      balance +=amount;
5.      victim = msg.sender;
6.    }
```

Fig. 4 (A) Pseudocode for DAO-like contract, instancing permission for fund withdrawal by the investor [21]. (B) Pseudocode for DAO-like exploit [21].

The above function in the code was maliciously misused on June 2016, to steal around $50 million funds from the DAO. The statement in Line 3 is a call to a function in the customer's contract. Fig. 4B shows the customer's code. Withdraw function instantly called by the customer's contract in Line 4. This call in Line 3 once again tests whether adequate funds are there for withdrawal. When the nested call is complete, the withdraw () function decreases the balance. The funds are transferred a second time when the test passes and so on. The loop stops when the call stack overflows. This form of attack might at first glimpse appear like a strange risk brought in by a novel fashion of programming, but if we alter our view slightly, we can distinguish an unexpected difficulty [21].

Herlihy [21] and Hoare [22] proposed a concurrent programming language called monitor. This programming language is an object with a built-in mutex lock. This lock is adopted automatically when this method is called and released when the method returns. A wait () call is also provisioned by Monitors which permits the thread to stop holding on to the monitor lock, suspend, awaken, and reacquire the lock. The monitor constant is the primary equipment for rationalizing the rightness of a monitor execution. When there is no thread executing in the monitor then this statement holds. When the thread is holding the monitor lock, the constant can be violated, but when the thread releases the lock, the monitor lock is restored either by returning from a method, or by suspending via wait ().

The vulnerability appears common then the smart contracts are seen via the monitors and monitor constants. The DAO-like contract assumed that the constant that each customer's entrance record in the balance table manifest its factual balance. The fault happened when the constant, which was not permanently violated, was not fixed before ceasing the monitor lock by attaining an external call.

3.2 Distributed ledger for DAO

Distributed ledger technology (DLT) enables organizations to share data and code. Novel distributed applications are emerging such as DAPPs which are developed by groups of common suspecting institutions which unitedly try to figure out the issues of decentralization in their individual organizations.

Distributed Ledger Technology cuts down the requirement for a single centralized party, to supervise transactions, contracts and other processes, which were earlier based on centralized governance. Initially, it appears beneficial but there are critical worries for parties which relies on centralized body such as banks. Current rules are difficult to enforce as distributed ledgers are not susceptible to a particular legislation.

It is difficult to organize groups of mutually distrusting organizations to develop distributed applications. The growth in the development of the DAPP institutions is inferior as they are hesitant to connect with the distrusting organizations, unless there is a mutual model to share code and data.

Many projects turned back to develop new centralized company to supervise the governance of their DAPP. This makes the usage of distributed ledger technology unsuccessful and difficult to actualize the advantages of distributed ledger technology [23].

3.3 Use case/applications of DAO

(a) Ethereum DAO

The Ethereum DAO functions on Smart contracts. Smart contracts are simple programs that execute the logic built by the developers. Smart contracts have several applications which not only include funds transfer but also other applications in waste management industry. For example, via smart contracts reward points can be automatically transferred to people who manage their waste in a good way [24].

(b) Democratizing ownership of organization through smart contract on blockchain

1. **DigixDAO**: **DigixDAO** platform was established in 2016, for **utilizing token on gold for immutability, transparency** and auditability**. It was based on Ethereum, gaining advantage from the Distributed Ledger**. Further details can be recovered at https://digix.global [15]
2. **The DAO**: The DAO also called as Genesis DAO and was originated in 2016. It is a kind of capitalist directed venture capital fund. DAO code is open source. Further details can be recovered at http://DAO.Link [25]
3. **MakerDAO**: Maker DAO believes in the rationale that balanced digital assets are essential to attain complete potency of the blockchain technology. To accomplish that, it presented Dai, which is endorsed by collateral. They act to reduce the instability of DAI. It is a stable token in contrast to the U.S. dollar. Further details can be recovered at https://makerdao.com [26]

(c) Use cases of DAO

DAO implementation on use cases described below can benefit them immensely.

1. **Pay as a service**: DAO can bring in conceptions such as "Pay by use," constructing applications around it. For example, ***Slock.it***. ***Slock.it*** is a DAO recorded company which combines blockchain technology with the Internet of things (IoT) by forming a secure and decentralized ubiquitous Sharing network for renters and owners for using renting services. There will be no requirement of third party in between to deal with the payment, such as Airbnb, PayPal or Visa.
2. **Crowdfunding/ICO**: DAO's are generally used as vehicles for crowdfunding. Tokens are provisioned in exchange for funds through Ethereum and Bitcoin crypto-currencies and deployed using the smart contract for automatic collection and allocation of tokens.
3. **Governance**: DAO's can be employed to each and every aspect of governance. Using DAO people can vote for any new law or policy given by the government in a transparent fashion. People will not need to rely on the intermediate party and they will have more power to create the changes for the improvement of the society.

4. Implementations of DAO

4.1 Ethereum

Ethereum's aim is to create an alternate protocol for constructing decentralized applications. These applications provision distinct tradeoffs which are very effective for extensive division of decentralized applications. Specific importance is given on conditions where applications require rapid development time, efficient interaction, and security for moderate and seldom utilized applications [27].

4.2 Implementations of DAO

This section introduces a prospective solution employing blockchain Ethereum, [28,29]) which incorporates a Turing complete programming language with smart contract computing functionality. A solution is elaborated that permits the formation of organizations where participants preserve straight real-time check of contributed collects and governance policies are formalized, automatized and imposed using software. Distinctly, basic code for smart contract [30,31] is composed to make a Decentralized Autonomous Organization (DAO) on the Ethereum blockchain. Next, we explain the working of DAOs code, centering on fundamental establishment and governance characteristics, which includes organization, formation and voting rights.

Ethereum blockchain constructs a conceptual structure layer with a built-in Turing-complete programming language. This layer permits any person to write smart contracts and decentralized applications where they can produce their own discretional laws for ownership, transaction formats and state transition functions [24].

(a) Ethereum accounts

In Ethereum, the state consists of objects known as "accounts," where every account has an address of 20-byte size and state transitions accounting for straight transfers of value and information between accounts.

An Ethereum account comprises of four values:

- A counter **nonce** to ensure every transaction is processed only once
- Up to date account's **ether balance**
- **Contract Code** of account
- **Storage** of account

"Ether" is the cryptocurrency of Ethereum, used for paying transaction fees. Generally, two kinds of accounts exist, one is account which is owned externally and other is contact account. The external accounts are operated by private keys and have no code whereas the **contract accounts** are operated by contract code.

(b) Messages and transactions

"Transaction" in Ethereum denote digitally signed data package holding a message. Transactions contain [24]:

- Receiver of the message
- Digital signature by the sender
- The total number of ether that need to transfer from the sender to the receiver
- Optional data field
- A *Startgas* value, denoting the upper limit of computational steps for executing the transaction.
- A *Gasprice* value, denoting the fee paid by sender per computational step.

Cryptocurrencies generally consists of the first three fields. The optional data field by default has no operation. However, there is an OPCODE in the virtual machine which is used by a contract to retrieve the data.

For Ethereum's anti-denial service model, the *Startgas* and *Gasprice* fields are essential.

Every transaction adjusts number of computational step, a code requires during execution to avoid computational loss. The computation is measured in gas unit. Generally, a single computation step costs a gas. Multiple gas fees are charged for several frames of data transaction.

(c) Messages

Contracts are able to send messages to other contracts. Messages are non-serializable virtual objects that subsist in the execution environment of Ethereum. A message comprises of:

- Message sender (absolute)
- Message receiver
- The total number of ether to transfer with the message
- Optional data field
- *Startgas* value

The message is generated by a contract during execution of CALL OPCODE during code execution. Contracts have relationships with other contracts.

The gas allowance allotted by a contract utilizes the total gas consumed by all transaction and all sub-executions.

For example, if an external agent X sends a transaction to Y with 1200 gas, and Y takes 700 gas before sending a transaction to Z, and the internal execution of Z takes 300 gas before repaying, then Y can expend another 200 gas before ceasing of gas.

The state transition function for Ethereum is $EMPLOY\,(R,\ TS) \rightarrow R'$. Function is defined as follows (Fig. 5):

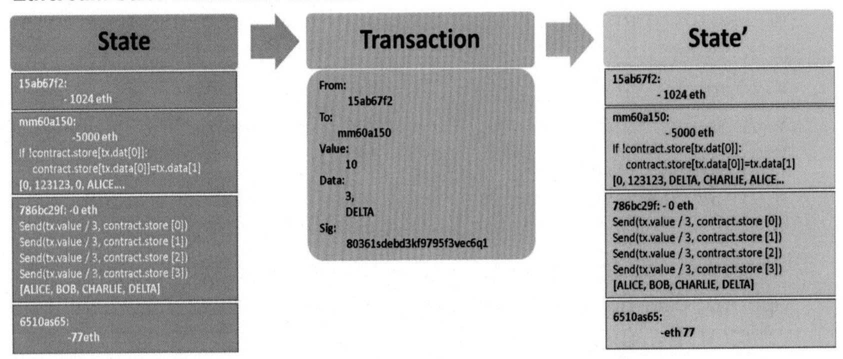

Fig. 5 An illustration of state transition of Ethereum smart contract.

1. Examine if the transaction is correct and the signature is legal, and the nonce corresponds the nonce in the sender's account. Return ERROR if the condition is not met.
2. Compute the transaction fee as *Startgas* * *Gasprice*, and from the signature find out the broadcasting address. Deduct the fees, maintain the balance of the sender's account and increase the sender's nonce. Return an ERROR message if adequate balance is not there.
3. Assign *Gas* = *Startgas* and deduct a definite amount of gas per byte to pay for the transaction bytes.
4. Move the transaction amount from the sender's account to the receiver's account. Make the receiver's account if it is not there. If the receiver's account is a contract, execute contract's code till end or till the execution consumes the gas.
5. If the transfer value fails due to sender not having adequate balance, or the code execution consumes the gas, then roll back all state alterations excluding the fees payment and add up the fees to the miner's account.
6. Else, repay the fees for all leftover gas to the sender, and transfer the fees to the miner for gas consumed.

For example, assume that the contract's code is:

if!auto.store[assigndatapayload(0)]:

auto.store[assigndatapayload(0)] = assigndatapayload (64).

This is high-level language Serpent code and this code can be compiled to EVM code. The contract code is scripted in the low-level EVM code.

Assume the contract's store kickoffs vacant, and a transaction is transmitted with 20 ether value, 3000 gas, 0.001 ether *Gasprice*, and 64 bytes of data, with bytes 0–31 expressing the number 2, bytes 32–63 expressing the string CHARLIE and bytes 64–127 expressing the string DELTA. The state transition function process in this situation is described as follows:

- Examine that the transaction is legal and correct.
- Examine that the transaction sender has at least 3000 * 0.001 = 3 ether. If true, then deduct three ethers from the sender's account.
- Assign gas = 3000; supposing the transaction is 200 bytes of length and the byte-fee is 6, deduct 1200 so remaining left is 1800 gas.
- Deduct 20 more ether from the sender's account and add up to the contract's account.

- Execute the code. The code examines if the contract's storage at index 3 is occupied, finds if it is not occupied, then it fixes the storage at index 3 to the value DELTA. Assume it takes 287 gas, so the remaining left amount of gas is $1800 - 287 = 1513$.
- Add 1513 * $0.001 = 1.513$ ether back to the sender's account and return the resulting state.

The total transaction fee is equal to the supplied *Gasprice* multiplied by the transaction length in bytes, in case if there is no contract at the receiver's side of the transaction.

The working of messages is equivalent to transactions with regard to returns. If some message execution ceases the gas, then execution of that message and other messages activated by that message returns, but parent executions don't require returning. It is harmless for one contract to call another contract. If X calls Y with G gas, then X's execution is assured to consume at most G gas. CREATE OPCODE creates a contract and its working is similar to CALL.

(d) Code execution

Ethereum Virtual Machine Code or EVM code is a low level, stack-based bytecode language. Ethereum contracts are written in EVM code. Every byte in the code expresses a function. When the code executes, it enters in an infinite loop that consists of several times execution of the function at the current program counter. The program counter starts at zero and increment by one till the code is terminated or in case of an error or if instructions like STOP or RETURN is noticed [32]. The functions can retrieve to three kinds of space to store data:

- **Stack**, LIFO (last-in-first-out) data structure in which data can be pushed and popped
- **Memory**, an array of bytes
- The contract's long-run reset key and value **storage**.

The conventional execution model of EVM code is amazingly simple. EVM is determined by a tuple having eight attributes such as tuple (block_state, transaction, message, code, memory, stack, pc, gas), where block_state is the state containing all accounts globally and permits balances and storage. In each round of execution, at the PCth byte of code we find the current instruction. For example, the ADD operation in stack pops two data values from the stack and pushes their sum in the stack. Further, the operation reduces gas by 1 and the PC is incremented by 1. The STORE operation pops two top data values from the stack and appends the second data value at the contract's storage at the address mentioned in the first data value in the stack.

(e) Blockchain and Mining

In many ways the Ethereum blockchain is alike the Bitcoin blockchain, but still there are few dissimilarities. With respect to the architecture of Blockchain, the blocks in Ethereum comprise a copy of transaction list and the latest state whereas Bitcoin only contains a copy of the transaction list [32].

The fundamental block validation algorithm in Ethereum is as follows:

- Examine that the previous block referenced is legal and is present.
- Examine that the timestamp of the block is more than the previously referenced block and less than 15 min into the future.
- Examine the legality of the block number, difficultness, transaction root, uncle root and gas limit.
- Examine that the Proof of Work (PoW) is legal on the block.
- Let state is S [0] at the end of the previous block.
- Let the block transaction list is TL, having n transactions. For all i in 0 ... n − 1, set S [i + 1] = APPLY(S[i], TL[i]). Return an ERROR if total gas spent in the block overreach the GASLIMIT or the application returns an error.
- Let the final state be S[n], after adding the block incentive point paid to the miner.
- To check the validity of the block, examine if the Merkle tree root of the final state S_FINAL is equal to the final state root provided in the block header otherwise, it is not valid.

As the complete state is required to store in each block, the approach may appear extremely ineffective at first glimpse. The efficiency of the approach ought to be Conforming to that of Bitcoin. This is because the state is stored in tree structure, and only a small portion of tree requires modification. Therefore, the tree structure are quite alike for two neighboring blocks and data can be stored once and referenced twice using pointers. To do so, a "Patricia tree" is used, considering an adjustment to the Merkle tree conception which allows for not only effective change of nodes but also insertion and deletion of nodes [32].

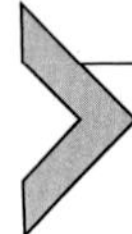

5. The future of organizations

5.1 Legacy organizations

Coordination of representatives enhances their skillfulness with regard to extrinsic contending market drivers. This is the fundamental source of the organization and the grounds companies desire to develop. However, cooperation of representative implicates enhancing cooperation monetary value as the company develops, and so companies cannot develop for an indefinite time. During development, companies require fixed structure in situ, and thus face a raising challenge to: (a) hold agility with regard to quickly varying considerations, and (b) maintain arrangement of concerns, faith and employment between their participants. In brief, the bigger a company is, the more internal hard work it needs to deal with and the smaller the company, more external competition is frequent. The true size of a company in general is the optimum equilibrium among these two pulls. From time to time, the foundation of a new technology or paradigm change revises the simplification of coordination cost, forcing up the scale and efficiency of companies to novel stages. The Internet allotted for an open, peer-to-peer and real-time information interchange on an international standard. Internet media in effect is more scalable than Traditional media.

5.2 Open organizations

The current economic system is established on win-lose game. The consequence of contest actuates growth for q execution, which is maximized with regard to local instead of global win. The noncritical Nash equilibrium has this problem. Nash equilibrium means that even if there is other beneficial collaborative balance for everybody, an individual actor does not have the influence to exclusively alter his/her scheme. Big-scale interaction has this trouble which forbids the conversion from contention to allegiance. Nash equilibrium is mostly frequent in production and development. This is depicted distinctly in the form of approach or non-clarity of knowledge and information and moreover by and large anti-competitor assets. Anti-competitor assets are infinite in consumption, and more they are used, more they become worthy.

For example, code is never exhausted, and the more it is looked upon it become finer. Simultaneously, enterprises have no bonus to open-source their code; else they would unsymmetrically devote reward to their challengers. If there are several contending companies which develop a similar product, they can gain more profit in co-developing the common components of their products rather independent production. Intellectual Property (IP) is the conventional intend to change anti-competitor assets into rare components, thus making them salable, although it has turned less reasonable and functional in today's fast pace of innovation. Beginning common assets is conflicting with the present economic system. Contrary, it is fundamental for big-scale, open collaboration, and DAOs. The DAO demands bonus and incentive for being effective while sharing of reusable components. The development and effectiveness of DAO will increase by present shareable components. Open organizations intend to change from the present non-collaborative Nash equilibrium to a prospective collaborative state. However, DAOs will replace the subsisting organized base not because they are better or more incorrupt, but instead simple as they will be much efficient.

5.3 Impact of DAO

There is prospective impact that the decentralized autonomous organizations can bring to the world irrespective of the ultimate failure of The DAO's. Because of this the "The DAO" was in the limelight with its vast short-term success. The technology is still growing as progressive intellectuals are inquiring about DAOs based projects and dApps [33]. This will guide us in a novel age of economic freedoms which is not controlled by third parties.

The supporters of **Dash** admits users to vote on teams and projects which will be financed as a part of its treasury model [34].

Finally, DAOs are novel and extremely observational gain to the technology world such like blockchain technology. According to the proponents, the history of DAO aids a great learning experience to the forthcoming project instead of rejecting their prospective due to the aftermath of the DAO.

5.4 The future of organizations

The power to coordinate well and organize an enormous amount of people is the most important force and drivers of society, which has become thru continuous development throughout thousands of years. In this section we discuss the issues of the legacy organization today, and a newly potential shape of web establishment: the DAO

(a) What DAO implementing can bring to the modern world?

The DAO completely control the technical way over the conventional ways of infrastructure and communications. However, there is a lack of hierarchyin the system. In contrast, when you are part of the

DAO, you are concerned in your work's results. You yourself bear the wholesome responsibility for your work, which makes you a better decision maker.

Every participant in the DAO project has crucial and equal rights. The success or failure entirely depends on each and every participant and it impacts every facet of DAO's work. An individual's impacts rest on the sum of money one has invested in the project. This viewpoint is more trustworthy than the obsolete system of power distribution. Anonymity is another astounding characteristic of the DAO. This means that the stakeholders do not need to prove their identity if they want to take part in the investment. Thus, it maintains privacy where there is no control of the government.

On the whole, decentralized autonomous organizations implementation is immensely beneficial and successful for our world than the conventional organizations, which is progressing toward a political, economic and business society. Another bright aspect of DAO is growth in the critical reasoning of large groups of people.

6. Summary

In this chapter we discussed Decentralized Autonomous Organization (DAO) based on smart contact of Ethereum blockchain, which enable the formation of organizations where participants preserve straight real-time check of contributed collects and governance policies are formalized, automatized and imposed using software. We provided the basic code for smart contract to form a Decentralized Autonomous Organization (DAO) on the Ethereum blockchain. We also explained the working of DAOs code, centering on fundamental establishment and governance characteristics, which includes organization, formation and their voting rights. DAOs are considered to agree to the expectation of the business work in the future.

Glossary

Smart Contracts The smart contract is a self-executing program with predefined contractual policy that provides additional security features such as control user data and monitor access rights in blockchain. The smart contract is executed autonomously when the set of predefined conditions are met. If the mutual contract between the two parties is breached, the smart contract handles the issue by removing the ability for peer-to-peer nodes to conduct transactions without the need for a broker.

Decentralized Autonomous Organization (DAO) Definition of Decentralized Autonomous Organization is an organization that can run on its own pre-defined protocols without having any hierarchical management. It operates based on the pre-defined smart contract code such as EVM in Ethereum platform.

Futarchy Futarchy was originally proposed by economist Robin Hanson as a futuristic form of government. Individuals would vote on a metric to determine how well their country (or organization) is doing, and then prediction markets would be used to determine the policies that best optimize the metric.

Ethereum Gas Gas is a unit used to calculate the amount of fees in order to execute certain operations in the Ethereum smart contract system. Gas means the amount of computational cost that it will take to execute certain transactions. There is no fixed price of conversion for Gas into Ether, usually the sender of a transaction specifies the highest gas price that one willing to pay.

References

[1] H.J. Leavitt, Applied organization and readings. Changes in industry: structural, technical and human approach, in: W.W. Cooper et al., (Ed.), New Perspectives in Organization Research, Wiley, New York, NY, 1962.
[2] D. Katz, R.L. Kahn, The Social Psychology of Organizations, Wiley, New York, NY, 1978.
[3] C. Jentzsch, Decentralized Autonomous Organization to Automated Governance, Founder, Slock, IT, White Paper, 2017.
[4] R.C. Merkle, DAOs, Democracy and Governance, Whitepaper, 2016.
[5] https://www.blockchain-council.org/blockchain/decentralized-autonomous-organization-dao-dao-works/.
[6] J. Achterbergh, D. Vriens, Managing viable knowledge, Syst. Res. Behav. Sci. 19 (2002) 223–241. https://doi.org/10.1002/sres.440.
[7] https://blockonomi.com/bitshares-guide/.
[8] https://bitsonline.com/vitalik-buterin-daico-new/.
[9] https://www.coindesk.com/information/what-is-a-dao-ethereum/.
[10] K. Wojdyło, J. Czarnecki, Blockchain, Smart Contracts and DAO, White Paper, 2016. http://www.codozasady.pl/wp-content/uploads/2016/11/Wardynski-and-Partners-Blockchain-smart-contracts-and-DAO.pdf.
[11] S. Beer, The viable system model: its provenance, development, methodology and pathology, J. Oper. Res. Soc. 35 (1) (1984) 7–25, https://doi.org/10.1057/jors.1984.2.
[12] R. Espejo, D. Bowling, P. Hoverstadt, The viable system model and the Viplan software, Kybernetes 28 (6/7) (1999) 661–678.
[13] R. Espejo, R. Harnden, The Viable Systems Model—Interpretations and Applications of Stafford Beer's VSM, Wiley, 1989. Chichester. https://searchworks.stanford.edu/view/1346153.
[14] V. Butrin's, https://vitalik.ca/general/2017/03/14/forks_and_markets.html", 2017.
[15] https://digix.global/.
[16] J. De Wilt, DAO, Can It Be Viable? An Exploratory Research on the Viability of a Blockchain Based Decentralized Autonomous Organization, Unpublished Master thesis, Radboud University, Nijmegen, Netherlands, 2018.
[17] White Paper, DAOStack: An Operating System for Collective Intelligence, White Paper, 2018. Work in Progress. https://daostack.io/wp/DAOstack-White-Paper-en.pdf.

[18] https://medium.com/ethex-market/how-to-use-the-alchemy-dapp-and-the-genesis-dao-6bfd91a357dc.
[19] S. Grossman, I. Abraham, G. Golan-Gueta, Y. Michalevsky, N. Rinetzky, M. Sagiv, Y. Zohar, Online detection of effectively callback free objects with applications to smart contracts, Proc. ACM Program. Lang. 2 (POPL) (2017) 28, Article no. 48. https://doi.org/10.1145/3158136.
[20] C. Smith, Blockchain, Smart Contracts and DAO, Wardyński & Partners, Whitepaper, 2016.
[21] M. Herlihy, Blockchains from a distributed computing perspective, Commun. ACM 62 (2) (2019) 78–85. https://doi.org/10.1145/3209623.
[22] C.A.R. Hoare, Monitors: an operating system structuring concept, Commun. ACM 17 (10) (1974) 549–557.
[23] https://medium.com/@rickcrook/the-case-for-de-centralisation-1ac14935a3fc.
[24] https://github.com/ethereum/wiki/wiki/White-Paper#decentralized-autonomous-organizations.
[25] http://DAO.Link.
[26] https://makerdao.com.
[27] F. Santos, V. Kostakis, The DAO: A Million Doller Lesson in Blockchain Governance, School of Business and Governance, Ragnar Nurkse Department of Innovation and Governance, 2018.
[28] V. Buterin's, A Prehistory of the Ethereum Protocol, https://vitalik.ca/2017-09-15-prehistory.html, 2017.
[29] G. Wood, Ethereum: A Secure Decentralised Generalised Transaction Ledger Byzantium Version 72dde18–2018-04-07, Yellow Paper, 2014. https://pdfs.semanticscholar.org/ee5f/d86e5210b2b59f932a131fda164f030f915e.pdf?_ga=2.69799320.968260507.1555470712-544867317.1555314991.
[30] N. Szabo, Smart Contracts: Building Blocks for Digital Markets, http://www.fon.hum.uva.nl/rob/Courses/InformationInSpeech/CDROM/Literature/LOTwinterschool2006/szabo.best.vwh.net/smart_contracts_2.html, 2000.
[31] K. Delmolino, M. Arnett, A.E. Kosba, A. Miller, E. Shi, Step by step towards creating a safe smart contract: lessons and insights from a cryptocurrency lab, in: J. Clark, S. Meiklejohn, P.Y.A. Ryan, D. Wallach, M. Brenner, K. Rohloff (Eds.), Financial Cryptography and Data Security—FC 2016 International Work-Shops, BITCOIN, VOTING, and WAHC, Christ Church, Springer Berlin Heidelberg, Barbados, 2016, pp. 79–94. https://link.springer.com/chapter/10.1007%2F978-3-662-53357-4_6.
[32] D.A.O. Ethereum, A Next-Generation Smart Contract and Decentralized Application Platform, White Paper. https://github.com/ethereum/wiki/wiki/White-Paper#decentralized-autonomous-organizations.
[33] https://news.coinsquare.com/learn-coinsquare/dapps-a-look-into-the-world-of-decentralized-applications.
[34] https://news.coinsquare.com/learn-coinsquare/dash-what-it-is-and-how-to-buy-dash-in-canada/.

Further reading

[35] P. Brinch Hansen, The programming language concurrent pascal, IEEE Trans. Softw. Eng. 2 (June) (1975) 199–206.

About the authors

Madhusudan Singh is currently Assistant Professor at School of Technology Studies, Endicott College of International Studies, Woosong University, Daejeon, South Korea and also Visiting Research Professor at Seamless Transportation Lab, Yonsei University, Songdo, Incheon, Korea. Prior, He was Research Professor at Yonsei Institute of Convergence Technology, Yonsei University, International Campus, Songdo, Korea (June 2016–February 2018). Dr. Singh has also worked as Senior Engineer at the Research and Development department of Samsung Display, Yongin-Si, South Korea (March 2012–March 2016). Currently, he is Senior Member of IEEE (SMIEEE) and member of ACM, IEEE Standard, IEEE Blockchain Working Group, and many more research and scientific organizations. He is also associated as Editor/reviewer/TPC member multiple International journals/conferences. He has published more than 50+ refereed research articles, 9+ national/International Patents, and delivered 15+ technical talks as speakers. His fields of research interests are Blockchain Technology, Machine Learning, Cyber Security, Software Engineering, Internet of Things, and Computer Vision.

Shiho Kim is a professor in the school of integrated technology at Yonsei University, Seoul, Korea. His previous assignments include, being a System on chip design engineer, at LG Semicon Ltd. (currently SK Hynix), Korea, Seoul [1995–1996], Director of RAVERS (Research center for Advanced hybrid electric Vehicle EnergyRecovery System), a government-supported IT research center. Associate Director of the Yonsei Institute of Convergence Technology (YICT) performing Korean National ICT consilience program, which is a Korea National program for cultivating talented engineers in the field of information and communication Technology, Korea [from 2011 to 2012], Director of Seamless Transportation Lab, at Yonsei University, Korea [since 2011 to present]. His main research interest includes the development of software and hardware technologies for intelligent vehicles, blockchain technology for intelligent transportation systems, and reinforcement learning for autonomous vehicles. He is a member of the editorial board and reviewer for various Journals and International conferences. So far he has organized two International Conference as Technical Chair/General Chair. He is a member of IEIE (Institute of Electronics and Information Engineers of Korea), KSAE (Korean Society of Automotive Engineers), vice president of KINGC (Korean Institute of Next Generation Computing), and a senior member of IEEE. He is the coauthor for over 100 papers and holding more than 50 patents in the field of information and communication technology.

CHAPTER FIVE

Blockchain applications in healthcare and the opportunities and the advancements due to the new information technology framework

Ramzi Abujamra, David Randall
American Research and Policy Institute, Washington, DC, United States

Contents

Abstract

The promise of blockchain extends into healthcare allowing a transformation of our current system and its use of information technology. The current use of information technology in healthcare can be advanced with blockchain. The fundamental values that blockchain will bring include decentralization due to its distributed ledger technology, interoperability and increased security and immutability. These benefits enable a healthcare system that is integrated, seamless and more secure and with an increased level of compliance. Blockchains offer the opportunity to deliver secure and highly efficient transaction processing between patient, provider and payer and other healthcare parties in a more efficient and seamless way. The use of smart contracts in healthcare is the catalyst by which blockchains can deliver these transactions in a transformative way. The application of blockchains combined with the Internet of Things (IoT) is also a catalyst that will enable these transactions to be processed to disrupt healthcare from its current state at this time. We explore ways that blockchain and its associated information technology will revolutionize our healthcare system that we believe would allow for a more efficient and fluid environment with improved cost efficiencies.

Advances in Computers, Volume 115
ISSN 0065-2458
https://doi.org/10.1016/bs.adcom.2018.12.002

1. Introduction

Blockchain technology applications have the potential to transform our current use of health information technology and the associated hardware and software infrastructure. The underlying technology and associated cryptocurrency with its decentralized architecture suggests a range of applications that can, we argue, bring cost savings and efficiencies versus traditional legacy systems currently in use in not only the public healthcare space but also in associated private market participants. We suggest that there are numerous applications that can be implemented in public programs in the US healthcare system using blockchain technology.

The fundamental promise of the blockchain is the underlying Information Technology (IT) architecture and its "unbreakable" chain of data entries that allow for secure and open transactions. The decentralized and distributed database of a blockchain that contains data allows for an auditable and distributed ledger that allows all to see every transaction. The open source attributes of the blockchain make the technology a natural fit for the requirements associated with the complexities of transaction laden systems associated with health information technology in the public and private sector.

We suggest that there are specific applications in public programs that include both the US Medicaid and Medicare systems that could benefit from the deployment of blockchain applications by replacing expensive hardware and software IT systems with a blockchain infrastructure. The Medicaid program spent $553 Billion in Federal Fiscal Year 2016 with that amount predicted to grow at 5–8% or more per year [1]. Administrative costs of the Medicaid program are estimated at 5–6% of total spend or nearly $30 Billion, which includes costs associated with health information technology deployment and maintenance [2]. In addition, we argue that the potential cost savings would be significant versus current costs and ongoing system maintenance of existing legacy IT systems. Blockchain is relatively easier to program and to implement system wide changes in comparison to changes made to legacy IT systems.

The advantages of blockchain are obvious, but with any new technology there are questions about efficacy and efficiency. In this paper we attempt to answer questions about the technology, issues of interoperability and specific applications related to the healthcare space and associated costs. We also address some of the privacy and data security concerns associated with healthcare that are inherent in any health IT system. Finally, we also examine the policy implications of deploying blockchain technology and suggest further areas of research.

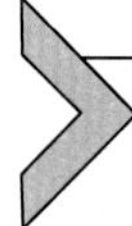

2. Technology description and framework for healthcare

The advent of blockchain technology can trace its beginnings to creation of Bitcoin as a digital cryptocurrency [3]. While the rise of digital only currencies in recent years has captured the attention of the financial industry, the public, and regulators it remains largely unknown to the general public. The technology that helped to launch Bitcoin or "blockchain" is of particular interest and potential use to a wide range of industries, including healthcare because of the open source and decentralized nature of the technology.

The fundamentals of blockchain technology are based on a decentralized distributed ledger. This consists of unbreakable chains that allow for secure and open transactions. These transactions are open for all to see, which is enabled by the decentralized nature of the blockchain [4]. In addition, transactions are secure and immutable; it is impossible to make any changes to the existing chains. In healthcare, the advantages of this type of distributed, secure and open architecture are clear as they address some of the core inefficiencies that are in existence today. One of the main benefits of blockchain is that it allows the elimination of a third party between transactions. This means that providers can interact directly with patients, for instance, and this type of transaction does not necessitate a third party or intermediary. With blockchains, all parties are equipped to add to the ledger and no need exists for a central intermediary to handle this type of transaction. Moreover, blockchains ensure a high degree of security and confidentiality due to the unbreakable nature of the chains. Legacy systems in existence today are thus no longer needed with the blockchain framework and transactions are performed immediately removing long delays and costs that exist with current legacy systems.

There are a variety of frameworks available to develop blockchains in healthcare. The most common and with the most applicability is the Ethereum blockchain. This blockchain framework is comparable to the Bitcoin blockchain that was the first development of blockchains in finance for cryptocurrency. Ethereum provides a framework to develop smart contracts in contrast to Bitcoin, due to its Turing complete capability [4]. Ethereum blockchains, like Bitcoin, are considered public blockchains, as anyone is allowed access to the blockchain. An alternative blockchain also popular in healthcare applications is IBM's Hyperledger blockchain. This blockchain is developed by the Linux foundation [5]. In addition, Hyperledger is considered a permissioned blockchain, which means that only designated parties are allowed access to the blockchain, in contrast to the public blockchain which is open to all. In healthcare, this type of blockchain design can be useful as it defines a defined user base consisting of providers, payers and other affiliated parties to only be allowed access in the blockchain. This type of permissioned access is conducive within the secure and confidential environment required in healthcare.

Many applications of blockchain in healthcare today leverage this type of architecture and there are many applications ranging from benefits, administration as well as clinical [6–8]. The basis of most of the applications is the use of smart contracts. These smart contracts are designed to adhere to certain rules that the blockchain follows. Transactions based on these smart contracts are executed in real time and with high degree of security between parties. These transactions are immediate and eliminate many of the inefficiencies that are present today in the legacy systems. These transactions are also more secure and enable a distributed architecture. Various healthcare stakeholders are able to interact in more complex relations and including with the patient as well. Integration of the electronic health record and other patient data becomes more easily integrated with other aspects of the health system, from providers to payers and pharmaceutical companies as well [9,10]. In this framework, the patient benefits greatly as they are added themselves as parties to the blockchain. This inclusion of the patient in the blockchain allows the distributed nature of the blockchain to capture data from the patient and enables that data to be integrated with the continuum of the care network. This can be revolutionary in our health system as the patient becomes central to the care system with this capability. This is further enabled with the strong security of blockchains to transmit the information with a high degree of confidentiality.

The interaction of stakeholders with the smart contract can be implemented with Decentralized apps (Dapps). These applications are similar to web applications in the realm of the Internet, but with the additional feature of operating on a decentralized blockchain network [4]. Dapps have many potential applications in healthcare [11,12]. Dapps are also conducive to working with a myriad of IoT technologies to deliver information across the healthcare continuum, allowing for a complete transformation of our current health system [13,14]. We see a way to revolutionize healthcare with an implementation of blockchains, ranging from providers, payers and government systems as well. The future of healthcare is a more robust and integrated system that is no longer beset by the inefficiencies of the past, the result due to the blockchain architecture.

3. Healthcare application

Blockchain has potential to improve healthcare in a number of innovative ways. Some of those examples include a Master Patient Identifier (MPI), autonomous automatic adjudication and interoperability [15,16]. MPIs offer a single person identifier that can follow the patient in various situations, enabling a more seamless and scalable health delivery across the continuum of care providers (as well as possibly beyond healthcare where data would also of relevance) [6].

Autonomous automatic adjudication would simplify and lead to significant efficiencies in how claims or other healthcare transactions are processed between parties. Essentially, blockchain could lead to elimination of the third party thus creating a more efficient process. The process would use smart contracts across parties that would enable automatic adjudication of claims. Enhancing claim adjudication in healthcare has the benefit of reduction in the occurrence of claim fraud that is currently prevalent [15]. All players in healthcare, from providers to payers, would benefit from this reduction in fraud.

Fraud detection and mitigation is increasingly becoming a policy issue for state Medicaid programs. Conservative estimates from the US Government Accountability Office (GAO) estimate that fraud in the Medicaid program is over $14 Billion per year or 4% of program spend and expected to grow as program spending increases [17]. Other research finds that Medicaid fraud could be as high as 15–22% of total spend as a result of over billing for services [18,19]. These findings suggest that the deployment of a blockchain application that utilizes a smart contract and verifiable ledger of all service and payment activities could reduce fraud and overpayment that is prevalent in the US Medicaid system.

Blockchain could also improve interoperability across systems and organizations. This is crucial for progress in our current health ecosystem which consists of a plethora of disparate IT legacy systems that have been amassed over the years and that do not communicate well with each other. Blockchain would provide the ability to replace these disparate systems with a single system that offers interoperability. With the use of smart contracts and fully auditable history, Blockchain would enable peer-to-peer interoperability among participants within transactions [16]. In addition to offering interoperability, blockchain transactions would also have the advantage of being cryptographically and irrevocable thus ensuring privacy across parties.

The patient would be able to designate by whom the data can be accessed (and at what level of access) by the use of keys that users would have access to (either private or public) [15].

The benefit of blockchain in healthcare would take place across the entire supply chain spectrum. Contractual agreements between payer and patient or provider and patient (or between provider and payer) would be implemented with the use of smart contracts within the blockchain [20]. Smart contracts lead to efficiencies as they enable a reduced number of intermediaries that exist today which lead to more streamlined transactions [21]. These transactions would enable a more holistic view of the patient's record for all parties involved and lead to an increase in transparency. Beyond contractual transactions, clinical transactions based on electronic health records would also occur on the blockchain. This would enable clinicians to have access to different components of the patient's data throughout the patient's lifetime with an increase in transparency (this access would be controlled so that providers have access to data only on a need-to-go basis) [22–25]. Improvements in the supply chain from blockchain would be extended to drug companies and manufacturers, as well as improvement to pharma clinical trials and longitudinal health research for the patient [15,26].

The federal government can also benefit from the improvement in blockchain with more streamlined transactions in the supply chain. For instance, one of the issues facing Medicaid recipients is high churn due to changing economic qualifications. Blockchain can be helpful to maintain a recipient's identity as they pivot between different governments systems in a more seamless way [27]. This would be especially valuable if account based plans from the private sector are implemented as an option for Medicaid recipients (with hope of reduction in costs and better outcomes) [28]. Other government programs would similarly benefit from a more streamlined delivery of care across all players within the supply chain, including with private prescription plans [20].

Overall, Blockchain applications have been surging in a number of different industries. According to Deloitte, 35% of health and life sciences companies surveyed plan to use blockchain, and 28% of respondents plan to invest $5 million or more (10% plan $10 million or more) [29]. In finance, it is expected that savings from reduced intermediaries (and slow payment networks) have potential savings of $15–20 billion by year 2022 [20,23]. The industries implementing blockchain are numerous and are led by real estate, supply chain and others. Within healthcare, a number of companies have begun implementations of Blockchain in various areas of applications including Gem (in collaboration with Philips Healthcare Blockchain Lab), PokitDok, Healthcoin, HashedHealth and many others [16]. The Hyperledger Healthcare Working Group is a consortium that provides an open source collaboration of member companies in healthcare with the goal to speed the development of commercial adoption of blockchain [5,26].

The applications of blockchain technology are numerous as well the potential to transform legacy health information technology. We focus our discussion on public health information technology because of not only its impact on the US healthcare system but also the complexities associated with integration and deployment that we believe deployment of blockchain technology can have not only ease of development and maintenance but also cost savings versus current IT infrastructure.

4. Records management

The promise of universal provider adoption of electronic records systems has been a "holy grail" for the health information technology market. Recent industry and government reports suggest that adoption of electronic health record standards outlined by the 2009 law that has funded much of the industry growth is lagging behind with a good deal of variation in provider adoption rates. These facts take on added significant as the Centers for Medicare and Medicaid Services (CMS) begin to tie future funding to provider adoption and "meaningful use" attestation by providers. Given these coming milestones and varying degrees of implementation success, serious questions remain about the future of wide-spread adoption of electronic health records systems and how the US healthcare system will look in the coming decade [30].

The US Recovery Act of 2009 provided an enormous financial boost toward adoption of electronic health records with over $34 Billion appropriated to fund provider information technology hardware and software solutions [31]. As implementation has progressed, there are numerous milestones established by both the authorizing statute, but also under CMS guidelines and regulations. The so called "meaningful use" guidelines are set in series of stages, with Stage 1 designed to facilitate data capture and sharing, which ended in 2012. Stage 2 of the program intended to advance clinical processes for all eligible healthcare professionals and hospitals and to report to CMS by the end of 2014. The final Stage 3 of the program was slated to be fully implemented by 2016 with goal of improved clinical outcomes as result of Electronic Health Records (EHR) use by providers, but has been delayed [32]. Each stage of the implementation process has had varying degrees of success with both hospitals and provider professionals as evidenced by both published research and various industry reports. The variation in adoption rates of EHR standards raises the greatest levels of concern among health IT professionals, policy makers and academics about the prospects for wide adoption of uniform electronic standards.

Adoption rates and the progress toward Stage 3 EHR standards continue to be of concern to not only providers, but policymakers as well. Recent reports suggest that while EHR use has grown from 20% of providers using the technology in 2002 to over 60% today, much of the increased use has been uneven across professional providers and hospitals. Rural hospitals have generally lagged their peers and older physicians (greater than age 55) have lower adoption rates [33]. Given these lower than expected adoption rates, many health IT firms have struggled to achieve CMS guidelines with their clients as many small providers struggle to implement Stage 2 standards, which leaves serious questions about how the majority of professional providers can achieve Stage 3 attestation and meaningful use standards.

Federal government agencies are not immune to the challenges of implementing EHR standards. Both the Veteran's Administration and the Department of Defense have had numerous implementation issues in their attempt to use EHR technology and to increase the interoperability of their respective systems [34]. Government records management in general and EHR specifically suggest that the deployment of a blockchain solution could potentially alleviate the issues discussed. Most notably the issues surrounding interoperability and secure data access can be achieved through APIs in a blockchain environment.

5. Medicaid management information systems

The Affordable Care Act authorized the US Department of Health and Human Services (HHS) and the Centers for Medicare and Medicaid Services (CMS) to create health information technology systems to assist in the implementation of the programs created under the law. Central to the architecture of the ACA is the coordination of diverse set of disparate programs (including Medicaid and various social programs) and accompanying legacy technology systems that are needed to verify eligibility, enrollment and ongoing programmatic support. Inherent in the implementation of these ACA programs are the varied nature of state policy choices and associated complexity [35].

In anticipation of the passage of the Affordable Care Act, CMS developed and codified a policy and financing structure designed to provide states with tools needed to achieve the immediate and substantial investment in information technology systems. In order to achieve this goal, federal funding is provided through a variety of venues to help states improve their eligibility and enrollment systems.

Central to the implementation process is the creation of a Medicaid Information Technology Architecture (MITA). MITA is a national framework to support improved systems development and healthcare management for the Medicaid enterprise. MITA has a number of goals, including development of seamless and integrated systems that communicate effectively through interoperability and common standards. The MITA standards are a critical standard for states in linking the complex systems associated with Medicaid Management Information Systems or MMIS.

The MMIS is an integrated group of procedures and computer processing operations (subsystems) developed at the general design level to meet principal objectives. For Title XIX purposes, "systems mechanization" and "mechanized claims processing and information retrieval systems" are identified in section 1903(a)(3) of the Act and defined in regulation at 42 CFR 433.111. The objectives of this system and its enhancements include the Title XIX program control and administrative costs; service to recipients, providers and inquiries; operations of claims control and computer capabilities; and management reporting for planning and control. States may receive 90% Federal Financial Participation (FFP) for design, development, or installation, and 75% FFP for operation of state mechanized claims processing and information retrieval systems approved by the secretary.

The MMIS program spent $3.7 Billion in 2015 and total administration and other technology spend on eligibility systems, electronic health records and technology associated with administration was over $25 Billion in 2015 [22]. Often, there are integration and interoperability issues that cause significant cost overruns and additional maintenance costs to both the state and federal government. The promise of blockchain can potentially solve these issues at a reduced cost due to relative ease of deployment versus traditional hardware and software infrastructure. We argue that the MMIS program is a candidate for the deployment of blockchain to only reduce costs but gain additional efficiency to deal with the inevitable changes in the program and technology.

6. Benefits administration

Blockchain technology has the ability to simplify and reduce the cost of benefits administration. As benefits design, enrollment, beneficiary engagement and provider payment systems have evolved overtime there are the inevitable and predictable issues associated with the myriad of health information technology systems designed to work together. Interoperability has become a key concern and challenge in the development and deployment of the health IT infrastructure. We suggest that the use of blockchain application can assist in dealing with the challenges and allow for potentially infinite modularity and allow multiple systems to work in greater efficiency. Benefits administration involves the use of disparate systems that are designed to gather and process data from numerous sources. This fact is especially true in public healthcare systems, including Medicaid and Medicare as well as Health Insurance Exchanges administered by the states and the US federal government. Integration complexity is a hallmark of these systems as previously discussed.

The challenges associated with blockchain deployment with any complex healthcare system not only include interoperability, but also issues associated with data access and privacy [24]. Data privacy and the ability to access sensitive patient specific information are some of the key challenges in the design of a healthcare blockchain application. Numerous researchers and software engineers suggest that addressing authentication of users can be achieved through design of systems that utilize a Proxy pattern to facilitate the transfer of data [25].

The promise of using blockchain applications is readily apparent when considering how the deployment of the technology can be flexible and address key issues of interoperability and data privacy. There are numerous processes and systems that potentially meet the criteria and challenges with the new technology; this includes the use of benefit incentive systems such as wellness programs, enrollment and eligibility systems and the previously discussed records management systems surrounding EHR deployment.

Current benefits administration systems suffer from several key limitations that we argue are a direct result of the centralized nature of health information technology today. We also suggest that existing large-scale health IT infrastructure is inflexible and costly to maintain and update to deal with programmatic technology changes. As an example, we argue that current eligibility and enrollment systems have rigid logic rules that often cannot deal with the inevitable issue with beneficiaries and their needs. This fact is largely a result of centralized networks that are often inflexible.

The centralized benefit administration is synonymous with data silos that are designed to keep people out and data secure, which are admirable attributes in today's cyber security environment but also prevent personalization of benefits and reduce timely coordination among various stakeholders. We suggest that the documented issues of duplication, fraud, and overutilization of service are traceable to the need for data security and data availability. The centralized nature of current systems leads to a limited and narrowly defined views of what data needs to be shared by whom to whom and who owns the data. We argue that the data belongs to the individual, not the system nor the corporate entities that build the system. We believe that a blockchain application environment utilizing smart contracts can remove many of the barriers and limitations inherent with centralized technology systems.

The potential impact of decentralized benefit and care management system could have numerous benefits to the healthcare system. We suggest that the impacts would include instant eligibility and recognition of the benefits through the use of a smart contract. Additionally, a blockchain application could provide an unbreakable chain that would allow for individualized care through smart care contracts that do not break administrative or clinical systems. We believe that the convergence of clinical and administrative systems that can be individualized through a blockchain smart contract would reduce the duplication inherent in centralized data silo systems.

The deployment of all the needed applications in a blockchain environment would naturally require the use of portals that would allow each of the needed systems to communicate. As discussed in addressing the use case for records management, interoperability becomes a prime challenge. The development of a so-called "smart benefit" design system we suggest can be achieved through blockchain applications that work together through the use of smart contract design similar to the Ethereum model with modifications that allows for the inclusion of HIPAA compliant privacy protocols that providers can modify data of patients via specific permissions.

The potential for the use of an integrated approach and associated challenges are issues that be solved through further software development and the use of a blockchain application protocol that allows for scalability, interoperability and unique patient and provider access. The ability for the applications to be easily modified we suggest will lead to significant cost reductions in not only IT system integration but also ongoing maintenance and modifications that offer the ability of blockchain to be far more flexible than current health information technology systems.

7. Policy implications and conclusions

We argue that blockchain technology has the ability to address the documented shortcomings of public and even private legacy health information technology systems. In each of the above addressable programs and markets core issues of interoperability, data access and privacy and the ability to adapt to changing programs and technology are potentially solved through the use of Blockchain applications. We suggest that deployment of blockchain applications be incremental in nature as with any emerging technology.

As discussed, federal and state agencies are relatively slow to adopt and adapt large scale information technology changes. This is evidenced by both the EHR rollout by the Veteran's Administration as well the processes that CMS utilizes for MMIS, which are understandable and somewhat rationale by policymakers given the scale and scope of populations being served. We argue as with Defense Advanced Research Projects Agency (DARPA) government examine selective system implementation of blockchain that can achieve elusive program goals of interoperability, flexibility and needed privacy considerations through experimental programs that focus on research that can quantify functionality of the technology and cost savings associated with deployment and ongoing maintenance costs.

We suggest that further research should focus on costs associated with deployment, ongoing maintenance of systems and the ability of systems integrators to adapt to changing programmatic mandates as well as technology innovation. Additionally, we suggest that use of multiple types of applications across varying program functions would be useful to examine as we have described in this paper. As with any new technology, there will be undoubtedly be issues in deployment and implementation but we believe that given the adaptive nature of the blockchain and the ability to make changes in applications that efficiency and efficacy will demonstrate the possibilities we discuss.

References

[1] http://www.kff.org/medicaid.
[2] http://www.kff.org/report-section/medicaid-financing-how-does-it-work-and-what-are-the-implications-issue-brief/#endnote_link_152625-6.
[3] http://www.bitcoin.org/bitcoin.pdf.
[4] J. Bambara, P.R. Allen, Blockchain a Practical Guide to Developing Business, Law, and Technology Solutions, McGraw-Hill Education, New York, 2018.
[5] http://www.hyperledger.org/.
[6] https://www.healthit.gov/sites/default/files/11-74-ablockchainforhealthcare.pdf.
[7] T.T. Kuo, H.E. Kim, L. Ohno-Machado, Blockchain distributed ledger technologies for biomedical and health care applications, J. Am. Med. Inform. Assoc. 24 (2017) 1211–1220.

[8] M. Carruthers, L. Bai, R. Shirra, Super secure: distributed ledger technology in life and health insurance, The Actuary Magazine 15 (2018) 13.
[9] https://www.healthcarefinancenews.com/news/blockchain-will-link-payer-provider-patient-data-never.
[10] https://healthitanalytics.com/news/five-blockchain-use-cases-for-healthcare-payers-providers.
[11] http://catai.net/blog/2018/04/decentralized-applications-dapps-healthcare/.
[12] https://medium.com/ethereum-dapp-builder/heres-how-dapps-can-revolutionize-the-healthcare-industry-6fce85346f8f.
[13] M. Simic, G. Sladic, B. Milosavljević, A case study IoT and blockchain powered healthcare, in: The 8th PSU-UNS International Conference on Engineering and Technology (ICET-2017), Novi Sad, Serbia, 2017.
[14] https://www.postscapes.com/blockchains-and-the-internet-of-things/.
[15] https://www.forbes.com/sites/reenitadas/2017/05/08/does-blockchain-have-a-place-in-healthcare/#6afb55a71c31.
[16] https://www.linkedin.com/pulse/how-blockchain-can-solve-real-problems-healthcare-tamara-stclaire.
[17] http://www.gao.gov/assets/670/668233.pdf.
[18] http://ehrintelligence.com/2012/04/25/ehr-adoption-depends-on-age-group-practice-size-and-setting-and-specialty/.
[19] S.T. Parente, S. Oberlin, L. Tomai, D. Randall, The potential savings of using predictive analytics to staunch medicaid fraud, J. Health Med. Econ. 2 (2016) 2.
[20] https://www.linkedin.com/pulse/blockchain-transformational-technology-health-care-bruce-broussard?trk=vsrp_people_res_infl_post_title.
[21] https://www.healthit.gov/sites/default/files/8-31-blockchain-ibm_ideation-challenge_aug8.pdf.
[22] MACPAC (Medicaid and CHIP Payment and Access Commission), MACStats: Medicaid and CHIP Program Statistics, MACPAC, Washington, DC, 2016, pp. 78–81.
[23] http://www.coindesk.com/santander-blockchain-tech-can-save-banks-20-billion-a-year/.
[24] http://www.healthcareitnews.com/news/blockchain-faces-tough-roadblocks-healthcare.
[25] P. Zhang, J. White, D.C. Schmidt, G. Lenz, Applying software patterns to address interoperability in blockchain-based healthcare apps, in: The 24th Pattern Languages of Programming Conference, October 22–25, 2017, Vancouver, Canada, 2017.
[26] http://www.healthcareitnews.com/node/53648.
[27] https://www.healthit.gov/sites/default/files/14-38-blockchain_medicaid_solution.8.8.15.pdf.
[28] D. Randall, S.T. Parente, R. Abujamra, Medicaid expansion and the use of account-based health plans, Health Care Curr. Rev. 3 (2015) 134.
[29] http://www.healthcareitnews.com/news/how-does-blockchain-actually-work-healthcare.
[30] M. Ramlet, D. Randall, S.T. Parente, Insurer payment lags to physician practices: an opportunity to finance EMR adoption, J. Health Med. Inform. 4 (2013) 136.
[31] http://www.gpo.gov/fdsys/pkg/PLAW-111publ5/html/PLAW-111publ5.htm.
[32] http://healthit.gov/providers-professionals/meaningful-use-definition-objectives.
[33] S. Decker, E. Jamoon, J. Sisk, Physicians in nonprimary care and small practices and those age 55 and older lag in adopting electronic health record systems, Health Aff. 31 (2012) 1108–1114.
[34] http://www.healthcareitnews.com/news/congress-demands-va-dod-interoperability-progress.
[35] D. Randall, S.T. Parente, US medicaid managed care markets: explaining state policy choice variation, Insur. Mark. Co. Anal. Actuar. Comput. 3 (2012) 35–49.

About the authors

Ramzi Abujamra is an informatics and actuary professional. He has extensive experiences working in health insurance plans in data analytics and as an actuary. His experience spans working with both Medicare and Medicaid government programs, with a number of publications. He has completed a PhD degree from the University of Minnesota, Minneapolis in Healthcare Informatics.

David Randall currently serves as the Senior Policy Advisor and Advisory Board Member to the Solve.Care Foundation (https://solve.care) and is also the Resident Scholar with the American Research and Policy Institute (www.arapi.org) in Washington, DC.

Dr. Randall has extensive experience as a former top insurance regulator, legislative staff member, healthcare lobbyist, consultant and executive with not only insurance companies, but also with several provider trade groups in Washington. He has testified before both US House and Senate committees of Congress on a variety of health policy issues. He also has over two dozen peer-reviewed and professional publications as well as book chapters on health information technology, healthcare entitlements and the policy process surrounding healthcare reform. His research includes study of Medicaid and Medicare entitlements and privatization, Health Insurance Exchanges, healthcare analytics and healthcare information technology. He received his PhD from Kent State University.

CHAPTER SIX

Testing at scale of IoT blockchain applications

Michael A. Walker, Douglas C. Schmidt, Abhishek Dubey
Institute for Software Integrated Systems, Vanderbilt University, Nashville, TN, United States

Contents

Advances in Computers, Volume 115
ISSN 0065-2458
https://doi.org/10.1016/bs.adcom.2019.07.008

Abstract

Due to the ever-increasing adaptation of Blockchain technologies in the private, public, and business domains, both the use of Distributed Systems and the increased demand for their reliability has exploded recently, especially with their desired integration with Internet-of-Things devices. This has resulted in a lot of work being done in the fields of distributed system analysis and design, specifically in the areas of blockchain smart contract design and formal verification. However, the focus on formal verification methodologies has meant that less attention has been given toward more traditional testing methodologies, such as unit testing and integration testing. This includes a lack of full support by most, if not all, the major blockchain implementations for testing at scale, except on fully public test networks. This has several drawbacks, such as: (1) The inability to do repeatable testing under identical scenarios, (2) reliance upon public mining of blocks, which introduces unreasonable amounts of delay for a test driven development scenario that a private network could reduce or eliminate, and (3) the inability to design scenarios where parts of the network go down. In this chapter we discuss design, testing methodologies, and tools to allow Testing at Scale of IoT Blockchain Applications.

1. Introduction

Blockchain deployments (and specifically Ethereum, which is the main focus of this chapter due to its large installed base and its powerful smart contract language) are generally managed via programs that have different modes in which they can operate. They broadly fall into Command-Line Interfaces (CLI), RPC APIs, or creating Graphical Interfaces via the use of HTML pages and JavaScript code. These interfaces provide standard means to either run Ethereum applications within the clients themselves, or to interface other applications with the Ethereum clients.

In practice, however, the existing blockchain deployment interfaces lack built-in fault tolerance, most notably for either network communication errors or application execution faults. Moreover, Ethereum clients are deployed manually since no official manager exists for them. As a result, developers can—and do—lose all of their Ether (Ethereum's digital currency) due to insecure client configurations. Addressing this problem requires patterns and tools that enable the deployment of blockchain clients in a repeatable and systematic way. This requirement becomes even more important when integrating IoT blockchain applications (ITBAs). The IoT component of ITBAs add other requirements atop traditional blockchain applications due to their interactions with the physical environment and increased privacy concerns, e.g., thus preventing leakage of personal data, such as energy usage that would reveal a user's activity patterns in their home. Additionally, ITBAs may not only communicate over the blockchain, but may also use off-blockchain communications via TCP/IP or other networking protocols for reasons related to their operation.

In this book chapter we present Best Practices for Testing-at-Scale of Blockchain Systems making use of the structure and functionality of PlaTIBART, which is a Platform for Transactive IoT Blockchain Applications with Repeatable Testing that provides a set of tools and techniques for enhancing the development, deployment, execution, management, and testing of blockchain systems and specifically ITBAs. Sections of this chapter originate from an article by the name of PlaTIBART: a platform for transactive IoT blockchain applications with repeatable testing, and is available at https://doi.org/10.1145/3152141.3152392. In particular, we describe a pattern for developing ITBAs, a Domain Specific Language (DSL) for defining a private blockchain deployment network, Actor components upon which the application can be deployed and tested, a tool using these DSL models to manage deployment networks in a reproducible test environment, and interfaces that provide fault tolerance via an application of the Observer pattern. The technology/technical terms used in the book chapter are explained wherever they appear or Section 11 and then References are provided.

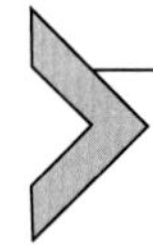

2. Introduction of distributed ledgers/blockchain testing concepts

Interest in—and commercial adoption of—blockchain technology has increased in recent years [1]. For example, blockchain adoption in the financial industry has yielded market capitalization surpassing $75 billion USD [2] for Bitcoin and $36 billion USD for Ethereum [3]. Blockchain's growth, at least partially, stems from its combination of existing technologies to enable the interoperation of non-trusted parties in a decentralized, cryptographically secure, and immutable ecosystem without the need of a trusted central authority. Blockchain, a specific type of Distributed Ledger, provides these features in different ways depending on implementation. However, generally blockchains work by creating a cryptographically signed chain of blocks, hence the name, that are decentralized via a consensus mechanism such as Proof-of-Work, that is not controlled by a central authority. Distributed Ledgers, which share many similarities to blockchain, do not necessarily require decentralized authority. However, for this chapter we discuss both but focus on blockchain versions of distributed ledgers due to the fully distributed non-central authority being easier to implement and manage, and therefore we assume more likely, for IoT manufacturers to integrate with. Blockchain deployments (and specifically Ethereum, which is the focus of this chapter due to its large installed base and its powerful, smart contract language) are generally managed via programs that have different modes in which they can operate. They broadly fall into Command-Line Interfaces (CLI), RPC APIs, or creating Graphical Interfaces via the use of HTML pages and JavaScript code [4]. These interfaces provide standard means to either run Ethereum applications within the clients themselves or to interface other applications with the Ethereum clients. In practice, however, the existing blockchain deployment interfaces lack built-in fault tolerance, most notably for either network communication errors or application execution faults. Moreover, Ethereum clients are deployed manually since no official manager exists for them. As a result, developers can—and do [5]—lose all their Ether (Ethereum's digital currency) due to unsecure client configurations. This problem is compounded by the fact that Ethereum's clients do not warn of this risk within their built-in help feature, and instead rely upon online documentation to warn developers. Addressing this problem requires patterns and tools that enable the deployment of blockchain clients in a repeatable and systematic way.

3. Testing analysis of blockchain and IoT systems

Blockchain systems can be subdivided into two broad categories: Turing Complete and Non-Turing Complete. This means the design of the system's contract language is either Turing Complete or it is not. The largest of each of these two categories is Bitcoin as a non-Turing Complete contract language and Ethereum as a Turing-Complete contract language. The reason this is important is because it describes the inherent design goal of language. Turing-Complete languages allow for theoretically any computation to be completed, whereas non-Turing Complete languages have a more limited instruction set that specifically limit the actions available in that language. The reason for adding these limitations to the language is to limit the functionality and therefore potential complexity of code put onto the blockchain's public ledger and executed distributedly. Non-Turing Complete contract languages are easier to analyze and predict runtime behavior, results, and potential faults. Additionally, there are blockchain/distributed ledger frameworks such as Hyperledger

Fabric which do not provide a specific public blockchain for use, but instead provide tools for developing customizable blockchain/distributed ledger applications or implementations modularly.

During roughly the same time as the growth of blockchain, the increased proliferation of IoT devices has motivated the need for transactional integrity due to the transition of IoT devices from just being smart-sensors to being active participants that impact their environment via communication, decision making, and physical actuation. These abilities require transactional integrity to provide auditing of actions made by potentially untrusted networked third party IoT devices. The demand for transactional integrity in IoT devices that simultaneously leverage blockchain features (such as decentralization, cryptographic security, and immutability) has motivated research on creating transactive IoT blockchain applications [6,7].

This requirement becomes even more important when integrating IoT blockchain applications (ITBAs). The IoT component of ITBAs add other requirements atop traditional blockchain applications due to their interactions with the physical environment and increased privacy concerns, e.g., thus preventing leakage of personal data, such as energy usage that would reveal a user's activity patterns in their home [8]. Moreover, ITBAs may not only communicate over the blockchain, but may also use off-blockchain communications via TCP/IP or other networking protocols for the following reasons:

- There are interactions with the physical environment that might require communication with sensors and/or actuators. For example, a user's smart-meter might communicate wirelessly with their smart-car's battery to activate charging based on current energy production/cost considerations.
- The distributed ledger (which makes an immutable record of transactions in blockchain) is public, so it is common to only include information within transactions that can safely be stored publicly. In particular, if some or all data from a transaction must be kept secret for privacy or any other reasons the transaction can, instead, contain the meta-data and a cryptographic hash of the secret data. Private information must, therefore, be communicated off-blockchain while still preserving integrity by storing meta-data and hash information on the blockchain ledger.
- Management tasks such as: updates, monitoring, calibration, debugging, or auditing may require off-blockchain communication (with possible on-blockchain components for logging). Currently, these management tasks are done manually in conventional blockchain ecosystems. Similar

to the need for a systematic means of deploying apps in a blockchain network, there is a need to systematically configure the network topology between all components of ITBAs.

4. Desired functionality of testing IoT blockchain systems

In this section we list desired functionality of Testing IoT Blockchain Systems. Specifically, what we believe is the simplest way to delineate progressive levels of increased testing of IoT blockchain systems. These stages, starting at the most easily achievable and becoming progressively more difficult, are: Unit Testing, Simple IoT Device Integration, Multiple IoT Device Integration, Test Driven Development, and Fully Automated Test-Driven Development.

Unit testing of software has become a standard requirement in well developed code. However, contract languages do not always include default unit-testing capability in the language or default build environment. However, the largest implementations for different categories of Blockchain solutions: Ethereum, Bitcoin, and Hyperledger Fabric all provide unit testing functionality, so any solution that does not do so should not be considered production level ready.

Beyond unit testing, the next level of desired testing of ITBAs is integration testing. Integration testing of purely software-based distributed systems provides a unique challenge due to coordination of multiple instances, networking and runtime configuration, etc. ITBAs compound this by requiring not only multiple software instances to be run for integration testing, but also require integration with the IoT component(s) of the system to verify runtime characteristics, hardware and software compatibility, etc. Therefore, we have decided to split the stage of testing with IoT devices into two sub-stages: one where integration testing is only done with one device, and then into a second stage where multiple devices are integrated into testing. This division provides a cleaner progression of desirability for analysis of testing progress.

The next level of desired testing ITBA systems is continuous integration. Continuous integration, like unit testing and integration testing, are commonplace in software development now. However, the adoption of these practices is less dependent upon the core blockchain, or even IoT, system being used and more about the support software designed to assist in development of that specific system. Therefore, like unit testing, we suggest considering any system that doesn't yet provide continuous integration support via support libraries, tools, etc. to be non-production level ready.

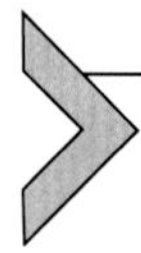

5. Existing shortcomings in testing IoT blockchain systems

This section reviews the state-of-the-art in IoT and blockchain integration, focusing on testing. Prior work [9] has shown that IoT and blockchain can be integrated, allowing peers to interact in a trustless, auditable manner via the use of blockchain as a resilient, decentralized, and peer-to-peer ledger. Work has also been done on the topics of security and privacy of IoT and Blockchain integrations [10,11]. Beyond that, work has focused on formal verification of smart contracts [12], and how to write smart contracts "defensively" [13] to avoid exceptions when multiple contracts interact. The current state-of-the-art with respect to testing, however, is lacking because blockchains are infrequently tested at scale in a systematic and repeatable manner, so we focus on that below.

5.1 Functional vs model-based declarations

Currently, as far as we can tell, PlaTIBART is the only model-based system for deploying Blockchain test networks, with Ethereum or otherwise. There are some tools, such as Nixos,[a] that provide for repeatable installation of their Linux distribution and therefore via use of the NixOps devops tool, can declaratively define deployments of private Ethereum networks. However, this still requires functional declaration of the instances to be created. The benefits of a model-based approach are that it allows much easier variation in the outputs, additionally, a model-based declaration can be modified to create the functional declaration inputs of other systems easily, thereby maintaining easy adaptability while also increasing interoperability with other tools, toolchains, and workflows.

5.2 Testing on live environments, non-repeatable

Blockchain systems, particularly Ethereum, focused extensively at the start on testing your smart contract code on a public, global, and non-modifiable instance of the Ethereum network they call the Test Network. Ethereum has at least added support for smart contract unit testing, testing smart contracts

[a] https://nixos.org/.

in an emulator, and calling that integration testing. However, these approaches lack robustness and repeatability.

The use of a public non-blockchain, even a testing one, for development poses several potential issues for developers. Firstly, the chance of publishing content to the blockchain that is intended to be secret is a high concern in a test environment. Secondly, reliance upon a public blockchain for testing removes the ability to control the frequency, latency, and predictability, or lack thereof, of your testing environment. This is important due to the common need for tests to be faster than real-time execution speed.

The use of an emulator to do integration testing of only the smart contract component of the system lacks robustness because of several reasons. First, it does not use the same client as production code would. Second, it ignores the need to include the client itself in the integration testing process. Third, it focuses on the HTML/JavaScript interface of the official client, while ignoring the other interfaces that geth provides, such as the JSON RPC API.

Therefore, we believe Ethereum has issues with the design philosophy of their testing mechanisms. Additionally, we have noted previously [14] that Ethereum's documentation was incomplete and spread across multiple pages for the same APIs, and as of the date of this publication the issue still exists.

5.3 Lack of defined integration/testing methodologies

Unfortunately, there is currently a severe lack of support for testing Blockchain systems and software when not using the precise scenarios envisioned by the Blockchain system's creators. For instance, Ethereum does not have any tools, testing or otherwise, that assist in integrating the official command line client of Ethereum: geth into applications. There is an official IDE, the Remix Solidity IDE, which enables unit testing but no support for integration testing at all currently. Their focus is on unit testing their smart contracts and "integration testing" their contracts inside a separate simulator, and not the geth client and private test networks. Other Blockchain and/or Distributed Ledger technologies, such as Hyperledger Fabric, at least have unit testing support and support integration testing, but at the time of writing, they have zero documentation on it.

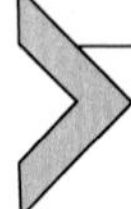

6. Platform for transactive IoT blockchain applications with repeatable testing (PlaTIBART)

The following sections will describe the PlaTIBART architecture, components, and components.

6.1 System design/rationale

PlaTIBART architecture for creating repeatable test network deployments of IoT/blockchain applications combines a Domain Specific Language (DSL) to define the network topology and settings, a Python program leveraging the Fabric API to manage the test network, and the RIAPS middleware [15] to facilitate communication between nodes on the network. Each of these components is described below.

6.2 Application platform

The *Resilient Information Architecture Platform* for Smart Grid (RIAPS) [15] is the application platform used by PlaTIBART to implement our case-study examples.]

RIAPS provides actor and component based abstraction, as well as support for deploying algorithms on devices across the network[b] and solves problems collaboratively by providing micro-second level time synchronization [15], failure based reconfiguration [7], and group creation and coordination services (still under active development), in addition to the services described in [18]. It is capable of handling different communications and running implemented algorithms in real-time.

6.3 Actor pattern

Each application client in the network is implemented as an actor with two main components: (1) a wrapper class specific to the role the actor is given and (2) a geth client, the reference client for Ethereum.[c] Fig. 1 shows a small network of five actors (indicated by an ellipse around a wrapper and geth client pair) and the networking connections between each actor's components. Geth clients communicate exclusively via on-blockchain means, i.e., the geth client of each actor communicates directly with its associated wrapper, and the wrapper communicates directly with other wrappers via an off-blockchain channel, such as TCP P2P communications.

6.4 Fault tolerance

A key benefit of decoupling the blockchain client and the wrapper into two components of an actor is enhanced fault tolerance around transaction loss, compared with tightly coupled solutions. Specifically, it allows the wrapper to not only monitor the blockchain client, but also shut down and restart the

[b] RIAPS uses ZeroMQ [16] and Cap'n Proto [17] to manage the communication layer.

[c] https://github.com/ethereum/go-ethereum/wiki/geth.

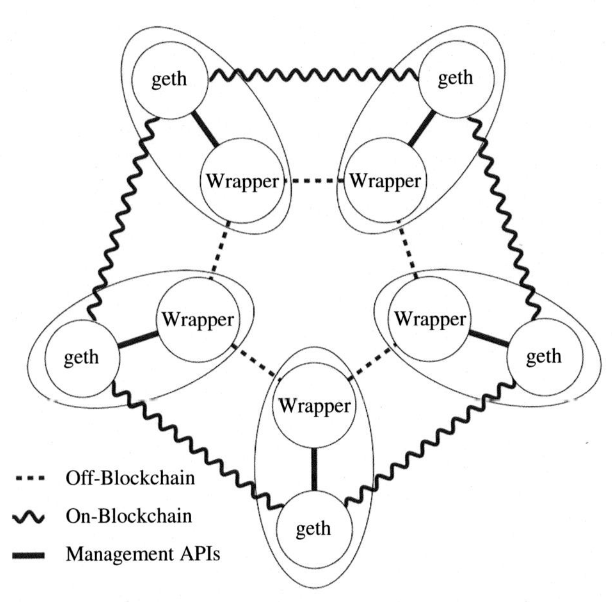

Fig. 1 Sample actor component network with an actor is a geth client and a wrapper.

client as needed. This design allows the wrapper component to ensure that if any known or discovered faults arise from defects in the blockchain software, the wrapper can at least attempt to recover.

For example, in our Ethereum test network described in Section 8, we have encountered faults where transactions are never mined [5] a client is restarted. These lost transactions are problematic since they prevent a client from being able to interact with the blockchain network. Other types of faults, such as those related to an actor's communication with other components of the network, are handled by other middleware solutions, such as RIAPS.

PlaTIBART applies the *Observer* pattern to notify the wrapper of the occurrence of events, such as faults and other blockchain-related conditions. This notification is accomplished by a separate thread within the wrapper that monitors its paired geth client for new events, such as completed transactions, or potential faults. This thread then notifies registered callback(s) when target events occur. For example, if the geth client becomes unresponsive or transactions appear to have stalled, then registered callback method(s) are called to notify the wrapper.

6.5 Domain specific language

PlaTIBART's DSL defines the roles that different clients in our network have, based on the Actor pattern. This DSL model implements a correct-by-construction design, thereby allowing for a verification stage on the model to check for internal consistency before any deployment is attempted. This verification prevents inconsistencies, such as two clients requesting the same port on the same host.

Fig. 1 shows an example of our DSL, which specifies a full network configuration file for a test network. The first two lines of the configuration file contain two unique identifiers for this test network and its current version, "configurationName" and "configurationVersion," respectively. Next, it contains values specific for the creation of an Ethereum private network's Genesis block.

A Genesis block in Ethereum is the first block in a blockchain and has special properties, such as not having a predecessor and being able to declare accounts that already have balances before any mining or transactions begin. The "chainID" is a unique positive integer identifying which blockchain the test network is using; 1 through 4 are public Ethereum blockchains of varying production/testing phases and should not be used for creation of private networks.

Next, "difficulty" indicates how computationally hard it is to mine a block, and "gasLimit" is the maximum difficulty of a transaction based on length in bytes of the data and other Ethereum runtime values. The "balance" is the starting balance that we allocate to each client's starting account upon creation of the network,[d] which eliminates the situation where clients cannot begin transactions to request assets before any mining has begun. Lastly, the "clients" represent the actual nodes in our network.

Fig. 2 shows how Clients are defined. Clients in the DSL represent the individual actors in our network, comprised of a geth client and a RIAPs instance using a wrapper interface. The geth client has two interface/TCP port pairs associated with it: one for incoming Blockchain connections, and one for administration and communication with RIAPs.

6.6 Network manager

Based on our experience developing decentralized apps (DApps) for blockchain ecosystems [14,19], three key capabilities are essential for DApps to function effectively in an ITBA ecosystem: traditional IoT computations and interactions should be supported, information should be robustly sorted in a distributed database, and a system-wide accepted sequential log of events

[d] "balance" applies only to accounts created before a new blockchain is created. Accounts created after the blockchain, be it public or private, is created will not receive any starting balance.

```
{" configurationNa me ":" test network a001 ",
" configurationVe rs io n ":"1" ,
" chainId ": 15 ,
" difficulty ": 100000 ,
" gasLimit ": 200000000000000000 ,
" balance ": 4000000000000000000000000 ,
" genesisBlockOutFil e ":" genesis - data . json ",
" clients ": {
        " startPort ": 9000 ,
        " prosumer ":{
                " count ": 15 ,
                " hosts ": [ "10.4.209.25" ,
                             "10.4.209.26" ,
                             "10.4.209.27" ,
                             "10.4.209.28" ]
                }
        , " dso ": {
                " count ": 1,
                " hosts ": [ "10.4.209.29" ]
                },
        " miner ": {
                " count ": 1,
                " hosts ": [ "10.4.209.30" ]
                }
        }
}
```


Fig. 2 Sample DSL model.

should be provided. Each requirement can be delegated to a separate layer in a three-tiered architecture. The first tier is the IoT middleware layer that facilitates communication between networked devices, which can be addressed by existing IoT middleware, such as RIAPS [15]. The second tier is a distributed database layer. The third tier is a sequential log of events layer, which can be solved by blockchain integration. PlaTIBART provides an architecture for coordinating all these layers in a fault tolerant manner, along with tools for repeatable testing at scale. It leverages the Actor model [20] to integrate these three layers.

Each layer is composed of components that accomplish their designated layer-dependent tasks. These components are then combined into a single actor that can interact with each layer and other actors in the network, as described in Section 8 Case Study: Transactive Energy System. Transactive Energy Systems (TES) have emerged in response to the shift in the power industry away from centralized, monolithic business models characterized by bulk generation and one-way delivery toward a decentralized model in which end users play a more active role in both production and consumption [21,22]. The GridWise Architecture Council defines TES as "a system of economic and control mechanisms that allows the dynamic balance of supply and demand across the entire electrical infrastructure, using value as a key operational parameter" [22]. In this paper, we consider a class of TES that operates in a gridconnected mode, meaning the local electric network is connected to a Distribution System Operator (DSO) that provides electricity when the demand is greater than what the local-network can generate. The main actors are the consumers, which are comprised primarily of residential loads, and prosumers who operate distributed energy resources, such as rooftop solar batteries or flexible loads capable of demand/response. Additionally, the DSO manages the grid connection of the network. Such installations are equipped with an advanced metering infrastructure consisting of TES-enabled smart meters. Examples of such installations include the Brooklyn Microgrid Project [23] and the Sterling Ranch learning community [24]. A key component of TES is a transaction management platform (TMP), which handles market clearing functions in a way that balances supply and demand in a local market.

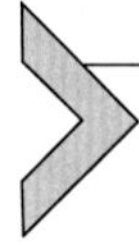

7. In-depth guided walkthrough of PlaTIBART network manager

The goal of PlaTIBAT is to use models to design and deploy repeatable testing networks for IoT Blockchain Applications. Therefore, we present a guided walkthrough of how PlaTIBART's Network Manager allows for simple command line creation of blockchain networks, currently only Ethereum but with more to come in future revisions of the software.

The Network Manager, having file name network-manager.py, is designed to be a command line tool for eventual integration into other systems, such as automated build systems, etc. Therefore, it follows best practices of command line tools and has a built-in help menu to assist users when learning the system. Additionally, it doesn't do anything to the system that can't already be done by a series of repetitive, and potentially complicated, command line instructions. Meaning that the purpose of the Network Manager is to simplify the process of creating blockchain test networks in a repeatable and model-driven manner. The use of a model allows the command line instructions to remain the same for almost all variations of supported network designs. Currently the Network Manager only supports networks designs where each blockchain node connects to each of the bootnodes, or to each of the first class of clients. Higher level of customization in network connections is a future area of research and development. Having the same series of instructions for the Network Manager enables a series of simple commands to be written that fully automate test network creation and testing. To show this, we'll be examining a complete cycle of creating, running, stopping and deleting a test network.

7.1 Guided walkthrough of creating new test network

The first step in creating a test network is to delete the temporary files that can be used. The following commands delete the /new-blockchain/ directory where the blockchain is created and saved to. The remaining files are possible files that may or may not be created depending on system design. The file static-nodes.json is a list of static nodes, a possible Ethereum discovery mechanism, that could have been previously created. The new miners and clients json files are lists of newly created miner clients and standard non-miner clients. Clients are separated into two categories due to the relatively massive memory requirements for miners, 4 GB minimum to even start mining and growing from there as the blockchain grows in length, versus the relatively minor approximately 250 MB of ram used by a non-miner client note these requirements being specific to the Ethereum network and Ethereum's client: geth. The genesis-data.json file is the traditional input file that geth uses to create a new blockchain network, the Network Manager creates this as an intermediary artifact during network creation (Figs. 3 and 4).

The first step in creating a new Ethereum test network is to create Bootnodes if your network is going to use them. We're going to assume a valid PlaTIBART model file is passed as a parameter to $1 in the following command line instructions both save space and to reinforce that the Network Manager's commands don't change based on input model.

```
rm -f ./static-nodes.json
rm -f ./new-miners.json
rm -f ./new-clients.json
rm -rf ./new-blockchain/
rm -f ./genesis-data.json
```

Fig. 3 Commands to delete network manager temporary files.

```
./network-manager.py bootnodes create --file $1 \
 --out=./static-nodes.json
```

Fig. 4 How to create bootnodes with the network manager.

The next step in creating a new Ethereum test network is to create the clients and miners defined in the input model file. The order of these commands isn't dependent upon one another (Fig. 5).

Next in creating a new Ethereum test network is to make the genesis-data.json input file that allows Ethereum's geth client to create a new blockchain network. This file contains all the meta-data about the network to be created, such as staring Ether for known clients, complexity of the beginning mining calculations, and the ChainID, which prevents unrelated chains from communicating with each other. This is then fed to the local copy of geth on the host machine and creates the genesis block (first block in a blockchain) (Fig. 6).

Here is the creation of the genesis block, which is done on the host machine's local geth client (Figs. 7 and 8).

This newly created block contains all the model's meta data and allows pre-mining distribution to each of the miners and clients. This pre-mining distribution helps prevent a potential race-condition where a client is always

```
./network-manager.py clients create --file $1 \
 --out=./new-clients.json

./network-manager.py miners create --file $1 \
 --out=./new-miners.json
```

Fig. 5 Creating miners and clients.

```
./network-manager.py blockchains make --file $1 \
 --clients ./new-clients.json
```

Fig. 6 Making genesis file for new test network.

```
./network-manager.py blockchains create \
 --file genesis-data.json --datadir ./new-blockchain/
```

Fig. 7 Creating the new blockchain genesis block.

```
./network-manager.py clients distribute --file $1 \
 --local ./new-blockchain/

./network-manager.py miners distribute --file $1 \
 --local ./new-blockchain/
```

Fig. 8 Distributing genesis block to clients and miners.

trying to "catch-up" to the newest created block and never does. If the mining difficulty is set too low compared to the processing power of the system(s) hosting the miner(s) this is a possibility that can occur.

At this point, localized logic and data files can also be distributed via the Network manager to the hosts for each one of the generated clients from the model. The specifics of these files will depend upon the ITBA(s) that you are testing. This example distributed the code used in our Use-Case 1. At this point the new Blockchain, Ethereum in this example, test network is now fully created and ready for use. Some use cases, such as our Use-Case 2, will archive this network for future use, while others will make use of it as is.

7.2 Guided walkthrough of starting and stopping test networks

Starting the network for mining and then the processing of data and requests start by starting the miners and clients. If the miners and clients can reach a single bootnode in the bootnode network, then they should eventually sync up if the network is moderately reliable. Otherwise, if not using bootnodes, the miners will need connected to the clients manually (Figs. 9–11).

Stopping the entire network can be done step-by-step by using the above commands, substituting "start" with "stop". Alternatively, you can use the Network Manager's network options to stop the entire network at once. Starting the entire network also works, but it was explained in detail above for clarity (Fig. 12).

```
./network-manager.py clients distribute --file $1 \
 --local ./components/ --subdir components/

./network-manager.py clients distribute --file $1 \
 --local ./data/ --subdir components/data/
```

Fig. 9 Distributing logic code and data to each client.

```
./network-manager.py miners connect --file $1
```

Fig. 10 Connecting miners to each client to connect the network.

```
./network-manager.py miners start --file $1

./network-manager.py clients start --file $1
```

Fig. 11 Starting miner and clients.

```
./network-manager.py network stop --file $1
```

Fig. 12 Network manager stopping entire network.

```
./network-manager.py network delete --file $1
```

Fig. 13 Network manager deleting entire network.

Deleting the network will delete all files created by the setup process on all clients, miners, and bootnodes in the network, but will not delete the host machine's generated files, if those are to be kept separately (Fig. 13).

8. Example use-case 1: Transactive energy

Transactive Energy Systems (TES) have emerged in response to the shift in the power industry away from centralized, monolithic business models characterized by bulk generation and one-way delivery toward a decentralized model in which end users play a more active role in both production and consumption [21,22]. The GridWise Architecture Council defines TES as "a system of economic and control mechanisms that allows the dynamic balance of supply and demand across the entire electrical infrastructure, using value as a key operational parameter" [22].

8.1 Sample problem

In this section, we consider a class of TES that operates in a gridconnected mode, meaning the local electric network is connected to a Distribution System Operator (DSO) that provides electricity when the demand is greater than what the local-network can generate. The main actors are the consumers, which are comprised primarily of residential loads, and prosumers who operate distributed energy resources, such as rooftop solar batteries or flexible loads capable of demand/response. Additionally, the DSO manages the grid connection of the network. Such installations are equipped with an advanced metering infrastructure consisting of TES-enabled smart meters. Examples of such installations include the Brooklyn Microgrid Project [23] and the Sterling Ranch learning community [24]. A key component of TES is a transaction management platform (TMP), which handles market clearing functions in a way that balances supply and demand in a local market.

8.2 In-depth guided walkthrough

To test PlaTIBART we implemented a solution to the Transactive Energy case study and deployed it to the test network defined in Fig. 2. This network was installed on a private cloud instance hosted at Vanderbilt University. We ran our tests on 6 virtual hosts, each with: 4GB RAM, 40GB hard drive space, running Ubuntu 16.04.02, and gigabit networking. For these tests we implemented a custom smart contract and wrappers for both Smart Grid distribution system operators (DSO) and prosumer clients in Python. Each wrapper had one geth client associated with it. We used PlaTIBART's network manager tool outlined in Section 6.6 and commands detailed in Sections 7.1 and 7.2 to create, start, shutdown, and delete the test network. We manually paired each wrapper with its geth client's IP address and port (in future work this is to be integrated and automated into the network manager's capabilities). Using our custom written wrappers, smart contract, and managed test network we simulated a day's worth of transactive energy trading between actors. Via the Linux "time" command we measured each step needed in the entire process to create a test network, including Clients Create, Miners Create, Blockchain Make, Blockchain Create, Distribute to Clients, and Distribute to Miners. We also measured the steps required to start and connect the geth instance for each "clients" ("prosumer" and "DSO") to the geth client of each "miner." Currently, this star-network is the only network topology supported by PlaTIBART, but we will expand the supported topologies in the future.

8.3 System output and analysis

After running our tests, described above, we found the standard deviation for each testing phase was small (the largest being 0.09% of the time taken). Likewise, the average time either remained relatively static, or scaled linearly, in relation to the number of clients (2, 5, 10, 15, 20 prosumers +1 DSO+ 1 miner).

The test phases that remained relatively static included: Miners Create, Blockchain Make, Blockchain Create, Distribute to Miners, Miners Start, and Network Delete. The test phases that scaled with increase in number of prosumers were: Clients Create, Distribute to Clients, Full Network Created, Clients Start, Network Connect, and Network Stop. The scaling increases were linear (Std Dev < 0.065) after dividing the average time increase by the difference in number of clients.

The results of our experiments indicate that there exists high consistency and predictability of managing PlaTIBART-managed blockchain test networks. These results help build confidence that PlaTIBART's approach to creating repeatable testing networks for IoT blockchain applications scales well, which is important to encourage adoption by IoT system developers.

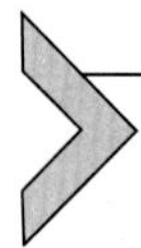

9. Example use-case 2: Blockchain/distributed systems education

As we discussed in An Elastic Platform for Large-scale Assessment of Software Assignments for MOOCs (EPLASAM), there are significant challenges presented when attempting to scale software assignments for use in MOOCs [25]. Attempting to scale distributed system, specifically ITBA, software assignments presents additional challenges beyond those of traditional software assignments. Even ignoring the need for physical IoT hardware to test code on, the need for: private repeatable Blockchain networks, easily adjustable network designs, and ease of use of network setup by both instructors and staff, but also learners, becomes crucial for individual assessment.

9.1 Sample problem

The use of PlaTIBART does not provide a complete solution, as described in EPLASM, for MOOC scalable assessment, or even just testing, of ITBAs. However, it does provide the design philosophy, tools, and methods that enable repeatable individual assessment of ITBAs and/or ITBA components. In our classes at both Vanderbilt University and Youngstown State University, we have made use of PlaTIBART to assist in the creation assignments that assessed the Blockchain components of ITBAs. Additionally, we were able to leverage PlaTIBART to provision 25 × IoT clusters (Cisco router, 4 × Raspberry Pi 3B +, ethernet cabling, and an IoT Electronics kit) for a series of IoT lectures and workshops held in partnership with Youngstown State University, the Youngstown Business Incubator, with support from Cisco.[e] These clusters included a custom PDF guide, all the software required to operate the electronics kits via Python, and all the software required to operate an Ethereum network on each of the clusters, including PlaTIBART being installed on each cluster's first Raspberry Pi device.

[e] https://oh-iot.com/.

```
{" configurationName ":" test network b001 ",
" configurationVersion ":"1",
" chainId ": 15 ,
" difficulty ": 100000 ,
" gasLimit ": 2000000000000000000 ,
" balance ": 4000000000000000000000000 ,
" genesisBlockOutFil e ":" genesis - data . json ",
" clients ": {
        " startPort ": 9000 ,
        " client ":{ " count ": 1,  " hosts ": [ "127.0.0.1" ] }
        " miner ": { " count ": 1,  " hosts ": [ "127.0.0.1" ] }
}}
```

Fig. 14 Blockchain assignment sample model json file.

9.2 In-depth guided walkthrough

The benefit to the design of PlaTIBART is that the only difference between creating a Dockerized container containing all the required code to allow students to easily start learning Blockchain and configuring IoT workshop clusters is slight modification of the input model file and changing the files distributed to each client. Fig. 14 shows the model input we used for creating the Docker image with a single client and a single miner, both on the same host, using localhost 127.0.0.1, but the network manager handles giving them ports in different ranges.

Now to support deploying to actual individual ITBA clusters the only requirement is that each system already have the host machine's public SSH key, have geth installed, and accept SSH connections.

Fig. 15 shows that the only change needed is adjusting the "client" section of the json. Increasing the number of clients and changing what IPs the clients will be installed on. Both scenarios were completed via the exact same command line instructions as discussed in Sections 7.1 and 7.2, the only difference being what files were distributed to each client.

9.3 System output and analysis

Both examples discussed in this section proved to be viable and were successful in creating ITBA component educational assessments. Both Docker images of Blockchain assignments for university courses, and IoT and Blockchain workshop preparation and instruction were successful. However, neither of these were studied in-depth for formal verification, but instead were used as a means of rapidly creating repeatable Blockchain test networks in the educational scenarios. Therefore, these examples show more the adaptability of the PlaTIBART design when crafting future ITBA assignments. Future work will need to verify the efficacy of this approach.

```
{" configurationName ":" test network c001 ",
" configurationVersion ":"1",
" chainId ": 15 ,
" difficulty ": 100000 ,
" gasLimit ": 2000000000000000000 ,
" balance ": 4000000000000000000000000000 ,
" genesisBlockOutFil e ":" genesis - data . json ",
" clients ": {
        " startPort ": 9000 ,
        " client ":{ " count ": 4,  " hosts ": [
                        "10.0.1.1","10.0.1.2","10.0.1.3","10.0.1.4" ] }
        " miner ": { " count ": 1,  " hosts ": [ "127.0.0.1" ] }
}}
```

Fig. 15 IoT/blockchain-cluster configuration sample model json file.

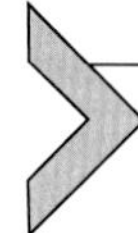

10. Research directions in testing at scale of IoT and distributed systems

Testing at scale of distributed systems is an ongoing research focus that will simultaneously have many different approaches. Formal validation of Blockchain related contract languages, systems, and tools, including PlaTIBART, is one area that will see continued focus. Additionally, verification of the efficacy of using PlaTIBART in an educational environment to teach ITBAs needs to be proven; this is a research area which we are currently pursuing. Expanding the network topology supported by PlaTIBART is future work that needs to be addressed, possibly with the integration with network simulation or management software. Work needs to be done on including private test Blockchain networks into both Unit and Integration Testing frameworks, and currently there don't appear to be tools, at least for Ethereum, for integration testing outside of Solidity IDE.

11. Key terminology and definitions

IoT Blockchain Applications (*ITBAs*): Blockchain Applications that run on an IoT system where Blockchain is leveraged for a wide range of potential uses ranging from distributed logging up to integration into the command and control decision making process of the IoT device. Both Blockchain and IoT use case scenarios can change drastically when they are used together in a system. Therefore, we use this term to describe the added complexity of Blockchain applications that interact with the physical world through IoT devices.

Testing-at-Scale: Testing distributed systems incurs a much heavier cost, both in complexity and required resources, when attempting to test a fully integrated distributed system versus traditional systems. This is because fully testing a distributed system requires a large amount of potentially heterogeneous devices running on potentially multiple platforms and/or architectures. This not only requires a system and methodology of testing that can be run easily with each new variation of the overall system, be it hardware or software, but also allows for consistent benchmarking, profiling, and analysis of performance, reliability, and other metrics.

References

[1] HBR, The Truth About Blockchain, HBR, (Accessed on 08/30/2017), January. https://hbr.org/2017/01/the-truth-about-blockchain, 2017.
[2] CoinMarketCap, Bitcoin (BTC) Price, Charts, Market Cap, and Other Metrics, CoinMarketCap, (Accessed on 08/30/2017), August. https://coinmarketcap.com/currencies/bitcoin/, 2017.
[3] CoinMarketCap, Ethereum (ETH) $381.84 (3.83%), CoinMarketCap, (Accessed on 08/30/2017), August. https://coinmarketcap.com/currencies/ethereum/, 2017.
[4] Ethereum Git Books, Interfaces | Ethereum Frontier Guide, (Accessed on 08/30/2017). https://ethereum.gitbooks.io/frontier-guide/content/interfaces.html, 2017.
[5] Github, Use RPC API Personal_SendTransaction Lost Coin Issue #14901 · Ethereum/Goethereum, Github, (Accessed on 08/30/2017), August. https://github.com/ethereum/go-ethereum/issues/14901, 2017.
[6] A. Bogner, M. Chanson, A. Meeuw, A decentralised sharing app running a smart contract on the Ethereum blockchain. in: Proceedings of the 6th International Conference on the Internet of Things, 177–178. IoT'16, ACM, Stuttgart, Germany, 2016. https://doi.org/10.1145/2991561.2998465. Isbn: 978-1-4503-48140.
[7] F. Buccafurri, G. Lax, S. Nicolazzo, A. Nocera, Overcoming limits of blockchain for IoT applications, in: Proceedings of the 12th International Conference on Availability, Reliability and Security, 26:1–26:6. ARES '17, ACM, Reggio Calabria, Italy, 2017. https://doi.org/10.1145/3098954.3098983. Isbn: 978-1-45035257-4.
[8] J. Gubbi, R. Buyya, S. Marusic, M. Palaniswami, Internet of things (IoT): a vision, architectural elements, and future directions, Futur. Gener. Comput. Syst. 29 (7) (2013) 1645–1660.
[9] K. Christidis, M. Devetsikiotis, Blockchains and smart contracts for the internet of things, IEEE Access 4 (2016) 2292–2303.
[10] A. Dorri, S.S. Kanhere, R. Jurdak, P. Gauravaram, Blockchain for IoT security and privacy: the case study of a smart home, in: IEEE International Conference on Pervasive Computing and Communications Workshops (PerCom Workshops), 2017, IEEE, 2017, pp. 618–623.
[11] A. Ouaddah, A. Abou Elkalam, A.A. Ouahman, Towards a novel privacy-preserving access control model based on blockchain technology in IoT, in: Europe and MENA Cooperation Advances in Information and Communication Technologies, Springer, 2017, pp. 523–533.
[12] R. Kumaresan, I. Bentov, How to use bitcoin to incentivize correct computations, in: Proceedings of the 2014 ACM SIGSAC Conference on Computer and Communications Security, ACM, 2014, pp. 30–41.

[13] K. Delmolino, M. Arnett, A. Kosba, A. Miller, E. Shi, Step by step towards creating a safe smart contract: lessons and insights from a cryptocurrency lab, in: International Conference on Financial Cryptography and Data Security, Springer, 2016, pp. 79–94.
[14] M.A. Walker, A. Dubey, A. Laszka, D.C. Schmidt, Platibart: a platform for transactive IoT blockchain applications with repeatable testing, in: Proceedings of the 4th Workshop on Middleware and Applications for the Internet of Things, ACM, 2017, pp. 17–22.
[15] S. Eisele, I. Mardari, A. Dubey, G. Karsai, RIAPS: resilient information architecture platform for decentralized smart systems, in: 2017 IEEE 20th International Symposium on Real-Time Distributed Computing (ISORC), 2017, pp. 125–132. May. https://doi.org/10.1109/ISORC.2017.22.
[16] ZeroMQ, P. Hintjens, ZeroMQ: The Guide, ZeroMQ, URL. http://zeromq.org, 2010.
[17] Cap'n Proto, K. Varda, Cap'n Proto, 2015. https://capnproto.org/.
[18] H. Lee, S. Niddodi, A. Srivastava, D. Bakken, Decentralized voltage stability monitoring and control in the smart grid using distributed computing architecture, in: 2016 IEEE Industry Applications Society Annual Meeting, 2016, pp. 1–9. October. https://doi.org/10.1109/IAS.2016.7731871.
[19] GitHub, JSON RPC—Ethereum/Wiki, Accessed on 08/28/2017. https://github.com/ethereum/wiki/wiki/JSON-RPC, 2017.
[20] E.A. Lee, S. Neuendorffer, M.J. Wirthlin, Actor-oriented design of embedded hardware and software systems, J. Circuit. Syst. Comp. 12 (03) (2003) 231–260.
[21] E. Cazalet, P. De Marini, J. Price, E. Woychik, J. Caldwell, Transactive Energy Models, Technical Report, National Institute of Standards Technology, 2016.
[22] R.B. Melton, Gridwise Transactive Energy Framework, Technical Report, Pacific Northwest National Laboratory, 2013.
[23] Brooklyn Microgrid, Brooklyn Microgrid - Home. http://brooklynmicrogrid.com/, 2017.
[24] Sterling Ranch Development Company, The Nature of Sterling Ranch, Sterling Ranch Development Company. http://sterlingranchcolorado.com/about/, 2017.
[25] M. Walker, D.C. Schmidt, J. White, An elastic platform for large-scale assessment of software assignments for MOOCs (EPLASAM), in: User-Centered Design Strategies for Massive Open Online Courses (MOOCs), IGI Global, 2016, pp. 187–206.

Further reading

[26] H. Agrawal, J.R. Horgan, E.W. Krauser, S.A. London, Incremental regression testing, in: Proceedings., Conference on Software Maintenance, 1993. CSM93, IEEE, 1993, pp. 348–357.
[27] A. Banafa, IoT and Blockchain Convergence: Benefits and Challenges—IEEE Internet of Things, IEEE, Accessed on 08/31/2017. https://iot.ieee.org/newsletter/january2017/iot-and-blockchain-convergence-benefits-and-challenges.html, 2017.
[28] R. Beck, J. Stenum Czepluch, N. Lollike, S. Malone, Blockchain-the Gateway to Trust-Free Cryptographic Transactions, ECIS, 2016, ResearchPaper153.
[29] A. Dubey, G. Karsai, S. Pradhan, Resilience at the edge in cyber-physical systems, in: 2017 Second International Conference on Fog and Mobile Edge Computing (FMEC), IEEE, 2017, pp. 139–146.
[30] H.K.N. Leung, L. White, A study of integration testing and software regression at the integration level, in: Proceedings, Conference on Software Maintenance, 1990, IEEE, 1990, pp. 290–301.
[31] J. Mirkovic, T. Benzel, Teaching cybersecurity with DeterLab, IEEE Secur. Priv. 10 (1) (2012) 73–76.

[32] G. Rothermel, R.H. Untch, C. Chu, M.J. Harrold, Prioritizing test cases for regression testing, IEEE Trans. Softw. Eng. 27 (10) (2001) 929–948.
[33] C. Siaterlis, A.P. Garcia, B. Genge, On the use of Emulab testbeds for scientifically rigorous experiments, IEEE Commun. Surv. Tutorials 15 (2) (2013) 929–942.
[34] M. Simić, G. Sladić, B. Milosavljević, A Case Study IoT and Blockchain Powered Healthcare, June, 2017. In: The 8th PSU-UNS International Conference on Engineering and Technology (ICET-2017).
[35] Github, Sometimes, Transactions Disappear From Txpool Rather Than Being Mined Into the Next Block—Issue #14893—Ethereum/Go-Ethereum, 2017, (Accessed 6 September 2017). https://github.com/ethereum/go-ethereum/issues/14893.
[36] F. Zhang, E. Cecchetti, K. Croman, A. Juels, E. Shi, Town crier: an authenticated data feed for smart contracts, in: Proceedings of the 2016 ACM SIGSAC Conference on Computer and Communications Security, ACM, Vienna, Austria, 2016, pp. 270–282. isbn: 978-1-45034139-4. CCS '16. https://doi.org/10.1145/2976749.2978326.

About the authors

Mr. Michael A. Walker is a Graduate Research Assistant pursuing his PhD in Computer Science at Vanderbilt University, Nashville, TN, USA. Currently he is an Instructor for the Computer Science & Information Systems department at Youngstown State University. He previously received his Masters in Science in Computer Science from Vanderbilt University [2011 – 2013], and obtained his Bachelors of Science in Computer Science from Youngstown State University [2006–2011].

Mr. Walker's research interests include Distributed Systems, Learning-at-Scale, Privacy, Security, and Software Design Patterns. He has published more than 11 research papers in various conferences, workshops and international journals of repute, including IEEE and ACM. He has been involved with 10 Massive Open Online Courses, acting as a Teaching Staff for three and an Instructor for four courses. He is a present and past board member of several non-profits directed toward outreach and education of Science, Technology, Engineering, and Mathematics, with a concentration on computer literacy and scientific understanding, specifically focused on benefiting young girls from disadvantaged backgrounds. Additionally, he has given several conference presentations on the subject of bridging the academic and industry divide for non-traditional students.

Dr. Douglas C. Schmidt is the Cornelius Vanderbilt Professor of Computer Science, Associate Provost for Research Development and Technologies, Co-Chair of the Data Sciences Institute, and a Senior Researcher at the Institute for Software Integrated Systems, all at Vanderbilt University.

His research covers a range of software-related topics, including patterns, optimization techniques, and empirical analyses of middleware frameworks for distributed real-time embedded systems and mobile cloud computing applications.

Dr. Schmidt has published 12 books and >600 technical papers covering a range of software-related topics, including patterns, optimization techniques, and empirical analyses of frameworks and model-driven engineering tools that facilitate the development of mission-critical middleware and mobile cloud computing applications running over wireless/wired networks and embedded system interconnects. For the past three decades, Dr. Schmidt has led the development of ACE and TAO, which are open-source middleware frameworks that constitute some of the most successful examples of software R&D ever transitioned from research to industry.

Dr. Schmidt received B.A. and M.A. degrees in Sociology from the College of William and Mary in Williamsburg, Virginia, and an M.S. and a Ph.D. in Computer Science from the University of California, Irvine in 1984, 1986, 1990, and 1994, respectively.

Dr. Abhishek Dubey is an Assistant Professor of Electrical Engineering and Computer Science at Vanderbilt University, Senior Research Scientist at the Institute for Software-Integrated Systems and co-lead for the Vanderbilt Initiative for Smart Cities Operations and Research (VISOR). His research interests include model-driven and data-driven techniques for dynamic and resilient human cyber physical systems. He directs the Smart computing laboratory (scope.isis.vanderbilt.edu) at the university. The lab conducts research at the intersection of Distributed Systems, Big Data, and Cyber Physical System, especially in the domain of transportation and electrical networks. Abhishek completed his PhD in Electrical Engineering from Vanderbilt University in 2009. He received his M.S. in Electrical Engineering from Vanderbilt University in August 2005 and completed his undergraduate studies in Electrical Engineering from the Indian Institute of Technology, Banaras Hindu University, India in May 2001. He is a senior member of IEEE.

CHAPTER SEVEN

Consensus mechanisms and information security technologies

Peng Zhang[a], Douglas C. Schmidt[b], Jules White[b], Abhishek Dubey[b]

[a]Department of Biomedical Informatics, Vanderbilt University Medical Center, Nashville, TN, United States
[b]Institute for Software Integrated Systems, Vanderbilt University, Nashville, TN, United States

Contents

Abstract

Distributed Ledger Technology (DLT) helps maintain and distribute predefined types of information and data in a decentralized manner. It removes the reliance on a third-party intermediary, while securing information exchange and creating shared truth via transaction records that are hard to tamper with. The successful operation of DLT stems

Advances in Computers, Volume 115
ISSN 0065-2458
https://doi.org/10.1016/bs.adcom.2019.05.001

largely from two computer science technologies: consensus mechanisms and information security protocols. Consensus mechanisms, such as Proof of Work (PoW) and Raft, ensure that the DLT network collectively agrees on contents stored in the ledger. Information security protocols, such as encryption and hashing, protect data integrity and safeguard data against unauthorized access.

The most popular incarnation of DLT has been used in cryptocurrencies, such as Bitcoin and Ethereum, through public blockchains, which requires the application of more robust consensus protocols across the entire network. An example is PoW, which has been employed by Bitcoin, but which is also highly energy inefficient. Other forms of DLT include consortium and private blockchains where networks are configured within federated entities or a single organization, in which case less energy intensive consensus protocols (such as Raft) would suffice. This chapter surveys existing consensus mechanisms and information security technologies used in DLT.

1. Introduction

Blockchain technologies alleviate the reliance on a centralized authority to certify information integrity and ownership, as well as mediate transactions and exchange of digital assets, while enabling secure and pseudo-anonymous transactions along with agreements directly between interacting parties. Since the introduction of Bitcoin by Satoshi Nakamoto [1], blockchain technology has been studied by researchers, engineers, and domain experts to evaluate its utility and improve its usability. Bitcoin is the first successful application of blockchain technology that is widely recognized for its revolutionary mechanisms that allow the secure direct transfer of digital assets between involved parties without the need for a trusted intermediary.

The concept and importance of digital assets are integral to the inception of blockchain technology. A digital asset is anything that exists in a binary format that comes with some right to exercise. One type of digital assets is native assets, which are assets that lack physical substance, but can be owned or controlled to produce some value. Examples of native assets are digital music, images, movies, electronic funds, and software. The other type of digital assets is digital representations of traditional assets, which are or historically used to exist in paper certificates or titles [2]. Assets like land property, gold, automobile title, and "paper" currency are examples of this type of digital assets.

The global economy increasingly depends on the effective management of digital assets [3] in nearly every domain and aspect of our lives. For example, the entertainment industry requires digital rights management for movies and music; the finance industry has experienced far more electronic fund transfers than cash exchanges; the energy sector is moving toward digital trading of energy and the adoption of smart grids; social media requires the management and protection of online users' reputation; and online elections cannot succeed without proper management of votes [4].

There are some common operations that can be performed on digital assets to augment their usability. For example, digital assets are transferable across different entities and users via atomic online transactions, such that they are either executed as one unit or not at all. These transactions can take place during a transfer between two bank accounts, a record of fund movement between a sender and a recipient, or a purchase of merchandise using a credit card. Likewise, the management of a special digital asset—digital identity—is important to match these identities in various occurrences to reduce the replication of data. Yet another operation that may be exercised on digital assets is provenance tracking, in which a digital history can be provided for physical products (such as supplies or hardware components) to trace and verify their origins, attributes, and ownership [5].

In recent years, Distributed Ledger Technology (DLT) has emerged as a means to comprehensively capture the advancements of blockchain technologies and variations that extend its core principles [6]. Blockchain technologies today typically refer to decentralized ecosystems managed by consensus mechanisms where the majority of parties (i.e., more than 50%) eventually agrees to the same reality. In such decentralized ecosystems, all the data (i.e., transactions of digital assets) are structured as a chain of blocks and replicated across all network maintainers (miners) [7], just like the Bitcoin blockchain [1].

DLT is an umbrella term that defines any shared ledger (regardless of its internal data structure) maintained in a decentralized network that replicates identical copies of the data across multiple nodes residing in various geographic locations. Nodes in the network simultaneously reconcile their copies of the data through consensus to achieve a shared truth, such that data in the shared ledger is verifiable and tamper-aware. Key to the success of DLT is the consensus process that helps order all valid transactions in a deterministic manner.

In distributed computing, consensus is a mechanism that helps a distributed network establish agreement on the value of some shared data [8]. A distributed ledger network can deliver a consistent and reliable state, even in the event that one or more nodes may be unreliable due to corruption or hardware failure. Unlike a typical centralized system—where decisions are dictated by the single governing authority—decisions regarding data stored in a distributed ledger are collaboratively made by majority votes. Due to the increasing interest in distributed ledger designs, various consensus mechanisms and information security protocols have surfaced as common configurations employed in these designs. This chapter provides a survey of popular technologies in use or proposed as part of many popular distributed ledgers.

The remainder of this chapter is organized as follows. Section 1 provides an overview of two main types of consensus mechanisms—Byzantine and Non-Byzantine consensus mechanisms—implemented by various distributed ledger systems; Section 2 describes four popular Byzantine consensus mechanisms implemented by public blockchain networks; Section 3 presents three Non-Byzantine consensus mechanisms used by other types of DLT that are permissioned; Section 4 provides an overview of information security protocols implemented in DLT that drive its successful operation; and Section 5 presents concluding remarks.

2. Consensus mechanisms overview

This section provides an overview of two types of mainstream consensus mechanisms—Byzantine and Non-Byzantine consensus mechanisms—implemented by classic distributed systems, which are the foundational technology underlying DLTs today.

2.1 Byzantine consensus mechanisms

Consensus mechanisms in distributed systems can be divided into two categories based on whether they assume maliciousness among the agents. In a classic 1982 paper [9], Lamport, Shostak and Pease introduced the problem of achieving consensus under malicious failure scenarios. They used the example of the Byzantine army and the problem of reaching agreement (consensus) to attack or retreat to explain the problem. Using a basic setup of three generals, they showed that if even one of the generals became malicious the other two generals will not be able to reach consensus. In a general setting, they showed that if at most N generals are traitors, then at least $3N + 1$ generals are needed to ensure that all non-malicious generals $(2N + 1)$

agree on the decision as to whether to attack or retreat. This result has several interesting applications in distributed systems, specially providing a lower bound on the minimum investment needed to ensure fault-tolerance.

2.2 Non-Byzantine consensus mechanisms

In non-malicious (i.e., Non-Byzantine) failure scenarios the problem is to ensure convergence on agreement of a stated value or a sequence of actions even if certain nodes fail. Regardless of whether failures are fail stop or fail stuck, the state is consistently observed by all participants and they agree that the node has failed with the correct stop or stuck semantics. Conversely, in malicious (i.e., Byzantine) cases, a failed node can deceive the other agents into different observations, i.e., some nodes might receive one message from the failed node, whereas other nodes might receive a different message from the failed node.

2.3 Taxonomy of consensus mechanisms

This chapter presents and evaluates several consensus mechanisms implemented by popular public blockchains and other types of permissioned distributed ledger systems. Our evaluation focuses on the following criteria:

- *The degree of decentralization*: the number of miners, maintainers, and/or members allowed in the network
- *Scalability*: transaction throughput, in terms of the time taken for network nodes in a distributed ledger system to reach consensus over a number of transactions grouped in a block or how much time it takes for a block to be produced and accepted by majority nodes
- *Randomness in block generation and miner selection*: dependencies in mining hardware, stakeholding, impact, and importance to the network
- *Consensus type*: whether the consensus mechanism employed by a distributed ledger network is Byzantine or Non-Byzantine consensus in terms of their resiliency to attacks from malicious nodes in the network.

3. Consensus mechanisms used in public blockchains

Public blockchains underlie the vast majority of cryptocurrency-based platforms, such as Bitcoin [1], Ethereum [10], and Litecoin [11]. These types of blockchains are permissionless, decentralized computing architectures open to the public and maintained by arbitrary users who possess Internet

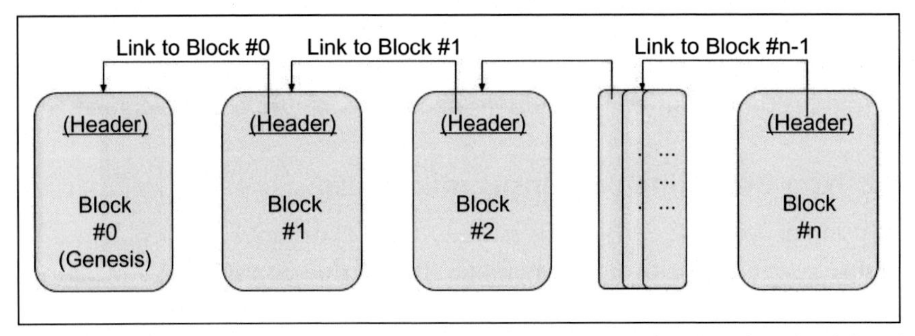

Fig. 1 Blockchain structure: A continuously growing chain of ordered and validated transactions.

access. Anyone with such access can participate in the exchange of digital assets in these platforms. Users are incentivized to contribute to the networks by validating transactions in the hope of being rewarded with digital tokens that may be used for commodity trading or in a shared market. Users are also attracted to public blockchains due to their "trustless" [1] nature, i.e., users can remain anonymous on-chain to protect personal identities and can feel confident that their transactions are carried out with integrity without having a trust relationship with any parties or brokers.

The most common public blockchains are structured as an append-only ledger of transactions that are continually reconciled and verified via a process called blockchain "mining." After a set of transactions is verified by the majority of the network maintainers, the transactions are then grouped into a structured block as being successfully mined. Consequently, the newly mined block is chained to the previous block in the sequence to persist a consistent and ordered transaction history, as shown in Fig. 1.

Public blockchains must guarantee that the shared ledger of transactions always provides the same snapshot to whoever accesses the chain at a given time to avoid incurring large volumes of digital asset exchanges. As a result, public blockchains typically implement the most robust mechanisms to reach consensus in highly decentralized global networks. These mechanisms may be more time consuming than others used in permissioned blockchains, but they more resilient to attacks from (minority) rogue players (in this context, we do not consider the 51% attack where the majority of network nodes collude to reverse blockchain transactions). Transactions in public ledgers are therefore immutable and transparent. Next, we present four consensus mechanisms that have been implemented by popular public blockchains.

3.1 Proof of Work (PoW)

Proof of Work (PoW) was the first prominent blockchain mining mechanism presented in the literature used by the Bitcoin blockchain [1]. With PoW, as new, unverified transactions become available or broadcast to the entire blockchain network, each node that maintains a copy of the ledger (also known as a "miner") verifies a set of those transactions by balancing the incoming and outgoing digital assets with previously validated transactions to prevent so called "double spending."

The miner next groups validated transactions into a tentative block. Each miner then competes in solving a computationally expensive algorithmic "puzzle" to ensure that their block is valid and that it follows in sequence from the last block in the current chain. Only the winner with a correct answer is privileged to append their block of transactions to the shared ledger and gains a mining reward in cryptocurrency (e.g., Bitcoins). This approach is also how native cryptocurrency tokens are minted.

The computationally expensive puzzle is at the heart of the PoW mechanism: it must be hard enough to solve to disincentivize attackers who intend to pollute the blockchain due to the high costs in obtaining a solution. Likewise, validating the proposed solution must be trivial so that it can be easily accepted by other nodes and the solution's correctness is transparent to the network, regardless of the computational power any network node possesses. The puzzle used by Bitcoin is to find a value called a "nonce." This nonce is created by combining the content in the proposed block to produce a new hash output that falls within a target range, such as a target hash prefixed with a number of 0's.

Due to the nature of hashing algorithms, the desirable output of a nonce can be computed only by brute force, i.e., guessing each nonce one by one until a solution is found. It is therefore highly unpredictable which node can successfully mine the next block, thereby protecting the validated transactions from tampering. The puzzle is adjusted regularly to maintain the same level of difficulty. For instance, the puzzle used by Bitcoin results in an average block formation time of 10 min. Fig. 2 illustrates the iterative process of solving the puzzle.

PoW has successfully sustained and secured the operations of two of the most popular public blockchains–Bitcoin and Ethereum–because it helps deter attackers with its requirement of expensive computational power and information transparency. In addition, the cryptocurrency reward incentives for competing to solve the puzzle makes it hard for a small group of rogue actors to manipulate and overpower the majority network nodes.

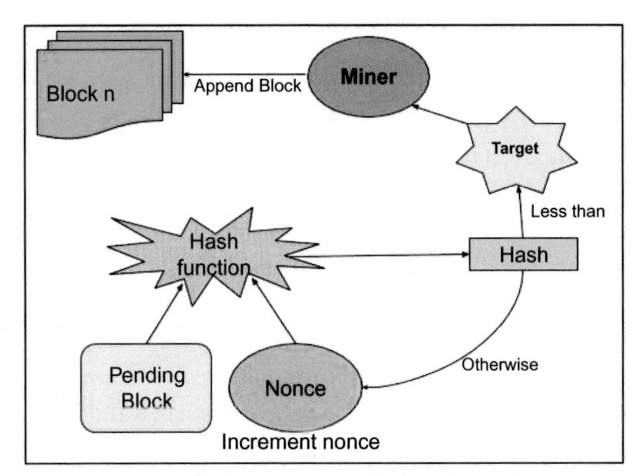

Fig. 2 The iterative process of solving the Proof of Work puzzle.

In practice, however, the computation power required to perform PoW is highly controversial because of its excessive energy consumption and wastage. Another major concern about PoW is its security vulnerability to 51% attack in small-scale public blockchain networks, where not many nodes compete in the mining process. An attacker could exploit those networks by much more easily obtaining a majority of the network's computation power and revert transactions.

3.2 Proof of Stake (PoS)

Another consensus mechanism that has frequently been compared with PoW is Proof of Stake (PoS). To reduce the energy consumption problem introduced by PoW's need for miners to solve a computationally expensive puzzle, PoS determines the next eligible block to append to the chain based on the current "stakes" held by the accounts, i.e., the total native cryptocurrency tokens they have. Stakeholders who are selected to maintain the PoS network are often required to lock in their stakes for a period of time during their service. This nature of PoS provides incentives for nodes to correctly create and validate blocks because by committing to network maintenance with their own shares of tokens, they could risk losing their stake and be deprived of their future privilege as a block producer if they are dishonest. All the locked shares are returnable to the good and fair stakeholders. Moreover, if a block is successfully appended to the blockchain its validator

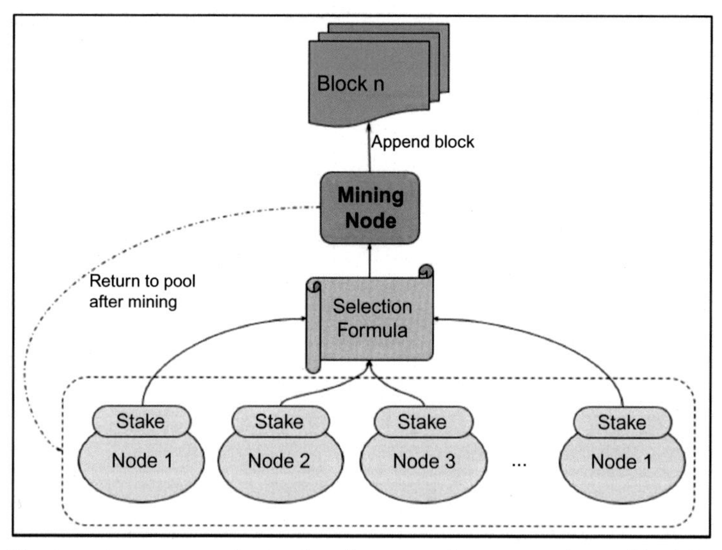

Fig. 3 The iterative process of selecting the next miner to produce a block based on node stakes in the network.

is rewarded with some transaction fees [12]. Fig. 3 summarizes the basic process of PoS-based mining.

Cryptocurrency tokens can typically only be created when the platform is initially launched and/or through the PoW mining process in the early phase (then switching to using PoS) as each new block of transactions is added to the blockchain. In PoS, networks nodes holding more stakes in the network are generally allowed to produce more blocks than others, though the percentage of blocks they are allowed to create are weighted according to the percentage of stakes they own over the entire network. PoS, however, does not simply select the next block based on its validator's cryptocurrency balance to avoid centralizing the network by permanently favoring those with more tokens. Instead, various methods have been proposed to select the next valid block in PoS.

One such method implemented by Nxt [13] and BlackCoin [14] uses randomization in the block selection process to create a formula that calculates the next block based on both the stake of a block's validator and the hash output from its validation. It is possible to predict the next block because all account balances in the shared ledger are available to the entire network. Another technique used in PeerCoin [15] leverages the concept of "coin age" to generate the next block according to both the amount and age of tokens available in the user accounts. To reduce the chance that one or a small group of users gains advantages due to high stakes in the system, after an account has generated a block, its coin age will be reset to 0 and the counting will restart until a predefined minimum age requirement is met again, e.g., 30 days in PeerCoin.

The main advantage of PoS is energy-efficiency because there is no need for block generators to perform computationally intensive tasks. Likewise, PoS incur lower requirements on computing hardware for users to partake in the block generation process. However, because PoS determines the block sequence according to the wealth of the network maintainers, stakes in cryptocurrency must be already established previously through other means–either minted in a PoW-based system prior to transitioning to PoS or acquired from other users with pre-established stakes. Moreover, because high stakes correspond to more rights to producing blocks, this model may discourage token distribution as users will likely want to keep their tokens instead of spending them.

3.3 Delegated Proof of Stake (DPoS)

The Delegated Proof of Stake (DPoS) consensus mechanism is an increasingly popular alternative to PoW and was first introduced by the founder of BitShares [16] to improve network efficiency and scalability over Bitcoin's PoW mining. DPoS essentially implements a reputation system with voting and an election process among stakeholders to reach network consensus in a digitally democratic manner. Generally, stakeholders in a DPoS-based blockchain system vote for super-representative roles, such as "Witnesses" and "Delegates", as a relatively small (compared to the total number of users with stakes in the network) and fixed number of people to perform critical tasks like validating transactions and generating blocks. Each stakeholder has a number of votes proportional to the tokens they own and may choose to delegate another stakeholder to cast their votes on their behalf [17].

The voting process selects the top *N* delegates, or also called "Witnesses" in Bitshares, (*N* being agreed upon by the network to ensure decentralization, e.g., 100 witnesses) based on the total votes received. Witnesses take turns producing and validating blocks and an even smaller number of Witnesses (such as 20) are rewarded with transaction fees for their service. This approach creates competitiveness of the role and deters fraudulent behavior. Voting is also an ongoing process such that any malicious or poorly-performing (due to missed blocks) witness can be voted out of their role at any given time.

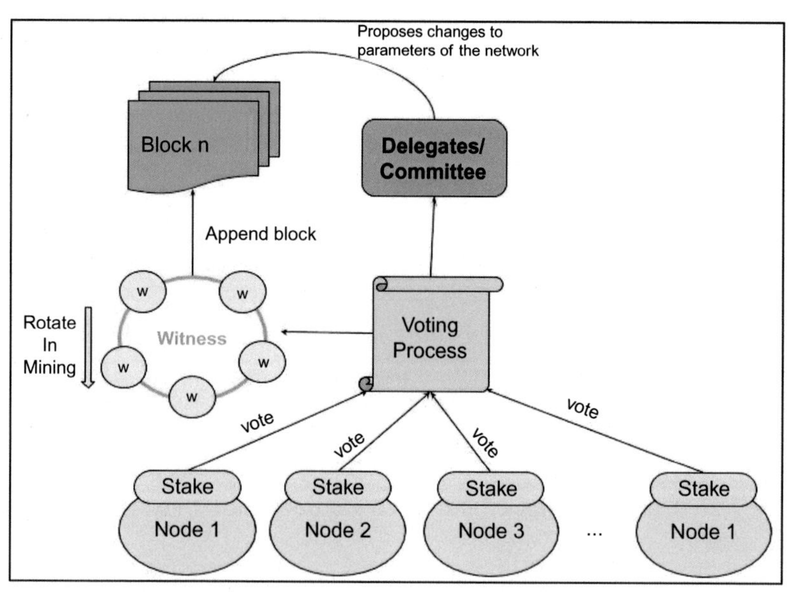

Fig. 4 Nodes use their stakes to vote for "Witnesses" to produce blocks and "Delegates/Committee" to propose changes to the network parameters.

Another group of delegates, also known as the "Committee" in Bitshares, is also elected to propose necessary changes to network parameters, such as fees paid to Witnesses, block sizes, block intervals, etc. When the majority in the Committee approves a change proposal, the stakeholders then review the proposal and vote to accept or nullify the change and/or vote out a Committee member. Ultimately, the administrative power is distributed to all the stakeholders; super-representatives are not meant to have direct authority to change the network by themselves. The architecture of DPoS is presented in Fig. 4.

A number of popular cryptocurrency networks have successfully implemented DPoS, such as BitShares [17], EOS [18], Steem [19], and Cardano [20]. Similar to PoS, DPoS offers more efficient transaction processing and a much lower requirement on computing hardware than PoW. In its existing deployments, DPoS has created more decentralized networks than its PoS predecessor, owing to its continuous election process that involves users even with the minimum token ownership. The downside to DPoS is its long-term sustainability as its governance relies heavily on users to actively elect a small set of delegates, which may be vulnerable to centralization overtime due to smaller stakeholders forfeit voting rights and/or tokens become poorly distributed.

3.4 Proof of Importance (PoI)

Proof of Importance (PoI) is a consensus mechanism introduced in the NEM blockchain platform [21]. It resembles the stake ownership consideration in PoS and DPoS since users are only eligible to forge (or "harvest" as in NEM) a block if they meet a minimum requirement of stakes in the network's native cryptocurrency. For instance, NEM currently only selects candidates from a pool of users with at least 10,000 XEM vested. PoI differs from all the aforementioned mechanisms, however, since it takes into account other factors than the amount of computation one puts into or the amount of stake one holds alone. In particular, PoI determines a user's eligibility of harvesting blocks according to their overall contribution to the network, rather than a single aspect, using a more holistic metric named the "importance score". Users having a higher importance score are more likely selected to harvest a block and are rewarded with transaction fees for their work.

Upon meeting the minimum stockholding requirement, PoI calculates an importance score for each user using a heuristic function based on the following three factors:

- *The user's token ownership*, where more tokens will correlate to a higher importance score as long as the tokens have been available in the user's account for a fixed period.
- *The net outbound token transactions*, which rewards the distribution of cryptocurrencies instead of accumulating or concentrating wealth. Calculating net transfers prevent accounts from gaming the system, such as transacting between different personal accounts.
- *The connectivity to other network nodes*, which promotes account interactions across the network by assigning more importance to more active users with diverse transaction history and larger transaction amount [22].

Fig. 5 presents an overview of the PoI architecture.

PoI was designed to minimize the centralization of wealth as a potential disadvantage of PoS. In particular, it uses the importance score to stimulate the spread of wealth by giving high scores to users with larger transactions and more/or connected transaction networks. PoI is intentionally resilient to the "nothing at stake" problem, where PoS valuators maintain and hold stakes in *every* fork encountered in the network because the costs to do so are trivial given that minimal external resources are needed, unlike in PoW. In this hypothetical scenario, a malicious attacker could exploit a double spend by using the current main chain and his/her fork. Since all forks are built

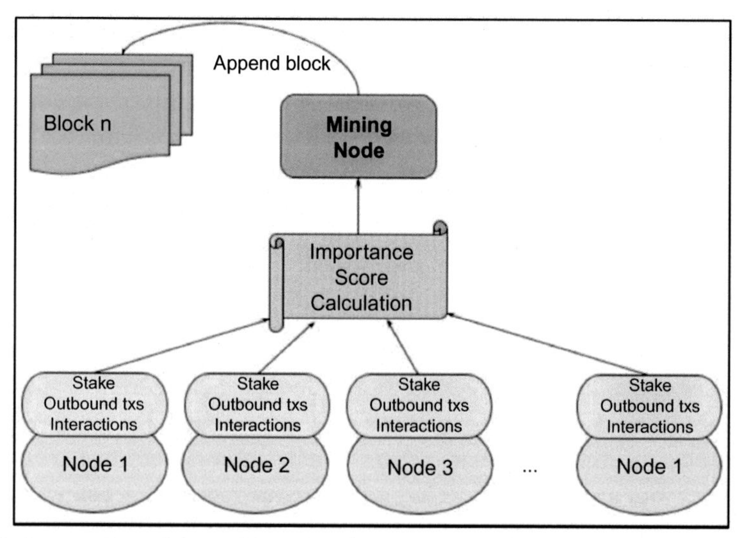

Fig. 5 An overview of the architecture of PoI consensus that calculates an importance score used to select the next mining node.

simultaneously an attacker could simply choose his/her fork containing the invalid transaction as the main chain and succeed in the attack when the stakeholding attacker is selected as the next block producer.

Despite the importance score taking into account several aspects of a user's involvement in and contribution to the network, it may be shifted towards more centralization in the longer term because the calculation still favors wealthy users with more stakes in the network and thus more flexibility to transact and interact with other users.

4. Consensus mechanisms used in other forms of distributed ledger technology

Public blockchains are designed to maximize the level of transparency and decentralization to provide a trustless environment for users interested in exploring the network and/or actively exchanging digital assets (such as cryptocurrencies) freely and anonymously. The openness and lack of restrictions on data access, however, may not be ideal for entities or functions that require sensitive data warehousing or exchange, such as enterprises or government agencies. Moreover, consensus mechanisms implemented in public blockchains must exercise strict orders to protect the networks, rendering transactions relatively slow compared to industry requirements.

To address these limitations, therefore, permissioned blockchains and other variants of distributed ledger technologies have surfaced as more closed ecosystems that restrict access and loosen some constraints of the consensus requirements of public blockchains. Permissioned blockchains are similar in structure to public blockchains, while permissioned distributed ledgers may store transactions in a single linear chain of blocks, multiple chains of blocks, or a directed acyclic graph (similar to a tree structure) that is non-linear [23].

Permissioned distributed ledgers allow their members to limit user access to the network by disclosing their identities prior to joining the consortium. They establish a much smaller and more controlled environment that has a modicum of trust among the member nodes. Consensus is therefore much easier and faster to achieve in permissioned networks compared to ones completely open to the public. Despite the lower degree of decentralization, consensus is still a crucial process to ensure that every member has equal rights to updating the shared ledger in the network. Below we describe three popular consensus mechanisms used in permissioned ledgers. Table 2 presents a summary that compares these mechanisms.

4.1 Proof of Elapsed Time (PoET)

In Proof of Elapsed Time (PoET), nodes are selected to produce blocks after waiting for a random period of time. This technique was developed by Intel and has been adopted by the Hyperledger Sawtooth project [24] as a much leaner alternative to PoW in a permissioned network [25]. Its core mechanism is based on Intel's Software Guard Extensions (SGX) technology [26] that has the ability to digitally attest that some code has been correctly set up in a so called "Trusted Execution Environment" [27]. In PoET, this code is a function that generates a random time period that must be waited out by each node.

When a participant joins the network, they download the time generator code and receive an attestation (in the form of a digital certificate) of the code setup from SGX that they announce to the network. Existing members can either approve or deny the join request. If the request is approved, the new node becomes an eligible candidate to produce blocks and participates in the random selection process. Whoever first completes the waiting period broadcasts a signed message to the network as the randomly chosen next block forger. The SGX is a critical component because (1) it warrants the integrity of the randomly generated wait period, preventing malicious nodes

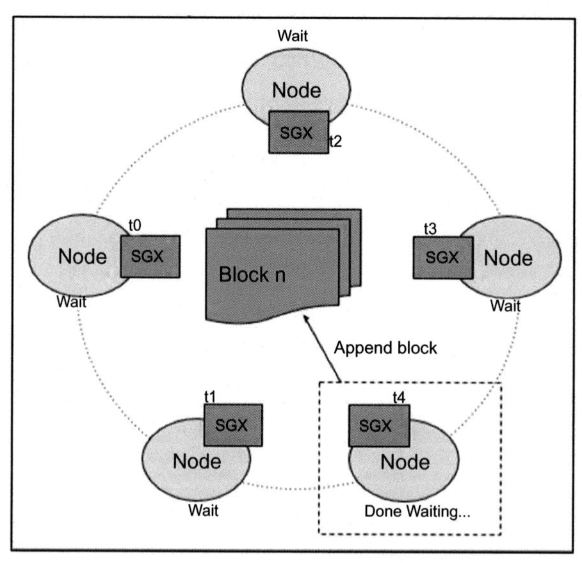

Fig. 6 Architecture of the PoET consensus mechanism that leverages the Software Guard Extension (SGX) technology to select the next miner based on a random wait period.

from altering the timer code to their advantage and (2) the attestations are an efficient method to verify the validity of wait time completion [25]. The PoET consensus mechanism is illustrated in Fig. 6.

PoET is an efficient and scalable mechanism, especially for permissioned network. It creates a randomized model for selecting block producers without resource-intensive computing as in PoW systems or complex calculations for determining miners used in consensus mechanisms involving stakes in the system as in PoS and PoI. However, PoET heavily relies on Intel's specialized, third-party hardware to operate, which creates entry barriers for participants without access to the SGX technology. It is also possible for nodes with more hardware available to gain a better chance of getting selected, but such nodes may likely be denied access to join the permissioned network.

4.2 Proof of Authority (PoA)

Proof of Authority (PoA) is designed to optimize the PoS mechanism and be used, ideally, in permissioned networks. Instead of choosing block miners on the basis of their stakes in cryptocurrency tokens, PoA selects a small group of authorities as transaction validators by their identity or reputation staked in the network [28]. To contend for validators, users go through a formal notarization process in which they provide documentation to prove their real identities and link them with their on-chain identities to establish their digital reputation. Existing validators can vote to add additional users into the authority group. A PoA-based system also rewards authorities for certifying and ordering transactions to incentivize honest behavior in providing service and moderating the network. PoA Network [29] and Ethereum's test net Kovan [30] are examples of public networks that use PoA consensus. An overview of the PoA mechanism is shown in Fig. 7 below.

PoA does not require intensive computation to complete hard tasks and only relies on a small number of validators to reach consensus. These features help improve transaction throughput and energy efficiency compared with PoW- and PoS-based systems. However, PoA also forgoes decentralization by concentrating mining power among a group of trusted authorities. As a result, this model can introduce censorship into the public network where one or more authorities may blacklist or deny all transactions from a particular user. On the other hand, a permissioned network established between different enterprise or large institutions can benefit from PoA because it offers a faster transaction processing speed and the identity-at-stake model aligns well with business operations that value trustworthiness and reputation.

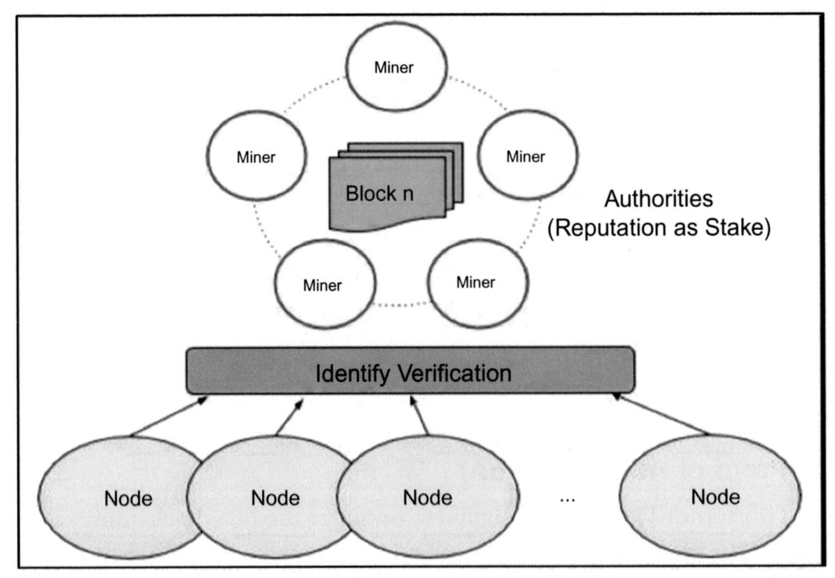

Fig. 7 Architecture of the PoA consensus mechanism in which nodes put their reputation at stake by verifying their identities before becoming miners in the network.

4.3 Ordering-based consensus

Ordering-based consensus is commonly applied in permissioned networks in which nodes are selected to participate in the network that supports a membership or identity service. A well-known permissioned ledger system, Hyperledger Fabric [31], implements a three-step process to reach consensus in this manner. First, the client application proposes a transaction to a set of nodes called "peer nodes" (who hosts of the shared ledger) for endorsement based on some predefined policy (e.g., requiring M out of N signatures from peer nodes). Peer nodes next return a response of the transaction proposal to the client, who then submits the endorsed transaction to a node called the "orderer." Orderer nodes may receive endorsed transactions in different orders, but they collectively determine the final, strict sequence of the transactions and package them into immutable blocks. Finally, the orderer nodes distribute transaction blocks to connected peer nodes for validation of the new transaction blocks. In this model, if a transaction is deemed invalid, it remains in the block, but is marked as invalid by the peer node because blocks created by orderer nodes are in their final states [32]. Fig. 8 presents the high-level architecture of order-based consensus.

The ordering service is key to reaching consensus regarding the states and sequence of transactions in the ledger. One of the consensus mechanisms from distributed computing that is implemented by the ordering service

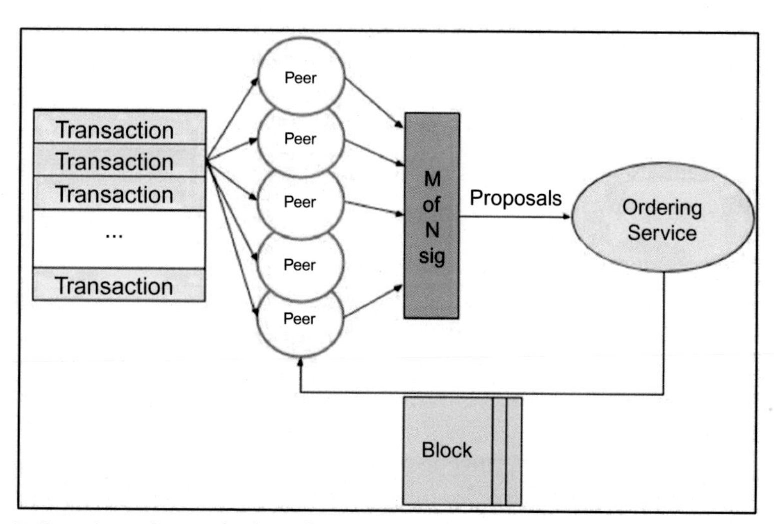

Fig. 8 Overview of an order-based consensus architecture that relies on an ordering service to determine sequences of transactions.

in Hyperledger Fabric is a leader-follower model called Raft [33]. In Raft, a single leader node is elected in each "term" of an arbitrary length to make decisions and propagate those decisions to the followers who then replicate them. Each Raft node maintains a log and is always in one of three states: follower, candidate, or leader. A node starts as a follower who accepts and replicates log entries from the leader. If a leader is not present or has not been responsive for a time period, the follower waits for a randomized timeout period and then moves into the candidate state. The candidate then requests votes from other nodes and becomes the leader if it receives majority votes from the network. All new transaction entries go through the leader, who appends the changes to its log and replicates the effective transaction sequence to the followers. The leader waits until a majority of followers have updated their local logs and then commits the updates and broadcasts the confirmation to follower nodes. The network is now in consensus of the transaction sequence and ledger states [34].

The ordering-based consensus process designates separate roles for verifying and ordering transactions. The leader-follower model orders transactions at a fast pace because the order is determined effectively by a single leader node at a time instead of being computed by every node. It does require configurations to set up the architecture but not specialized hardware or extensive computing resources, making it a cost-effective model for permissioned ledger systems. The leader election process can quickly detect a faulty leader in the network and replaces it so that the ledger can be updated continuously as new transactions are received. Raft-based ordering service alone, however, is not resilient to attacks from malicious nodes that may exist within the network, which requires other components in place to validate transactions added by leader nodes, as are present in the Hyperledger Fabric system.

5. Information security technologies

The success of distributed ledger technology is driven by a combination of

- core computer science concepts and principles from distributed systems, which are key to the development of consensus mechanisms, and

- information security, which ensures the security and integrity during transactions of digital assets.

Below we present five important concepts from information security commonly integrated into distributed ledgers.

5.1 Public key cryptography (encryption and signing)

Public key cryptography [35] is a critical component of distributed ledger technology. It is composed of a pair of mathematically related public and private keys generated from one-way cryptographic function. While the public key can be freely distributed, the private key must be protected by the owner of the key pair. It is computationally infeasible to acquire the private key given its paired public key in the cryptographic keys commonly used in information security today.

A private key is used commonly in public key encryption to safeguard the secrecy of messages using the intended recipient's public key to encrypt data packets, ensuring that only owners of the corresponding private key can decrypt them. Another standard use case is for the sender of data to create digital signatures with their private key, with the signatures being easily verifiable using their public key, to prevent data from being tampered with in transit. Through public key cryptography, public blockchains like Bitcoin and Ethereum provide so-called "pseudo-anonymity" [1], meaning that as long as users' real-life, personal identity is not linked with their public keys on-chain, their activities can remain anonymous.

5.2 Hashing

Hashing is a critical process employed by distributed ledgers to map data of an arbitrary length (e.g., different types/sizes of digital assets exchanged) to data of a fixed size (e.g., transaction hashes stored in the shared ledger) using cryptographic functions [36]. These functions are one-way, meaning that it is trivial to generate a hash value from a given input but impractical (given computation capabilities today) to reverse engineer and calculate the input value. Given the same input, hash functions will always produce the same output. Moreover, even the slightest change in the input will completely alter the output. Hashing is commonly used in blockchains to protect transacted data against tampering and link new validated transactions with the existing ledger to create a network-wide non-refutable history.

5.3 Multi-signature

Multi-Signature (multisig) is a joint digital signature created by more than one party to improve the protection and integrity of the original content [37]. The goal of this approach is to more securely authenticate transactions than the traditional single public-private key pair. Multisig is often employed by cryptocurrency wallets or public blockchain networks.

In early cryptocurrency-based networks, a user whose private key has been lost or stolen could permanently lose access to the ownership of their digital assets. Moreover, cryptocurrency accounts created for business operations are extremely prone to insider attacks because anyone with access to the shared secret key can withdraw or transfer the balance without being traceable. Multisig is designed to overcome these issues by requiring more than a single key pair to authenticate transactions. It requires M keys out of a set of N ($N >= M$) key pairs to perform a transaction, with 2 out of 3 keys ($M = 2$, $N = 3$) being the most common scheme today.

5.4 Ring signature

Despite the use of public key cryptography, logs in the shared ledger can be traced to identify certain patterns in users especially when users execute repeated transactions or interact with the same set of other users. The traceability of transactions is not ideal for privacy-focused blockchains that seek to obfuscate the identity of users originating the exchange of digital assets [4]. Ring signature is one method applied in Bytecoin [38] and Monero [39] to protect on-chain privacy for the sender of a transaction.

In ring signature, the actual signer of a message forms a group with an arbitrary number of other users or decoys, each of which has a public-private key pair. The signer then produces a ring signature and a one-time random ring key pair using a series of mathematic calculations based on a combination of the message, his own secret key, and all other users' public keys. The signature is verifiable using all the public keys from the ring. To an outside observer, the actual signer is computationally indistinguishable from other parties in the ring, and therefore, the identity of the signer is no longer traceable [40].

5.5 Zero-Knowledge Proof (ZKP)

Concerns about data privacy in shared environments are arising as distributed ledger technology is increasingly touted as a decentralized data transaction infrastructure that removes centralized control, in popular domains, such as finance [41,42], supply chain [43,44], and healthcare [45,46]. Vital information that could be used to identify an individual, such as date of birth, social security numbers (in the U.S.), employment information, and bank statements, is paramount to the safety and financial well-being of the identity owner. To safeguard sensitive information, initial applications of zero-knowledge proof (ZKP) techniques have surfaced in DLT projects like the zk-SNAKRS [47,48] protocol in ZCash.

ZKP is a complex scheme designed to incorporate encryption techniques to enable a prover to certify the truthfulness of a statement to a verifier without disclosing any more specifics other than the statement itself. A true ZKP must possess the following three key properties:

- *Completeness*—if the statement is true, an honest prover will convince the verifier,
- *Soundness*—if the statement is false, verifier will find out the prover is dishonest with very high probability, and
- *Zero knowledge*—if the statement is true, no extra information is revealed to the verifier other than the statement being true [49].

6. Concluding remarks

This chapter provided a survey of various consensus mechanisms and key information security concepts employed in public and permissioned distributed ledgers for exchanging and distributing digital assets. Each consensus mechanism can optimize at most two of the three attributes in a DLT network described in Section 2.3: *degree of decentralization* (number of network miners/maintainers/members), *scalability* (transaction throughput; number of transactions per second), and *randomness in block generation and miner selection* (dependencies in mining hardware, stakeholding, impact and importance to the network). Tables 1 and 2 provide an overview and summary of the consensus mechanisms described and compared in this chapter.

Generally, public blockchains require Byzantine consensus to maintain a robust and resilient decentralized network since any node, including malicious node, can theoretically become a miner. Permissioned networks, in contrast, are often less concerned with malicious nodes because strict rules are in place to onboard members to ensure that nodes in the network can be trusted with their reputation stake in the network.

Table 1 A comparison of consensus mechanisms used in public blockchains.

Consensus mechanism	Degree of decentralization	Scalability	Randomness in miner selection	Consensus type
Proof of Work (PoW)	High—allows any node to join the network and become a miner in the network; although as hardware requirement increases, mining power may become less decentralized	Low—transaction throughput is low due to difficulty in solving the cryptographic puzzle	High—cryptographic puzzle can only be solved by brute force; difficult to know which miner will solve it first despite nodes with more computational power having higher chances of winning	Byzantine consensus
Proof of Stake (PoS)	Medium—allows any node to join the network but only node with higher stakes in the network can become a miner; the stake holding requirement may eventually regionalize or centralize power	High—no requirement on solving a difficult puzzle to reach consensus	Medium—nodes with higher stakes in the network are more likely chosen as miners	Byzantine consensus
Delegated Proof of Stake (DPoS)	Medium—allows any node to join the network, but only a small subset of nodes is selected as miners	High—no requirement on solving a difficult puzzle to reach consensus	Low—although the election of miners is fairly random by majority votes, chosen miners take turn to generate blocks	Byzantine consensus
Proof of Importance (PoI)	High—allows any node to join the network and become a miner in the network, but the barrier of entry into the miner pool is high as initial stakes are required	High—no requirement on solving a difficult puzzle to reach consensus	Low—mining eligibility is highly dependent upon the calculation of an importance score, which can be highly predictable	Byzantine consensus

Table 2 A comparison of consensus mechanisms used in other permissioned distributed ledger networks.

Consensus mechanism	Degree of decentralization	Scalability	Randomness in miner selection	Consensus type
Proof of Elapsed Time (PoET)	Low—requires miners to possess specialized hardware	High in terms of transaction throughput—can produce transactions at a very fast speed without the need to solve a hard puzzle	High—specialized hardware used generates a random timer used to determine the next miner; the process is also verifiable	Non-Byzantine consensus
Proof of Authority (PoA)	Low—theoretically, anyone who is willing to go through an identity verification process could be appointed as an authority to generate blocks; in reality, a relatively small number of people will be appointed	High in terms of transaction throughput—a small set of authorized nodes are responsible for processing transactions and generating blocks	Low—the identity verification process most likely selects miners with established reputation	Non-Byzantine consensus
Ordering-based consensus	Low—resilient to faulty nodes but not to attacks from malicious nodes that may exist in the network; a tighter control would therefore be needed to prevent against those attacks	High in terms of transaction throughput—a single leader orders transactions at a time and replicates the changes to all follower nodes	High—the leader selection process is based on a randomized timeout period	Non-Byzantine consensus

Key terminology and definitions

Consensus A mechanism to achieve overall reliability in a distributed ledger network in the presence of a number of faulty maintainer and mining nodes. It requires the majority (51%) of nodes to agree on content stored in the shared ledger.

Delegated Proof of Stake A scalable consensus mechanism that uses a democratic voting and election process to select delegates who create and validate blocks of transactions; the number of votes a user can cast is proportional to the tokens they own; delegates are incentivized to be honest with rewards and also the risk of being voted out.

Hashing A cryptographic process that maps an input data of variable length to an output with a fixed size using a one-way function such that it is computationally infeasible to calculate the input based on the output; even the slightest changes to the input data will alter the data output; it is used to create transactions that are stored in shared ledgers to protect the content of transactions from being exposed or tampered with.

Multi-Signature (Multisig) A digital signature jointly created by more than one party to provide more security and integrity protection during an exchange of digital assets; it usually follows a scheme that requires M out of N signatures, where $M<=N$ and $N>=2$ such that a single party alone cannot sign a message or authorize a transaction; it is used by many cryptocurrency wallets to prevent frozen funds due to a single private key being lost or stolen.

Proof of Elapsed Time An efficient consensus protocol designed for permissioned distributed ledger networks to randomize the selection of block miners. It is based on Intel's Software Guard Executions technology that attests for the integrity of some trusted code used to generate a random waiting period for each node. The first node to finish waiting is given the privilege to mint the next block.

Proof of Importance A consensus mechanism introduced in the NEM blockchain that uses an importance score to select block generators, based on stake ownership, the spread of cryptocurrency, and interactions with other nodes to incentivize the distribution and transactions of native tokens.

Proof of Stake An energy-efficient consensus mechanism in which block producers are selected randomly based on their stakes of cryptocurrencies in the blockchain; users can mine a percentage of blocks in proportion to their token balance; malicious behavior is disincentivized because of the risk of losing stakes and privilege to produce blocks.

Proof of Work A robust consensus mechanism often used in public blockchains in which network nodes compete to solve a computationally expensive puzzle, whose solution is trivial to verify, to ensure the validity and integrity of ordered transactions.

Public Key Cryptography Also known as asymmetric cryptography, which is an encryption scheme that uses a mathematically related key pair, a public key and a private key, to secure information. The public key is used to encrypt data, while and the private key is used to decrypt cipher text to obtain the original data. It is computationally infeasible to calculate the private key based on the public key. As a result, public keys can be freely distributed for encrypting content and verifying digital signatures; however, private keys are kept secret with their owners to use for decrypting content and creating digital signatures.

Raft A consensus mechanism that follows a leader-follower model; A single leader is randomly elected to decide upon shared states of the network and broadcasts the changes to follower nodes; the election process that is based on randomized timeout settings occurs when a leader is not present or has not been responsive for a pre-defined time period.

Ring Signature A digital signature created by a signer who forms a group with other arbitrary users or decoys, each with a public-private key pair; the signature obfuscates the identity of the actual signer by generating a one-time signing key from all members of the group; its application in distributed ledger transactions protects the identity of the sender.

Zero-Knowledge Proof A cryptographic scheme that allows a certifying party to prove to a verifier that a statement is true without disclosing any other information about what the statement is; it allows a secret (such as sensitive information) to be used for verification purposes without the exact detail or specifics to be known to others.

References

[1] S. Nakamoto, Bitcoin: A Peer-to-Peer Electronic Cash System, 2008, Bitcoin.org.
[2] N. Cahn, Postmortem life on-line, Prob. Prop. 25 (2011) 36.
[3] E. Keathley, Digital Asset Management: Content Architectures, Project Management, and Creating Order Out of Media Chaos, Apress, 2014Retrieved 04-30-2019.
[4] J.W. TEP, S. Solicitors, Digital assets; a legal minefield, in: Zürich: STEP Verein & Basel Conference, 2014. URL: https://stoanalytics.com/article/digital-assets-a-legal-minefield/. Retrieved 04-30-2019.
[5] T. Regli, Digital and marketing asset management: the real story about DAM technology and practices, Rosenfeld Media, 2016.
[6] S. Davidson, P. De Filippi, J. Potts, Disrupting governance: the new institutional economics of distributed ledger technology, Available at SSRN 2811995, SSRN https://www.ssrn.com/index.cfm/en/, 2016.
[7] M. Crosby, P. Pattanayak, S. Verma, V. Kalyanaraman, Blockchain technology: beyond bitcoin, Appl. Innov. 2 (6–10) (2016) 71.
[8] M.J. Fischer, N.A. Lynch, M.S. Paterson, Impossibility of distributed consensus with one faulty process, Massachusetts Inst of Tech Cambridge lab for Computer Science, 1982.
[9] L. Lamport, R. Shostak, M. Pease, The byzantine generals problem, ACM Trans. Program. Lang. Syst. 4 (3) (1982) 382–401.
[10] G. Wood, Ethereum: a secure decentralised generalised transaction ledger, vol. 151, Ethereum Project Yellow Paper, 2014, pp. 1–32.
[11] "Litecoin v0.16.3," Sept. 29, 2018, URL: https://litecoin.org/. Retrieved 04-30-2019.
[12] V. Buterin, What Proof of Stake is And Why it Matters, vol. 26, Bitcoin Magazine, 2013. August.
[13] "Whitepaper:Nxt(Blocks)", URL: https://nxtwiki.org/wiki/Whitepaper:Nxt, 2018. Retrieved 04-30-2019.
[14] P. Vasin, Blackcoin's Proof-of-Stake Protocol v2, URL: https://blackcoin.org/blackcoin-pos-protocol-v2-whitepaper.pdf,vol. 71, 2014. Retrieved 04-30-2019.
[15] S. King, S. Nadal, Ppcoin: Peer-to-Peer Crypto-Currency With Proof-of-Stake, vol. 19, Self-Published Paper, 2012.
[16] D. Larimer, Delegated Proof-of-Stake (Dpos), Bitshare Whitepaper, 2014.
[17] "Delegated Proof-of-Stake Consensus" URL: https://bitshares.org/technology/delegated-proof-of-stake-consensus/. Retrieved 04-30-2019, 2019, bitshares.org.
[18] Eos.io, EOS.IO Technical White Paper v2, March 16, 2018, https://github.com/EOSIO/Documentation/blob/master/TechnicalWhitePaper.md.

[19] Steem.com, Steem An incentivized, blockchain-based, public content platform. August 2017, URL: https://steem.com/SteemWhitePaper.pdf.
[20] "Cardano", URL: https://www.cardano.org/en/home/. Retrieved 04-30-2019.
[21] NEM, February 23, 2018, "NEM—Distributed Ledger Technology (Blockchain) Technology," URL: https://nem.io/technology/.
[22] H. Lombardo, NEM Q&A—Original, Tested Blockchain Platform, Proof-of-Importance, "Change the World, Forever" Tech, 2015. URL: http://allcoinsnews.com/2015/04/07/nem-qa/, Retrieved 04-30-2019.
[23] N. Kannengießer, S. Lins, T. Dehling, A. Sunyaev, What does not fit can be made to fit! trade-offs in distributed ledger technology designs, in: Trade-Offs in Distributed Ledger Technology Designs, 2019 (January 10, 2019).
[24] Hyperledger Sawtooth, 2018, URL: https://www.hyperledger.org/projects/sawtooth. Retrieved 04-30-2019.
[25] K. Olson, M. Bowman, J. Mitchell, S. Amundson, D. Middleton, C. Montgomery, Sawtooth: An Introduction, The Linux Foundation, 2018.
[26] F. McKeen, et al., Intel® software guard extensions (intel® sgx) support for dynamic memory management inside an enclave, in: Proceedings of the Hardware and Architectural Support for Security and Privacy 2016, ACM, 2016, p. 10.
[27] M. Sabt, M. Achemlal, A. Bouabdallah, Trusted execution environment: what it is, and what it is not, in: 2015 IEEE Trustcom/BigDataSE/ISPA, vol. 1, IEEE, 2015, pp. 57–64.
[28] W. Gavin, PoA Private Chains, URL: https://github.com/ethereum/guide/blob/master/poa.md, 2015. Retrieved 04-30-2019.
[29] "POA", URL: https://poa.network/. Retrieved 04-30-2019.
[30] "Kovan PoA Testnet Proposal", URL: https://github.com/kovan-testnet/proposal, 2017. Retrieved 04-30-2019.
[31] E. Androulaki, et al., Hyperledger fabric: a distributed operating system for permissioned blockchains, in: Proceedings of the Thirteenth EuroSys Conference, ACM, 2018, p. 30.
[32] C. Cachin, Architecture of the hyperledger blockchain fabric, in: Workshop on distributed cryptocurrencies and consensus ledgers, vol. 310, 2016.
[33] The Raft Consensus Algorithm, URL: https://raft.github.io/. Retrieved 04-30-2019.
[34] D. Ongaro, J. Ousterhout, In search of an understandable consensus algorithm, in: 2014 {USENIX} Annual Technical Conference ({USENIX}{ATC} 14), 2014, pp. 305–319.
[35] W. Stallings, Cryptography and Network Security: Principles and Practice, Pearson, Upper Saddle River, 2017.
[36] A.G. Konheim, Hashing in Computer Science: Fifty Years of Slicing and Dicing, John Wiley & Sons, 2010.
[37] M. Bellare, G. Neven, Identity-based multi-signatures from RSA, in: Cryptographers' Track at the RSA Conference, Springer, 2007, pp. 145–162.
[38] Bytecoin, Cryptography behind Bytecoin, 2018URL: https://bytecoin.org/blog/cryptography-behind-bytecoin, 2018. Retrieved 04-30-2019.
[39] Moneropedia, In: Ring Signature, URL: https://www.getmonero.org/resources/moneropedia/ringsignatures.html. Retrieved 04-30-2019.
[40] R.L. Rivest, A. Shamir, Y. Tauman, How to leak a secret, in: International Conference on the Theory and Application of Cryptology and Information Security, Springer, 2001, pp. 552–565.
[41] B. Scott, How can cryptocurrency and blockchain technology Play a role in building social and solidarity finance? UNRISD Working Paper, 2016.
[42] Y. Guo, C. Liang, Blockchain application and outlook in the banking industry, Financ. Innov. 2 (1) (2016) 24.
[43] K. Korpela, J. Hallikas, T. Dahlberg, Digital supply chain transformation toward blockchain integration, in: Proceedings of the 50th Hawaii International Conference on System Sciences, 2017.

[44] S.A. Saveen, R.P. Monfared, Blockchain Ready Manufacturing Supply Chain Using Distributed Ledger, eSAT, 2016. https://dspace.lboro.ac.uk/dspace-jspui/handle/2134/22625.
[45] P. Zhang, D.C. Schmidt, J. White, G. Lenz, Blockchain technology use cases in healthcare, in: Advances in Computers, vol. 111, Elsevier, 2018, pp. 1–41.
[46] P. Zhang, J. White, D.C. Schmidt, G. Lenz, S.T. Rosenbloom, Fhirchain: applying blockchain to securely and scalably share clinical data, Comput. Struct. Biotechnol. J. 16 (2018) 267–278.
[47] C. Rackoff, D.R. Simon, Non-interactive zero-knowledge proof of knowledge and chosen ciphertext attack, in: Annual International Cryptology Conference, Springer, 1991, pp. 433–444.
[48] What are zk-SNARKs?, URL: https://z.cash/technology/zksnarks/. Retrieved 04-30-2019.
[49] O. Goldreich, Y. Oren, Definitions and properties of zero-knowledge proof systems, Journal of Cryptol. 7 (1) (1994) 1–32.

About the authors

Dr. Peng Zhang recently received her M.S. and Ph.D. in Computer Science from Vanderbilt University, Nashville, TN. She previously received her B.S. degree in Computer Engineering from Lipscomb University [2010–2013] also in Nashville. Her research interests include model-driven design for engineering and healthcare IT systems, intelligent model constructions using machine and deep learning, decentralized algorithms and protocols for facilitating and securing clinical communications, and application and enhancement of Blockchain technologies for moving towards patient-centered care. She has interned with industry companies such as HiTactics, Varian Medical Systems, and Center for Medical Interoperability to lead various machine learning and blockchain-related research projects. Dr. Zhang's work on FHIRChain, a blockchain-based architecture for enabling secure and scalable healthcare data sharing has been recently covered by HealthDataManagement, BeckersHospitalReview, HCANews, and several other media outlets.

Dr. Douglas C. Schmidt is the Cornelius Vanderbilt Professor of Computer Science, Associate Provost for Research Development and Technologies, Co-Chair of the Data Sciences Institute, and a Senior Researcher at the Institute for Software Integrated Systems, all at Vanderbilt University. His research covers a range of software-related topics, including patterns, optimization techniques, and empirical analyses of middleware frameworks for distributed real-time embedded systems and mobile cloud computing applications.

Dr. Schmidt has published 12 books and more than 600 technical papers covering a range of software-related topics, including patterns, optimization techniques, and empirical analyses of frameworks and model-driven engineering tools that facilitate the development of mission-critical middleware and mobile cloud computing applications running over wireless/wired networks and embedded system interconnects. For the past three decades, Dr. Schmidt has led the development of ACE and TAO, which are open-source middleware frameworks that constitute some of the most successful examples of software R&D ever transitioned from research to industry.

Dr. Schmidt received B.A. and M.A. degrees in Sociology from the College of William and Mary in Williamsburg, Virginia, and an M.S. and a Ph.D. in Computer Science from the University of California, Irvine in 1984, 1986, 1990, and 1994, respectively.

Dr. Jules White is an Assistant Professor in the Department of Electrical Engineering and Computer Science at Vanderbilt University. He was previously a faculty member in Electrical and Computer Engineering and won the Outstanding New Assistant Professor Award both at Virginia Tech. His research has produced over 85 papers and won 5 Best Paper and Best Student Paper Awards. Dr. White's research focuses on securing, optimizing, and leveraging data from mobile cyber-physical systems. His mobile cyber-physical systems research spans focus on: (1) mobile security and data collection, (2) high-precision mobile augmented reality, (3) mobile device and supporting cloud infrastructure power and configuration optimization, and (4) applications of mobile cyber-physical systems in multi-disciplinary domains, including energy-optimized cloud computing, smart grid systems, healthcare/manufacturing security, next-generation construction technologies, and citizen science. His research has been licensed and transitioned to industry, where it won an Innovation Award at CES 2013, attended by over 150,000 people, was a finalist for the Technical Achievement at Award at SXSW Interactive, and was a top 3 for mobile in the Accelerator Awards at SXSW 2013. His research is conducted through the Mobile Application computinG, optimizatoN, and secUrity Methods (MAGNUM) Group at Vanderbilt, which he directs.

Dr. Abhishek Dubey is an Assistant Professor of Electrical Engineering and Computer Science at Vanderbilt University, Senior Research Scientist at the Institute for Software-Integrated Systems and co-lead for the Vanderbilt Initiative for Smart Cities Operations and Research (VISOR). His research interests include model-driven and data-driven techniques for dynamic and resilient human cyber physical systems. He directs the Smart computing laboratory (scope.isis.vanderbilt.edu) at the university. The lab conducts research at the intersection of Distributed Systems, Big Data, and Cyber Physical System, especially in the domain of transportation and electrical networks. Abhishek completed his Ph.D. in Electrical Engineering from Vanderbilt University in 2009. He received his M.S. in Electrical Engineering from Vanderbilt University in August 2005 and completed his undergraduate studies in Electrical Engineering from the Indian Institute of Technology, Banaras Hindu University, India in May 2001. He is a senior member of IEEE.

CHAPTER EIGHT

A blockchain based access control framework for the security and privacy of IoT with strong anonymity unlinkability and intractability guarantees

Aafaf Ouaddah
Mchain, Admirals Way, Canary Wharf E14 9UH, London, United Kingdom

Contents

Advances in Computers, Volume 115
ISSN 0065-2458
https://doi.org/10.1016/bs.adcom.2018.11.001

Abstract

Motivated by the recent explosion of interest around the blockchain, we examine whether they make a good fit to build a lightweight and robust access control framework to address security and privacy issues in the Internet of Things (IoT) sector. In this direction, this chapter discusses the limitations of the centralized model to secure IoT and proposes the blockchain approach as example of successful distributed system to bring security and privacy to IoT devices. In this direction, we introduce FairAccess and PPDAC, as a lightweight and privacy-preserving access control framework based on the emergent blockchain technology, mainly the permissionless and public type, to ensure fine-grained access control functions for IoT devices with strong anonymity guarantee for IoT end-users. The proposed framework retains the benefits of the blockchain to meet IoT security and privacy arising needs while overcoming the challenges in integrating the blockchain to IoT.

1. Introduction

Two decades ago, a world where everyday objects could sense, analyze, store, or exchange information existed only in science-fiction novels [1]. Today, with the rapid grow of hardware technologies, such scenarios are increasingly becoming reality. In fact, Gartner identifies IoT as one of the top 10 strategic technology trends,[a] Cisco forecasts 50 billion devices connected by 2020[b] a potential market in excess of $14 trillion.[c] IoT is actually already here. Things are extending the world we live in by enabling a whole new range of applications. Being able to put a bunch of powerful, tiny, and cheap computers everywhere around us making it possible to monitor and interact with the physical world with a much finer spatial and temporal resolution than ever before. And yet the application scenarios and market opportunities offered by objects communicating actively and autonomously extend far beyond the foreseeable horizon [2].

IoT is both a global physical infrastructure and an umbrella term for conceiving many existing and evolving interoperable Information and Communication Technologies (ICT)'s, interconnected, devices, objects and services [3], increasingly deployed in various range of domains such as industrial control, health care, aviation, home automation, retail, transport, wearable, and more [4,5]. This revolution, also known as web of things [6], Internet of

[a] http://www.gartner.com/newsroom/id/2209615.
[b] http://share.cisco.com/internet-of-things.html.
[c] http://iotevent.eu/cisco-sees-14-trillion-opportunity-in-iot/.

Everything or Fog networks [7] where smart things are communicating among each other and with computers in the Internet in an intelligent [8] and a machine-to-machine way [9], aims at redefining the whole relationship of humans, work, and technology [10].

However, as quoted by the French theorist and urbanist Paul Virilio: "When you invent the ship, you also invent the shipwreck ... Every technology carries its own negativity, which is invented at the same time as technical progress" [11]. This wise adage gives us an eloquent understanding of how IoT will be a revolutionary technology if we can overcome its drawbacks mainly its security and privacy issues. Actually, in one hand, IoT is arising new security challenges due to the use of simple processors and operating systems that are not supporting sophisticated security approaches by the majority of IoT devices. In addition, the lack of authentication and authorization standards for IoT edge devices gives birth to malicious attacks on secrecy and authentication, silent attacks on service integrity, or attacks on network availability such as denial-of-sleep attacks that drain batteries, or denial-of-service (DoS) attacks [12,13].

On the other hand, privacy is no less serious issue. Actually, IoT devices are, by nature, "collectors and distributors of information" [8], so they constitute a huge challenge to individual privacy [14,15], especially, the omnipresence of IoT devices in user's everyday life and his ubiquitous interaction with smart objects. In addition to the uncontrolled, automated and unseen collection of fine-grained data by third parties lacking in transparency may consistently exhibit users to several threats, such as: identification, localization and tracking, monitoring, surveillance, manipulation, profiling, data linkage, privacy-violating interaction and presentation and even social engineering [16,17]. *Privacy-by-policy* and *privacy-by- design* are two emergent approaches, elicited by the academic literature [18–20] to enhance privacy in IoT. Privacy-by-policy aims to protect data from accidental disclosure or misuses, also promoting informed costumer choice [21,22], while privacy-by-design prompts to implement privacy throughout the engineering process, with a proactive and preventative approach, rather than ex post [23,24]. Combining both approaches has proven crucial to conceptualize and alleviate potential risks associated with the IoT, which nevertheless still a highly controversial issue.

Unfortunately, current solutions can hardly keep up with those new arisen challenges. Therefore, new security and privacy-engineering practices and distributed architectures are increasingly urgently needed to properly address the IoT major challenges. Thus, our research is concerned about how decentralization through blockchain technology, mainlythe permissionless and public type, can be applied to provide a privacy-preserving access control solution to IoT objects and what are the strengths and weaknesses in doing so. For that purpose, we introduced FairAccess [25,26] as a novel Distributed Privacy Preserving Access Control framework in IoT scenario that combines access control models and cryptocurrency blockchain mechanisms. In FairAccess, we propose the use of SmartContract to express fine-grained access control policies to make authorization decisions. We opt for authorization tokens as access control mechanism, delivered through emergent cryptocurrency solutions. We use the public blockchain first to ensure evaluating access policies in distributed environments where there is no central authority/administrator, and guarantee that policies will be properly enforced by all interacting entities and second to ensure token reuse detection. However, the public aspect of the blockchain ledger might initially appear to be at odds with the private aspect of some access control policies. To tackle this issue, this position chapter proposes a privacy-preserving distributed access control (PPDAC) scheme to be integrated with FairAccess. PPDAC uses a white box version of Distributed Cipher-text Policy Attribute Based Encryption "DCP-ABE" technique to add a privacy layer hiding sensitive attributes and private access control policies in the public blockchain while keeping the verification process transparent and public. In addition, the use of recent protocol in cryptography such as zero proof of knowledge protocol "zk-SNARK" enables our framework to guarantee high anonymity and intractability to IoT end users.

The reminder of this chapter is organized as follow: Section 2 begins with the definition of IoT's specific requirements in security and privacy. It discusses the drawbacks of the centralized approach and the advantages of the distributed one in IoT context. Section 3 proposes the blockchain approach as a promising distributed technology to enable security in IoT. This will be followed by some brief overview of the blockchain technology. Section 4 investigates our proposed approach to ensure smart objects security through a decentralized and user-driven access control framework based on the blockchain named FairAccess and discusses the limitations we have faced using the public blockchain mainly the privacy and traceability problems. Section 5 overcomes the defined problems by adding a privacy layer on top of the blockchain through PPDAC scheme. Finally, Section 6 concludes the chapter.

2. Problem statement and research questions

2.1 IoT security and privacy requirement

IoT is exposed to significant privacy and security risks due to its heterogeneous devices, dynamics, multiple domain application and undefined perimeters. As a result, it becomes more difficult for security researchers to find comprehensive solutions to the current security challenges. Therefore, the importance of understanding, defining and analyzing those requirement becomes paramount. In fact, each domain application presents peculiarities that often depend on the context and the devices interacting in that field, and that must be considered when dealing with security and privacy requirements. To meet this end, we have conducted a deep analysis that supports the specification of security and privacy requirements in a structured form in ref. [27]. During our study, we began by defining those requirements into six groups: Privacy, Confidentiality and Integrity, Reliability and availability, social and economic aspects, technologies constraints and usability. The six main security and privacy requirements along with their subcomponents are shown in Fig. 1.

Afterward, we listed different application fields of IoT and specify the characteristics and security requirements of each one. Indeed, we have categorized these applications into three domains: (1) Personal and home: at the scale of individual, home and healthcare. (2) Government and utilities: at the scale of community nation and region. (3) Enterprise and industry: at the scale of industries and big companies. Moreover, we gave an overview of different taxonomies and classification of constrained devices in IoT.

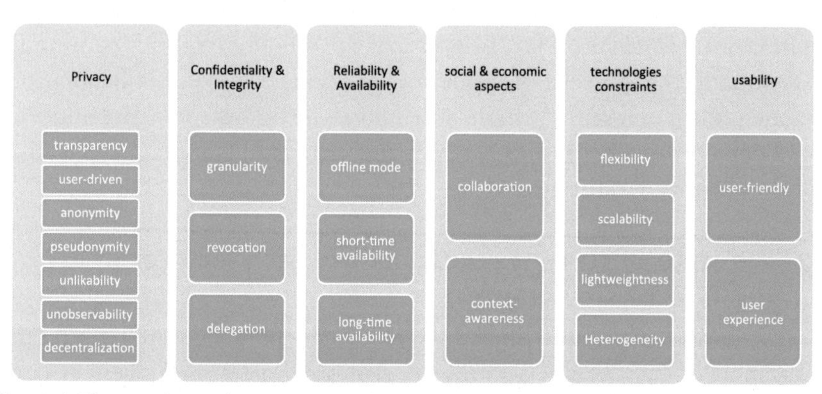

Fig. 1 IoT's security and privacy requirements [27].

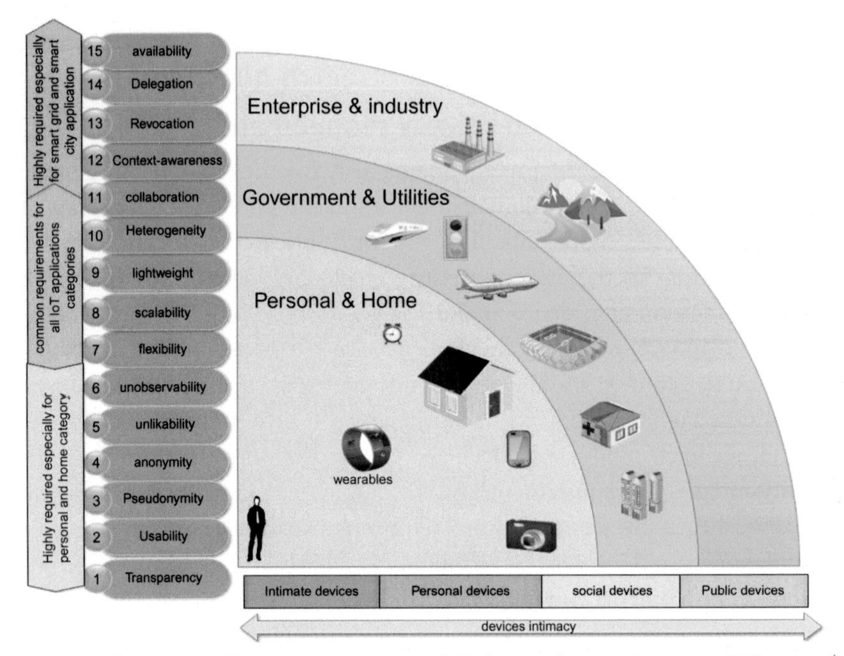

Fig. 2 IoT domain application taxonomy and their security requirements [27].

The study conducted to the following findings: In a glance, it is shown that with regard to personal and home category, access control solutions targeting this category are required to preserve the privacy of end users in a proprietary, primordial but user-friendly way. In addition, the availability and reliability needs are the outstanding objectives of security solutions targeting enterprise and industry category, while the confidentiality and integrity needs overwhelm when it comes to government and utilities applications. However, it is observed that the whole categories commonly share the following needs on principles of cooperation and collaboration, openness/interoperability, high scalability, flexibility and distribution. Fig. 2 depicts the relationship between the various IoT domain application and the security requirements.

2.2 The state of current access control architectures in IoT and related challenges

The integration of physical objects in the internet infrastructure requires the application of lightweight security mechanisms to be used even in constrained environments. However, current security standards and access control solutions were not designed with such aspects in mind. They are not sufficient to meet the needs of these nascent ecosystems regarding scalability, interoperability, lightness and end-to-end security [27]. These challenges have attracted more and more attention from research community and recently several efforts are starting to emerge in this direction. However, those solutions could be categorized within three approaches as follow (see Fig. 3):

- *The Centralized architecture*

  This approach consists in outsourcing IoT device's access control related operations to a trusted third entity. This entity, also known as a Policy Decision Point (PDP), could be instantiated by a back-end server or gateway directly connected to the device that it manages. It is responsible for analyzing access requests based on the stored access control policies. Therefore, requesters wishing to get access to the data provided by end devices are asked to pass by those trusted third parties. The bright side of this architecture consists in relieving constrained devices (i.e., sensors, actuators) from the burden of processing heavy access control functionalities, which enables the use of standard access control technologies such as SAML and HTTPS (to transport in a secure way authentication information), XACML (to define complex access control policies) among many others. However, this architecture presents major disadvantages, in the context of IoT, listed as follow: First, end to end security is dropped by the introduction of a trusted third party. Second, the role of IoT devices is strictly limited, within this architecture, in the decision making process. As a result, the elaboration of smart authorization policies, where access control decision are based on contextual information instantaneously collected from the environment of IoT end devices, is hardly challenging. Third, the Resource Owner (RO) access control policies as well as the users' authorization requests are revealed to the trusted party. As a result, the privacy of either the resource owner or the requester is corrupted. Fourth, the PDP remains a bottleneck and point of failure that can disrupt the entire network. This is particularly important when it is directly tied to critical IoT services.

- *The Decentralized architecture with trustful entity*

  In this approach, the device participates partially in the elaboration of the access control decision. Its main role consists in gathering the contextual information from its surrounding environment (location, temperature, humidity, power level, etc.) and send it to a trusted third party. This trusted third party receives the access control requests and

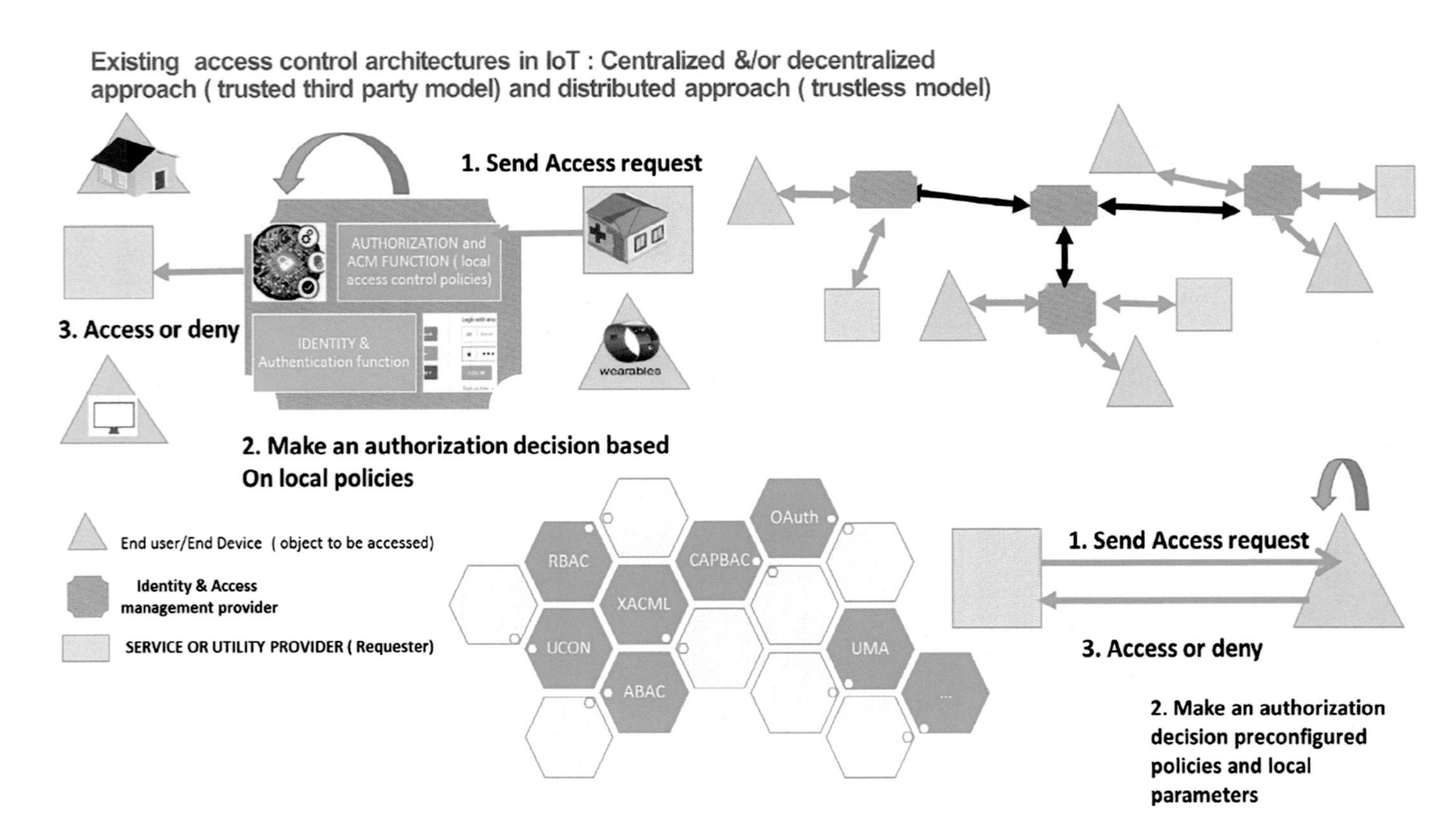

Fig. 3 Existing access control architectures in IoT.

makes the decision based on a pre-defined policies and the contextual information received from the smart object. Like the centralized approach, this architecture enables the use of already existing authorization technologies with the need to elaborate a connection between the trusted third party and the connected device to transfer the contextual information. However, in this scenario, additional security measures have to be taken to secure the communication channel between the trusted third party and the end-device to protect the transferred information. In addition, end-device has to be configured to providing or not its collected data. Moreover, the transfer of contextual information to the trusted party could not be achieved in instantaneous way, meaning that this architecture is not recommended in use cases where real time decisions access is required such as health-care scenario, or with SCADA systems. Finally, the privacy of the resource owner and the requester is not considered.

- *The Distributed architecture*

  The distributed approach consists in locating and embedding the intelligence of processing an access control decision in the device side. This approach matches perfectly with the real essence of IoT where intelligence is located in the edge of the network. It presents many impressive and promising advantages regarding the privacy of the resource owner and the requester as no trusted third party is involved. With the edge-intelligence principal, end-users are more empowered to control access over their own devices by defining their own policies. Furthermore, the possibility of making a smart access control decision in a real time is given. Moreover, the cost management of data generated by IoT devices is less expensive than the one in the two precedent approaches where providing a cloud back end for each connected smart object is required. In the distributed approach, devices are authorized to send information just when it is necessary. Finally end-to-end security could be achieved.

However, the most challenging hurdle in this approach arises from the inherent features of existing access control technologies such as RBAC [28] and ABAC [29] and OAUTH [30] that make their implementation unfeasible in resource-constrained devices [31].

Consequently, much effort has to be conducted to deeply analyze the viability of adapting existing access control models or defining new proposals that meet the requirements of a distributed access control approach.

2.3 Summary

The centralized and decentralized with trustful entity approaches where all devices are identified, authenticated and connected through cloud-based servers support huge processing and storage capacities. While this model has been used for decades to connect standards computing devices and will continue to fit small-scale IoT networks [32], it severely struggles to respond to the growing needs of the huge IoT ecosystems of tomorrow [33–35] for the following reasons.

- Cost: Existing IoT solutions are expensive due to two main reasons: (1) high maintenance cost: from the manufacturer's side, the centralized clouds, large server farms, and networking equipment have a high maintenance cost considering the distribution of software updates to millions of devices for years after they have been long discontinued [36]. (2) High infrastructure cost: the sheer amount of communications that will have to be handled when there are tens of billions of IoT devices needs to cater to a very high volume of messages (communication costs), data generated by the devices (storage costs), and analytical processes (server costs).
- Bottleneck and single point of failure: cloud servers and farms will remain a bottleneck and point of failure that can disrupt the entire network. This is particularly important when it is directly tied to critical IoT services such as healthcare services.
- Scalability: within the centralized paradigm, cloud-based IoT application platforms acquire information from entities located in data acquisition networks, and provide raw data and services to other entities. These application platforms control the reception of the whole information flow. This enforcement creates a bottleneck to scaling the IoT solutions to the exponentially growing number of devices and the amount of data generated and processed by those devices (i.e., the concept of "Big Data").
- Insufficient security: The tremendous amount of data collected from millions of devices raises information security and privacy concerns for either individuals, corporations, or governments. As proven by recent denial-of-service attacks on IoT devices [37], the huge number of low-cost and insecure devices connected to the internet is proving to be a major challenge in assuring IoT security.
- Privacy breaches and Lack of transparency: in the centralized models, from the consumer's side, there is an undebatable lack of a trust in service providers getting access to data collected by billions of entities

creating information. There is a need for a "security through transparency" approach allowing users to retain their anonymity in this super connected world.

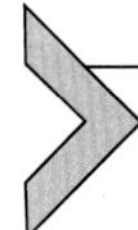

3. The solution: Decentralizing IoT networks in trustless way through the blockchain

A decentralized approach to IoT networking would solve many of the centralized related issues described above. The concept of a distributed IoT is not novel. In fact, many academic and industrial documents consider it as one of the most promising approaches that can push the dream of the IoT into the real world. Panikkar et al. [36] point out the need for a shift toward a decentralized architecture for the ever-expanding IoT device ecosystem to be sustainable. It has been explicitly mentioned that the development of decentralized autonomic architectures and the location of intelligence at the very edge of the networks are issues that need to be addressed. We argue that the blockchain provides an elegant solution to the aforementioned problems.

3.1 The blockchain phenomenon: A new wave of decentralization

Originally introduced by Satoshi Nakamoto in 2008 [38] to underpin the Bitcoin cryptocurrency network, the blockchain has taken up the bulk of technology industry and financial world attention. As a secure and decentralized computational infrastructure, it is widely acknowledged as a disruptive and efficient solution for the problems of centralization, privacy and security when recording tracking, monitoring, managing and sharing not only financial transactions but also any other value such as birth and death certificates, marriage licenses, deeds and titles of ownership, educational degrees, financial accounts, medical procedures, insurance claims, votes, provenance of food, and anything else that can be expressed in code [39–41].

In this chapter, we focus our study on the public and permissionless type of the blockcain. Then, we define this later as a distributed database for transaction processing. All transactions in a blockchain are stored into a single ledger. The blockchain technology is built on top of four fundamental building blocks, each building block has key properties, and each property is achieved through specific mechanisms.

We consider that the blockchain technology relies on the following building blocks as described below:

(1) Identifying the source and destination of a transaction: in a blockchain based ecosystem, users serve from digital identities called "addresses" to send and receive transactions. Those addresses should have the following key features:

- **Self-issued and independent**: users should be able to self-generate anonymous (e.g., using software) transaction identities (e.g., a cryptographic hash of the public-key, in Bitcoin system) in an independent way from any given authority (e.g., government, businesses, etc.).
- **Anonymous**: transaction identities should be anonymous in the sense that it reveals nothing about the real identity of its owner to fulfill the needs of user privacy, then true anonymity in digital identities requires more than self-issuance feature. It requires also the possession of the **unlikability** and **intractability** features.
- **Privacy-preserving verifiable**: any entities in the system should be able to verify the security of the digital identity. A publicly verification algorithms to validate the source of trust for any given (anonymous) transaction-identity is required while preserving the privacy of the owner of the identity.

(2) Transactions: A transaction records the transfer of a value (altcoin) from some source address to destination addresses. Transactions are generated by the sender and broadcasted the network of peers. Transactions are invalid unless they have been recorded in the public history of transactions, the blockchain. Ultimately, transaction processing with blockchain technology should satisfy the following properties:

- **Proprietary**: Only the owner who is authorized to perform transactions using her own identifying addresses.
- **Irreversible**: Once transaction has entered the ledger, it should be impossible to modify its information, or delete it, which would effectively reverse the transaction.
- **Publically verifiable**: The verification algorithm of a transaction should enable any node in the network to verify the validity of the transaction.
- **Immunity**: Once a transaction is recorded in the blockchain it cannot be altered without that alteration being detected and rejected by the other nodes in the network. In addition, if a

transaction conforms to a ledger protocol, it should be eventually added to the ledger.

(3) Condition for auto-processing a transaction: The transfer of any value (e.g., altcoins, tokens) with the blockchain or the execution of any function through the blockchain should be locked by logic conditions (e.g., low, contract) that must be written as a code and automatically executed by nodes in the network. **This condition should be self-executed**.

(4) Consensus: Every user or node in the network relies on algorithmically enforced rules to process transactions with no human interaction required to verify in an independent way the correct execution of the protocol and obtains the same results. Each node has exactly the same ledger as all of the other users or nodes in the network. This ensures a complete consensus from all users or nodes in the corresponding currencies blockchain.

The blockchain Mechanisms: To satisfy the properties cited above, a suit of cryptographic mechanisms has been introduced as explained below and described in Fig. 4:

Public key cryptography: enables a **self-issued**, **pseudonymous** and a **privacy-preserving** identification. **Digital** *signature*: is used to satisfy the feature of **transaction proprietary and publically verifiable properties**: Each user in the system possesses at least a pair of private and public keys; the public key is published publically to determine the digital identity, while the

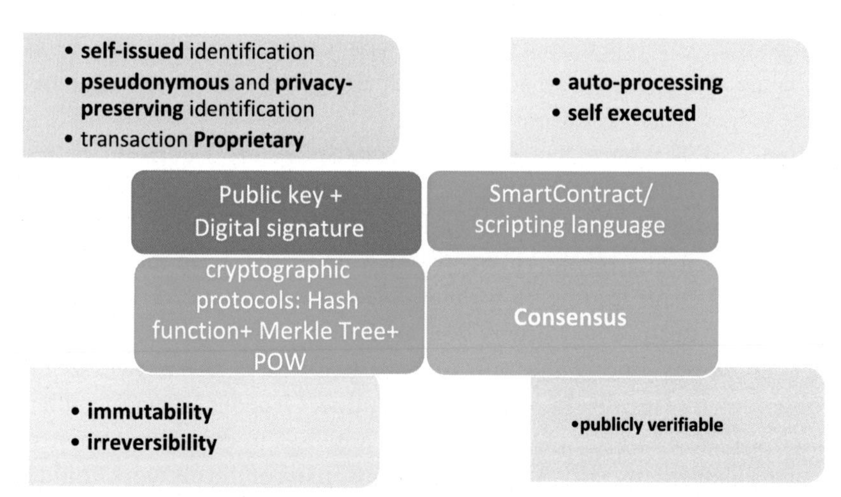

Fig. 4 Blockchain's building blocks and mechanisms.

private key is used to sign the corresponding transaction by it owner (or a trusted party on her behalf).

- The validity of the signature is tied to the knowledge of the sender's private key **(proprietary)**
- A signature can be verified by anyone knowing the sender's public key **(publically verifiable)**
- A signature is invalidated if any parameters of the transaction are changed.

Hash function [42] **and Merkle tree** [43]: are used to solve the problem of **Immutability, irreversibility** and **public verification** of transactions in blockchain systems. Those features are achieved by distributing transactions into time-ordered blocks and time stamping each of these blocks by its cryptographic hash.

Transactions are summarized in each block using a Merkle tree, also known as a binary hash tree. It is an efficient data structure that serves to summarize and verify the integrity of large sets of data. Organizing blocks in an ordered chain called (blockchain), where each block refers to the previous one, is a clever technique introduced by Nakamoto to make it infeasible to delete or replace the whole blocks. To simplify blockchain verification, key block parameters (such as a Merkle root) are collected in a block header. Thus, providing immutability of transactions is equivalent to providing immutability of block headers.

The **Immunity** feature of transactions is achieved through the properties of hashing functions. Actually, any alteration in the transaction results in changes in the Merkle root of the block, which is a part of its header. Hence, to change a single block, an attacker must also change all succeeding blocks in the ledger which is infeasible.

Proof of X where X $\in$ (Work/stack/Space etc.) consensus protocols: Once the transactions have been created, it is broadcasted in the network and each node verifies independently the received transaction and includes it in a block. However, each node may have different view of the blockchain leading to a divergence of branches of a blockchain. To mitigate this problem, there is a need of a distributed mechanism to reach consensus among the untrusted participants in the network. This problem is historically known as "Byzantine Generals" (BG) Problem, which was raised in ref. [44]. Reaching consensus in distributed environment is a challenging task. Actually, the **consensus** of the network could be achieved in a variety of ways, including proof of work (e.g., as used in Bitcoin), proof of stake (e.g., Nxt), delegated proof of stake (e.g., BitShares) and Practical byzantine fault tolerance (hyper ledger project). For more details, we refer the reader to a comprehensive survey of those four approaches in refs. [45, 46].

SmartContract/scripting language: are used to satisfy the auto-processing properties. The scripting language describes the execution of a certain program on a stack machine. Each transaction contains a script which locks, or encumbers, the value transferred by this transaction. A script is a part of each input and each output of a transaction. When we generalize this scripting language computation to arbitrary Turing complete logic, we obtain an expressive smart contract system. The interest in smart contract applications steadily risen since 2014 due to the appearance of Bitcoin-like technologies, such as Ethereum [47] and many other works designed specifically to decentralized smart contract system. SmartContracts are self-executed cryptographic "boxes" that are stored in the blockchain and contain value only unlocked if certain conditions are met.

3.2 Why the blockchain as a solution

The Blockchain technology and IoT offer a new world of promise and fascinating possibilities. Actually, the decentralized, autonomous, and trustless inherent capabilities of the blockchain make it an ideal component to become a foundational element of IoT solutions. It can potentially improve the IoT sector [48], and significantly help in achieving the vision of decentralized [49] and private-by-design IoT [50].

The following salient features of the blockchain, illustrated in Fig. 5, make it an attractive technology for addressing the aforementioned security and privacy challenges in IoT:

- Decentralization and trustlessness: The lack of central control ensures scalability and robustness by using resources of all participating nodes and eliminating many-to-one traffic flows, which in turn decreases delay and overcomes the problem of a single point of failure.
- Append-only. The major design goal of the blockchain is to make it such that deleting information from the blockchain (i.e., reversing transactions) is impossible or at least prohibitively expensive.
- Pseudo-anonymity: The inherent anonymity afforded is well-suited for most IoT use cases where the identity of the users must be kept private.
- Transparency: Each node in the network can independently verify the current state of the ledger and arrive to the same conclusion as the rest of the network.

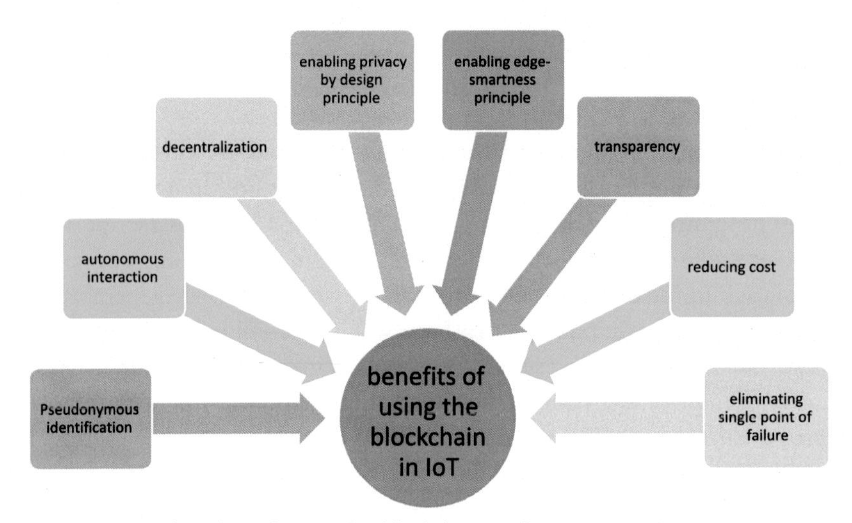

Fig. 5 The benefits of combining the blockchain with IoT.

- Cryptography model: The trust in the blockchain does not arise from an authority maintaining the ledger, but rather from the mathematical soundness of the cryptographic protocols used in the system and a prohibitive economic cost of an attack.

However, adopting the blockchain to handle authorization functions in IoT is not straightforward and will require addressing the following critical challenges:

- The public nature of the blockchain versus the private aspect of access control policies
- Tractability problem
- Consensus process is particularly computationally intensive, time consuming while the majority of IoT devices are resource restricted and most IoT applications low latency is desirable.

The previously defined problem statement leads to the following research questions:

1. How to solve access control challenges in IoT by leveraging the public and permissionless blockchain technology?
2. How to provide a lightweight access control framework for constrained IoT devices despite the intensive computation power required by the consensus protocol adopted in the public blockchain?
3. How to enable a privacy preserving access control solution that hides user's access control policy and overcomes the problem of traceability and profiling commonly known in the public blockchain?

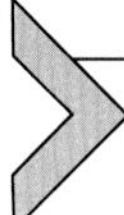

4. FairAccess: Using blockchain technology as access control infrastructure

To answer the aforementioned questions, we propose a framework composed of two main modules that are: FairAccess and PPDAC as shown in Fig. 6. In the next section we will give an overview and brief description of the both protocols.

4.1 FairAccess: Using blockchain technology as an access control infrastructure

FairAccess [25, 51] is a new distributed access control framework based on Blockchain technology that has combined, for the first time, access control models and cryptocurrency blockchain mechanisms. In FairAccess, we propose the use of SmartContract [47] to express fine-grained and contextual access control policies to make authorization decisions. We opt for authorization tokens as an access control mechanism, delivered through emergent cryptocurrency solutions. We use blockchain to ensure enforcing access policies in distributed environments where there is no central authority/administrator, and guarantee that policies will be properly implemented by all interacting entities (Fig. 7).

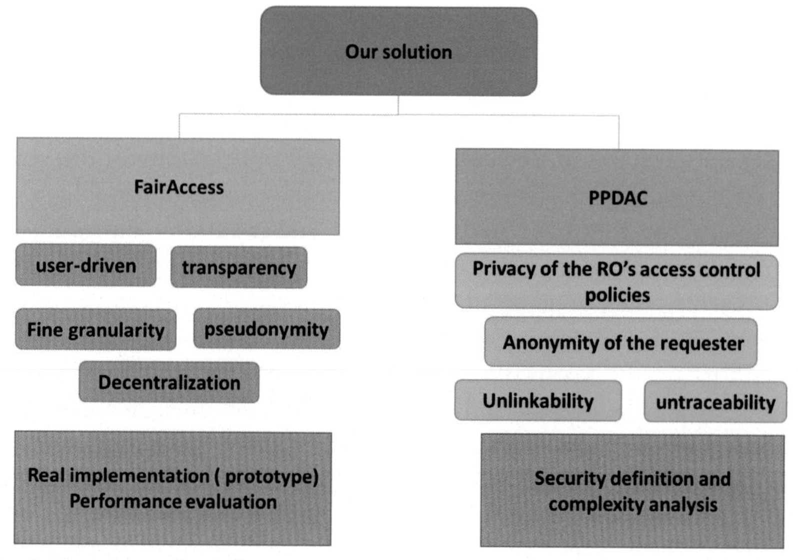

Fig. 6 Our proposed solution.

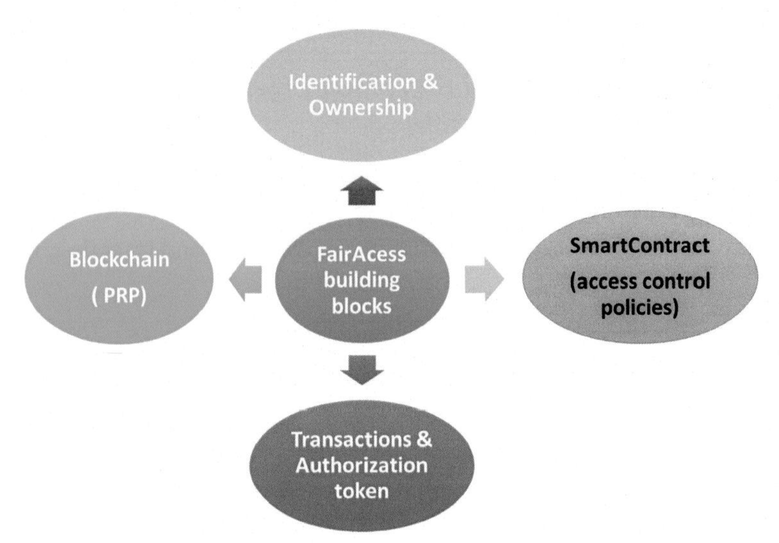

Fig. 7 FairAccess's building blocks.

Authorization token: In our FairAccess framework, we define an Authorization Token as a data structure that represents the access right or the entitlement defined by the creator of the SmartContract, generating the authorization token, to the entity interacting with this SmartContract in order to access a specific resource identified by its address. If the authorization token presented by the device was delivered by a SmartContract that matches with the one that is associated with the other device or service that manages that device, then the access is granted. Actually, the token-based access control enforced by the blockchain technology provides many advantages in IoT context. Actually, having the token securely held on the blockchain means smart devices can easily verify the validity of the access token relieving IoT constrained devices from the burden of handling a vast amount of access control-related information and at the same time mitigates the need for outsourcing these functionalities to a trusted powerful entity that prevents end-to-end security to be achieved. Moreover, it reduces communication cost, since no further authentication mechanisms are required to get the token since only signature is sufficient. Similarly, the resource owner can manage and update access to his numerous and heterogeneous devices by placing a resource hash of the requested data on the blockchain. Then the device can read the hash and check the update access control configurations. This hash can be updated with every new connection to the blockchain. This could enable easier management and updating of access control policies if the access control logics would be located in the device side, especially for devices that are placed in unreachable or hardly attended location such as a smart parking systems, where devices may be embedded directly in the asphalt. In addition, the token could be used for many access control operations such as getting, delegating, revoking and even updating access in an easier and flexible way.

Blockchain: FairAccess provides several useful mechanisms using the blockchain. In fact, in FairAccess, the blockchain is considered as a database or a policy retrieval point where all access control policies are stored in form of transactions; it serves also as logging databases that ensure auditing functions. Furthermore, it prevents forgery of token through transactions integrity checks and detects token reuse through the double spending detection mechanism.

SmartContract: The SmartContract has its own account on the blockchain, and the blockchain supports an *account-based model* [47]. It allows us to express logic functions in code. It operates as *autonomous actors*, whose behavior is completely predictable. As such they can be trusted to drive forward any on-chain logic that can be expressed as a function of on-chain data inputs, provided that the data they need to manage are within their own reach (in the example above, the contract would not be able to trade assets that it did not own). In addition, its code can be inspected by every network participant since all the interactions with a contract occur via *signed* messages on the blockchain, all the network participants get a cryptographically verifiable trace of the contract's operations.

This enables our framework to express fine grained access control policies. Actually, a policy is a set of rules and conditions (based on a specific context or attribute, etc.) that a requester entity has to fulfill in order to obtain the Access Token and gets access to the specific resource. This rules could be expressed by any access control model but must be transformed to a script language considered as locking script placed on the output of a transaction. Fortunately, new blockchain protocols are being developed including full Turing completeness capability, allowing anyone to write smart contracts and decentralized applications with their own arbitrary rules for ownership, transaction formats and state transition functions [47]. The use of those advanced languages, such as Ethereum, will certainly enable our FairAccess framework to express fine-grained context-aware access control policies. We believe that SmartContracts are a promising emergent field to express, granular, contextual and contractual access control model in general and in IoT in particular.

4.2 FairAccess's architecture

The vision of our decentralized access control framework is a system of autonomous organizations hinged around one or many Resource Owners in possession of one or many resources identified with addresses and

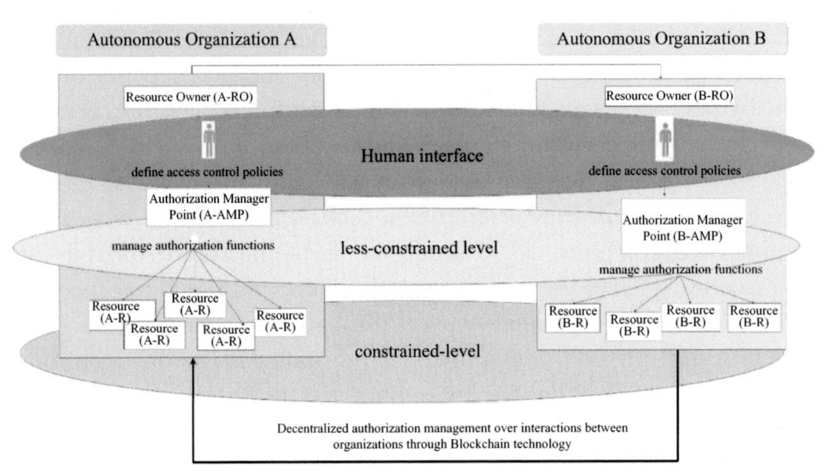

Fig. 8 FairAccess's architecture overview.

interacting between each other through transactions (requesting, granting, delegating and revoking access) under the control of their RO. The blockchain is a ledger keeping track and ensures the validity of access transaction among interacting organization. Each manages its own access policy, under only the control of his Resource Owner, resulting in an "Internet of Decentralized, Autonomous Organization" and thus the fairness of our access control framework (Fig. 8).

Considering the following scenario, see Fig. 9, where a subject (e.g., a device A, identified with the address rq) wants to perform an action (e.g., modify) on a protected resource (e.g., Device B temperature, identified with address rs). We assume that the requester already knows the access control policy regulating access to Device B. FairAccess workflow is as follow: the Rq fulfills the conditions specified in the access control policy and submits his request through his wallet in form of a RequestAccess Transaction. Afterword, the wallet broadcasts this transaction to the network nodes till it reaches miners. Those later act as distributed Policy Decision Point (dPDP), and evaluate the transaction and check the request with the defined policy, by executing a PolicyContract already deployed, by the owner of Device B, in the blockchain through a previous transaction called GrantAccess. The execution of PolicyContract determines whether the request should be permitted or denied. Finally, if it is being successfully executed, the PolicyContract generates and assigns an authorization token to the requester address through an AllowAccess transaction. Then the authorization token is recorded in the blockchain and appears in the requester's authorization token list. Finally, the requester presents the authorization token

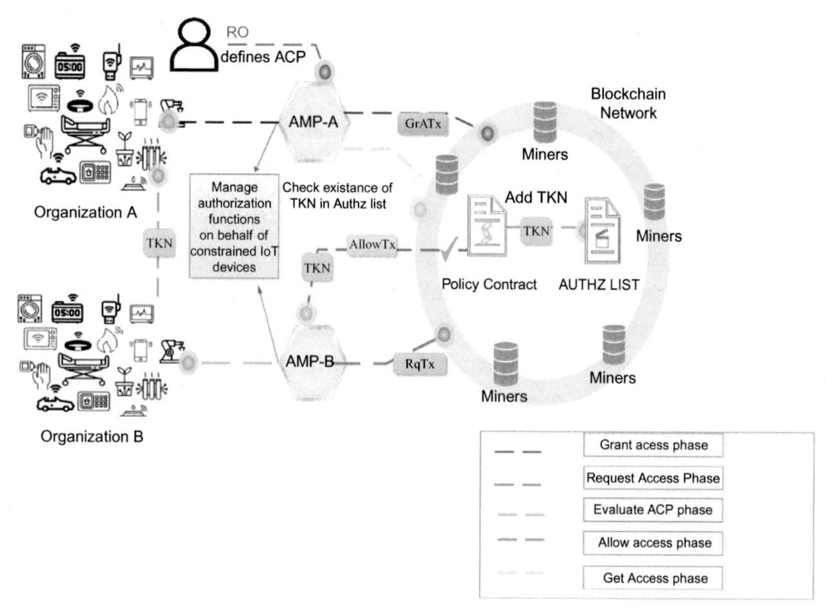

Fig. 9 FairAccess workflow.

to the end device through a GetAccess Transaction. The end device B checks the validity of the authorization token by referring to the blockchain; if this authorization token was delivered by the SmartContract corresponding to the device B, it allows access else it denies.

4.3 The faced challenges

FairAccess has successfully achieved the following IoT's security and privacy-preserving requirements, fixed in the beginning of this chapter, that are: (1) Distributed nature and the lack of a central authority. User-driven and transparency. (2) Lightweightness. (3) Fine-granularity. (4) Identification enabling thing to thing interaction. (5) Pseudonymity and Unlikability. However, adopting the blockchain technology to handle access control functions is not straightforward and additional critical issues emerge that are:

(1) The public and transparent aspect of the blockchain comes at odds with the private aspect of access control policies publically recorded in the blockchain: Often times, the policies for determining who can access the resources are sensitive also and need protection as well.

(2) Traceability issue such as the structure of the transaction graph [52,53] as well as the value and dates of transactions may conduct to learning devices authorization functionality pattern, as well as if a particular device is trying to communicate with other devices

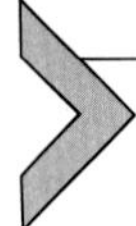

5. Toward privacy through transparency: A privacy-preserving distributed access control (PPDAC) scheme to enhance privacy and anonymity in FairAccess

In this section we will show how to overcome those two identified challenges by introducing PPDAC as a fully anonymous Privacy-Preserving Distributed Access Control scheme (PPDAC) to be integrated over FairAccess. PPDAC provides a strong privacy guarantees access control scheme that preserves the resource owner's access control policies and the requester's sensitive attributes and then preserves in a strong way the anonymity of both the requester and the Resource owner, see Fig. 10. The proposed scheme aims to maintain Blockchain transparency features while ensuring strong privacy guarantees for users. We develop a policy-hiding access control scheme that protects both sensitive attributes and sensitive policies. That is, nodes in the public blockchain can decide whether Alice's certified attribute values satisfy Bob's policy, without learning any other information about Alice's attribute values nor Bob's policy. To enable policy-hiding access control, and untraceability of authorization tokens, our construction for PPDAC uses a novel technique that combines a white box distributed multi-authority CP-ABE [54,55], zk-SNARKs [56] protocols and SmartContract (Fig. 11).

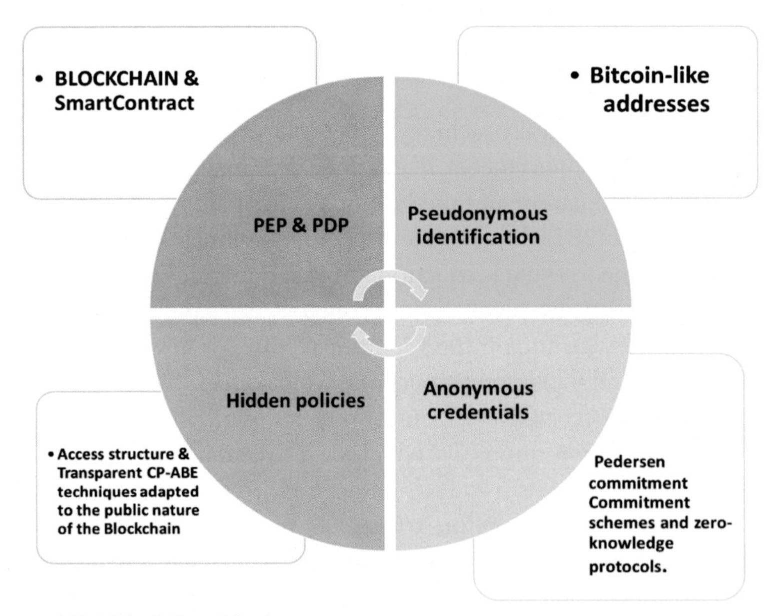

Fig. 10 PPDAC buildings blocks.

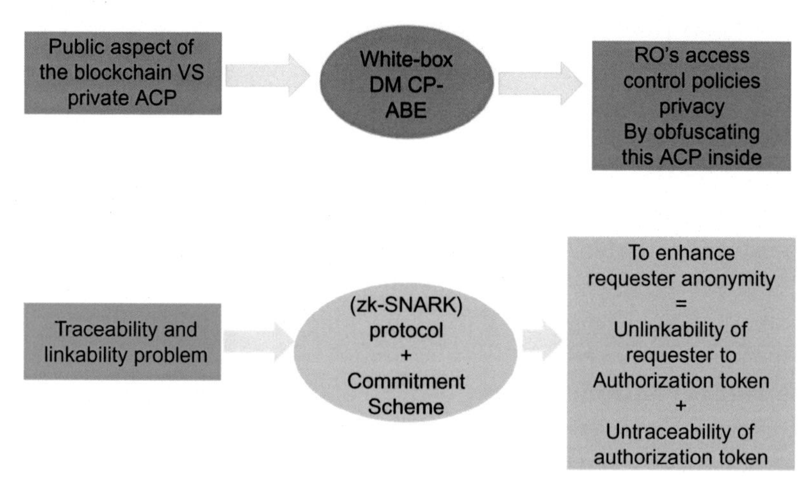

Fig. 11 PPDAC based technologies.

5.1 PPDAC building blocks

In a traditional CP-ABE scheme, an encryptor can encrypt data with a hidden access structure. A decryptor obtains his secret key associated with his attributes from a certifying authority in advance and if the attributes associated with the decryptor's secret key do not satisfy the access structure associated with the encrypted data, the decryptor cannot decrypt the data or even guess what access structure was specified by the encryptor. In our setting, due to the fact that our framework is built on a public blockchain, the aim to use the CP-ABE technique is not to decrypt data as it was introduced in the literature, but instead to ensure two properties:

1. Hiding the RO access control policies in the blockchain network
2. Enabling the requester to prove, in a public way to the network (miners), that his hidden attributes fit the hidden policies. For this purpose, we introduce two new notions:

A challenge: is an arbitrary string chosen at random by the resource owner and encrypted to get a ciphertext CT. It is considered as a proof of access control policy fulfillment used by the requester to prove, in a public way, to the blockchain network, his fulfillment of access control policy without revealing his attribute nor the access control policy, since the evaluation of access control policy is reduced to the ability of the requester to decrypt the ciphertext CT and obtain the publically known challenge. Actually, the requester will be able to decrypt the cipher text CT if and only if his attributes match with the hidden access structure.

An identity key (IDkey): in our setting, we introduce the notion of IDkey which plays the same role of a private key, but it is public. This IDkey holds the features of a cryptocurrency coin. The IDkey should fulfill three important properties:

- **Anonymous**: in the sense that it reveals no information about the real identity of his owner. Untraceable: we cannot guess the origin who sends the token to the blockchain. We acheive the anonymity and untraceability of IDkey using the zero-knowledge Succinct Non-interactive ARguments of Knowledge (zk-SNARKs) by hiding the origin of transaction spending the IDkey.
- **Usage is controlled by the owner**: We argue that the identity token IDkey, which is cryptographically secured by the blockchain as a coin, could not be secret, since the usage of this identity token is restricted to his owner or ceded under his consentement. To achieve this property, we will "tie" the usage of an IDkey to spending an associated IDcoin which is, at its turn, controlled by the possession of a private key only known by the owner.

Our protocol runs between the following entities: a Resource owner (RO) who owns IoT device (D), one or many requesters that could be service providers the (RO) wants to collaborate with to get smart services over his device. A set of authorities responsible for certifying a set of attributes

- Certifying authorities: the certifying authorities are trusted and manage their attributes set in an independent way.
- FairAccess node: A FairAccess node is equivalent to the role of miners. We suppose that those nodes are honest but curious. Hence, not only the identity of interacting entities should be anonymous but also the access control policy held in SmartContract should be hidden.
- Ro: the resource owner is responsible for defining an access control policy over his devices and obfuscating this access control policy inside a SmartContract using decentralized multi-authority CP-ABE with hidden policy. Then he deploys the SmartContract to the network.
- Requester: the certifying authorities generates relevant IDkeys for each requester. Furthermore, only requesters who's IDkeys satisfy the access control policy get an authorization token.

5.2 PPDAC phases and functionalities

The proposed protocol divides the interaction between parties into five phases as follows: a set up phase, a Grant Access phase, a RequestAccess phase, a GetAccess phase, an evaluating access control policy phase and an AllowAccess phase (Figs. 12–14).

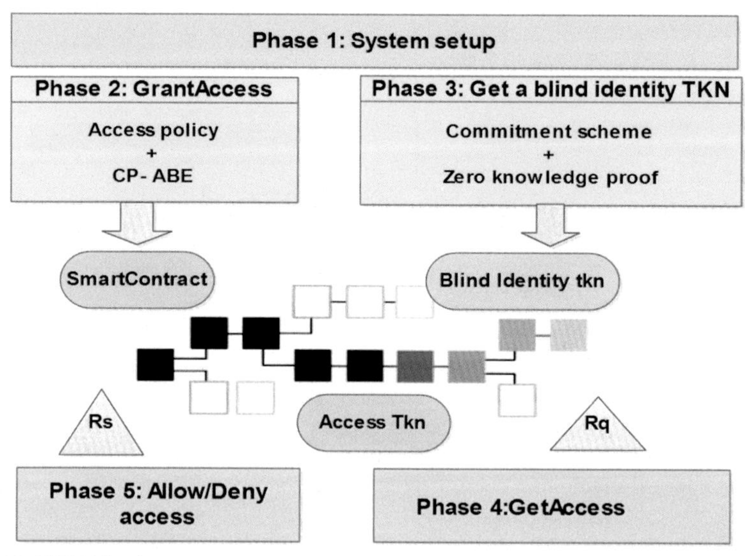

Fig. 12 PPDAC's phases.

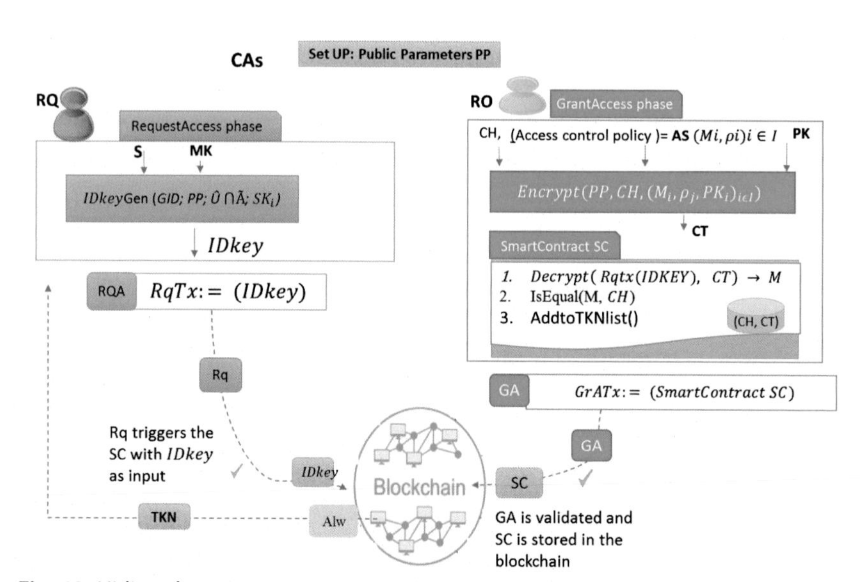

Fig. 13 Hiding the private access control policies inside the public blockchain.

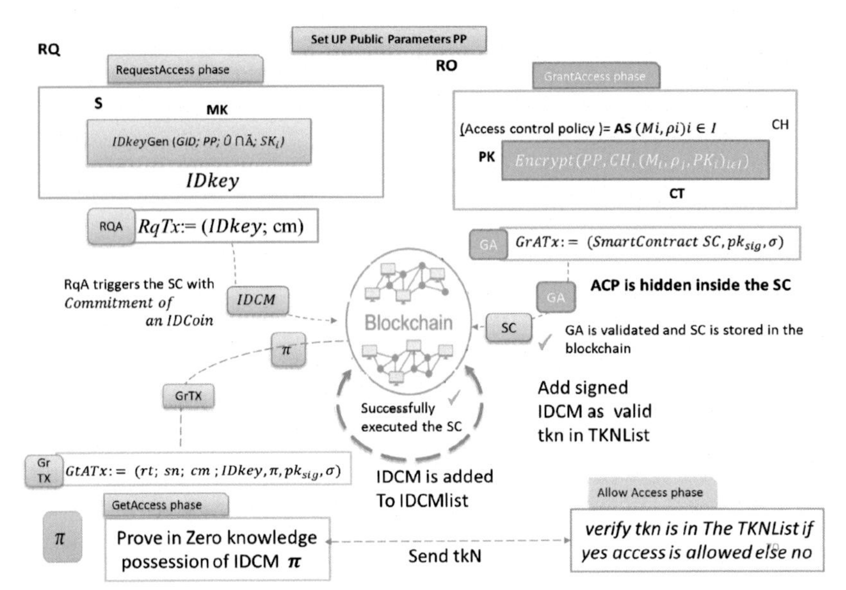

Fig. 14 Unlinkability and intractability of authorization tokens.

5.3 Background and definitions

To construct the PPDAC scheme, the following building blocks and definitions are adopted.

5.3.1 Pseudorandom function

Definition 1. Collision-resistant hashing. *We use a collision-resistant hash function*

$$\mathrm{CRH}: \{0,1\}^* \rightarrow \{0,1\}^{\mathrm{O}^{\Upsilon}} \tag{1}$$

Definition 2. Pseudorandom functions. *We use a pseudorandom function family PRF*

$$\mathrm{PRF} = \left\{ \mathrm{PRF}_{\lambda} : \{0,1\}^* \rightarrow \{0,1\}^{\mathrm{O}^{\Upsilon}} \right\}_{\lambda} \tag{2}$$

where λ denotes the seed. We derive, in arbitrary way, three pseudo random functions that are:

$$PRF_{\lambda}^{address}(x) := PRF_{\lambda}(00 \,\|\, x) \tag{3}$$

$$PRF_{\lambda}^{apk}(x) := PRF_{\lambda}(01 \,\|\, x) \tag{4}$$

We assume that PRF_{λ}^{Tsn} is also collision-resistant in the sense that it is infeasible to find

$(\lambda, x) \neq (\lambda', x')$ *such that* $PRF_{\lambda}^{Tsn}(x) = PRF_{\lambda'}^{Tsn}(x')$.

5.3.2 One-time strongly-unforgeable digital signatures

We use a digital signature scheme Sign $=(\mathcal{G}en, \mathcal{KS}i\mathcal{G}, sig, check)$ that works as follows:

$\boldsymbol{\mathcal{G}en(1^k) \rightarrow param_{sig}}$. Given a security parameter $\mathbf{1}^k$, $\mathcal{G}en$ samples public parameters $\boldsymbol{param_{sig}}$ for the signature scheme

$\boldsymbol{\mathcal{KS}i\mathcal{G}\left(param_{sig}\right) \rightarrow \left(pk_{sig}, sk_{sig}\right)}$. Given public parameters $\boldsymbol{param_{sig}}$, $\mathcal{KS}i\mathcal{G}$ samples a public key and a secret key for a single user

$\boldsymbol{sig_{sk_{sig}}(m) \rightarrow \sigma}$. Given a secret key sk_{sig} and a message m, $sig_{sk_{sig}}$ signs m to obtain a signature σ

$\boldsymbol{check_{pk_{sig}}(m, \sigma) \rightarrow b}$. Given a public key pk_{sig}, message m, and signature σ, $check_{pk_{sig}}$ outputs

b $=1$ if the signature σ is valid for message m; else it outputs b $=0$.

The signature scheme Sign satisfies the security property of one-time strong unforgeability against chosen-message attacks (SUF-1CMA security).

5.3.3 Access structure

Definition 3. (Access Structure [57]). *Let $\{P_1, P_2, \ldots, P_n\}$ be a set of parties, a collection $\mathcal{A} \subseteq 2^{\{P_1, P_2, \ldots, P_n\}}$ is monotone if $\forall B, C$: if $B \in \mathcal{A}$ and $B \subseteq C$ then $C \in \mathcal{A}$. An access structure (respectively, monotone access structure) is a collection (respectively, monotone collection) $\mathcal{A}$ of non-empty subsets of $\{P_1, P_2, \ldots, P_n\}$, i.e. $\mathcal{A} \subseteq 2^{\{P_1, P_2, \ldots, P_n\}} \setminus \{\phi\}$, the sets in $\mathcal{A}$ are called the authorized sets, and the sets not in $\mathcal{A}$ are called the unauthorized sets.*

5.3.4 Linear secret-sharing schemes (LSSS)

In our construction, we will use linear secret-sharing schemes (LSSS). Thus, we use the definition adapted from ref. [57].

Definition 4. (Linear Secret-Sharing Schemes (LSSS)). *A secret sharing scheme $\prod$ over a set of parties P is called linear (over $\mathbb{Z}_p$) if the following properties can be satisfied:*

(1) *The shares for each party form a vector over $\mathbb{Z}_p$.*

(2) *For $\prod$, there exists a matrix M with ℓ rows and n columns called the share-generating matrix. For $i=1, 2, \ldots, \ell$, the ith row is labeled with a party $\rho(i)$ where ρ: $\{1, 2, \ldots, \ell\} \rightarrow \mathbb{Z}_p$. To share a secret $s \in \mathbb{Z}_p$, a vector $\vec{v}=(s, v2, \ldots, vn)$ is selected, where $v2, \ldots, vn$ are randomly selected from $\mathbb{Z}_p$. $M\vec{v}$ is the vector of the ℓ shares according to $\prod$. The share $M_i\vec{v}$ belongs to the party $\rho(i)$, where M_i is the ith row of M.*

5.3.5 Distributed multi-authority CP-ABE scheme

In our setting we use a distributed multi-authority CP-ABE scheme based on the one introduced in [58].

Definition 5. *DCP-ABE scheme comprises four fundamental algorithms as follows: system Setup, Authority setup, Encrypt, KeyGen, and Decrypt.*

Syetem Setup $(\mathbf{1^k}) \rightarrow (\mathbf{PP})$*: The setup algorithm takes in a security parameter* $\mathbf{1^k}$*. It outputs the public parameters PP.*

Suppose that there are N Certifying Authorities $\{\mathring{CA}_1, \mathring{CA}_2, \ldots, \mathring{CA}_N\}$*, and each authority* $\mathring{CA}_i$ *monitors a set of attributes* $\tilde{A}_i$*. Each user U has a unique global identifier* GID_U *and holds a set of attributes* $\hat{U}$*.*

Authority setup$(1^k) \rightarrow (SK_i, PK_i)$*. Taking as input the security parameter*1^k*, and it outputs secret key and public key pair* $(SK_i; PK_i)$ *for each authority* $\tilde{A}$*, where* $KeyGen(1^k) \rightarrow (SK_i, PK_i)$*.*

Encrypt$\left(PP, \mathcal{M}, \left(M_i, \rho_j, PK_i\right)_{i \in I}\right) \rightarrow CT$*. It takes as inputs: the public parameters PP, the plaintext message*$\mathcal{M}$*, a set of access structures* $(Mi, \rho i)$ $i \in I$ *and a set of public keys* $(PKi) \in I$ *It outputs the cipher-text CT.*

keyGen $(GID; PP; \hat{U} \cap \tilde{A}; SK_i) \rightarrow idKey_{GID,i}$ *Taking as input the public parameter PP, the secret key*SK_i*, the requester's global identifier* GID_U *and a set of attributes* $\hat{U} \cap \tilde{A}$*, the IDkeyGen generation algorithm outputs an* $idKey_{GID,i}$*.*

Decrypt$(PP, (idKey_{GID,i})_{i \in I}, CT) \rightarrow M$*.The decrypt algorithm takes as input the public parameter PP, the cipher-text CT and the secret keys* $idKey_{GID,i}$*. If the set of attributes satisfies the access structure A then the algorithm will decrypt the cipher-text and return a message m else it returns* $\perp$*.*

Definition 6. *The decentralized cipher text-policy attribute-based encryption (DCP-ABE) is correct if*

$$Pr\left|\begin{array}{c|r} Decrypt\left(PP, \left(idKey_{GID,i}\right)_{i \in I}, CT\right) & Syetem\ Setup\left(1^k\right) \rightarrow (PP) \\ & Authority\ setup\left(1^k\right) \rightarrow (SK_i, PK_i) \\ \rightarrow M & Encrypt\left(PP, \mathcal{M}, \left(M_i, \rho_j, PK_i\right)_{i \in I}\right) \rightarrow CT \end{array}\right| = 1$$

where the probability is token over the random bits used by all the algorithms in the scheme.

Definition 7. (Selective-Access Structure Secure DCPABE (IND-sAS-CPA)). *A decentralized cipher text-policy attribute-based encryption (DCP-ABE) scheme is (T, q,* $\epsilon(k)$*) secure in the selective-access structure model if no* *probably polynomial-time adversary A. Making q secret key queries can win the above game with the advantage*

$$Adv_A^{ACP-ABE} = \left| Pr[b' = b] - \frac{1}{2} \right| > \epsilon(k)$$

where the probability is token over all the bits consumed by the challenger and the adversary.

5.3.6 zk-SNARKs protocol

Definition 8. *We informally define zk-SNARKs for arithmetic circuit satisfiability. We refer the reader to, e.g., ref.* [56] *for a formal definition.*

For a field $\mathbb{F}$, *an* $\mathbb{F}$*-arithmetic circuit takes inputs that are elements in* $\mathbb{F}$, *and its gates output elements in* $\mathbb{F}$. *We naturally associate a circuit with the function it computes. To model non-determinism we consider circuits that have an input* $x \in F^n$ *and an auxiliary input* $a \in \mathbb{F}^p$, *called a witness. The circuits we consider only have bilinear gates. Arithmetic circuit satisfiability is defined analogously to the boolean case, as follows:*

Definition 9. *The arithmetic circuit satisfiability problem of an* $\mathbb{F}$*-arithmetic circuit:*

C: $\mathbb{F}^n \times \mathbb{F}^p \rightarrow \mathbb{F}^l$ is *captured by the relation* $\mathbb{R}_C = \{(x, a) \in \mathbb{F}^n \times \mathbb{F}^p$ $C(x, a) = 0^l\}$; *its language is* $\mathcal{L}_C = \{x \in \mathbb{F}^n : \exists\, a \in \mathbb{F}^p$ *such that* $C(x, a) = 0^l\}$

Given a field $\mathbb{F}$, *a (publicly-verifiable preprocessing)*

zk-SNARK *for* $\mathbb{F}$*-arithmetic circuit satisfiability is a triple of polynomial-time algorithms (KeyGen; Prove; Verify):*

keyGen$(1^k, C) \rightarrow (pk, vk)$ *Takes as Input a security parameter k and an* $\mathbb{F}$*-arithmetic circuit C, the key generator keyGen probabilistically samples a proving key pk and a verification key vk. Both keys are published as public parameters and can be used, any number of times, to prove/verify membership in* $\mathcal{L}_C$.

Prove$(pk, x, a) \rightarrow \pi$ *Takes on input a proving key pk and any* $(x, a) \in \mathbb{R}_C$, *the prover Prove outputs a non-interactive proof* π *for the statement* $x \in \mathcal{L}_C$

Verify$(vk, x, \pi) \rightarrow b$ *Takes on input a verification key vk, an input x, and a proof* π, *the verifier Verify outputs* $b = 1$ *if he is convinced that* $x \in \mathcal{L}_C$.

A **zk-SNARK** satisfies the following properties:

Completeness. For every security parameter *k*, any $\mathbb{F}$-arithmetic circuit C, and any $(x, a) \in \mathbb{R}_C$, the honest prover can convince the verifier. Namely, b = 1 with probability $1 - negl\ (k)$ in the following experiment: $(pk, vk) \leftarrow$ ***keyGen***$(1^k, C)$, $\pi \leftarrow$ ***Prove***(pk, x, a), $b \leftarrow Verify(\ vk, x, \pi)$

Succinctness. An honestly-generated proof π has $\mathcal{O}_k(1)$ bit and *Verify* (vk, x, π) runs in time $\mathcal{O}_k(|\ x|\ )$. (Here, $\mathcal{O}_k$ hides a fixed polynomial factor in k.)

Proof of knowledge (and soundness). If the verifier accepts a proof output by a bounded prover, then the prover "knows" a witness for the given instance. (In particular, soundness holds against bounded provers.) Namely, for every *poly*(*k*)-size adversary $\mathcal{A}$, there is a *poly*(*k*)-size extractor $\mathcal{E}$ such that $Verify(\ vk, x, \pi) = 1$ and $(x, a) \in \mathbb{R}_C$ with probability $negl\ (k)$ in the following experiment: $(pk, vk) \leftarrow$ ***keyGen***$(1^k, C)$; $(x, \pi) \leftarrow \mathcal{A}(pk, vk)$; $a \leftarrow \mathcal{E}(pk, vk)$

Perfect zero knowledge. An honestly-generated proof is perfect zero knowledge. Namely, there is a *poly*(*k*)-size simulator $\mathcal{S}im$ such that for all stateful *poly*(*k*)-size distinguishers $\mathcal{D}$ the following two probabilities are equal:

1. The probability that $\mathcal{D}(\pi) = 1$ on an honest proof.

$$Pr\left[(x, a) \in \mathbb{R}_C\ \mathcal{D}(\pi) = 1 \left| \begin{array}{r} (pk, vk) \leftarrow \textbf{\textit{keyGen}}(\ C) \\ (x, a) \leftarrow \mathcal{D}(pk, vk) \\ \pi \leftarrow \textbf{\textit{Prove}}(pk, x, a) \end{array} \right. \right]$$

2. The probability that $\mathcal{D}(\pi) = 1$ on a simulated proof.

$$Pr\left[(x, a) \in \mathbb{R}_C\ \mathcal{D}(\pi) = 1 \left| \begin{array}{r} (pk, vk, trap) \leftarrow \textbf{\textit{Sim}}(\ C) \\ (x, a) \leftarrow \mathcal{D}(pk, vk) \\ \pi \leftarrow \textbf{\textit{Sim}}(pk, x, trap) \end{array} \right. \right]$$

5.4 Formal definition of privacy-preserving distributed access control scheme (PPDAC)

We begin by describing, and giving intuition about, the data structures used by a **PPDAC** scheme, the algorithms that use and let the details of the construction of each one in Section 5.5.

5.4.1 Data structure

Public Parameters (PP): A list of public parameters PP is available to all users in the system. These are generated by a trusted party at the "launch time" of the system and are used by the system's algorithms.

Addresses: As in FairAccess, each user in the system generates as many addresses as he wants. Each address corresponds to a pair of key (a_{pk}; a_{sk}). The public key a_{pk} is published and enables others to interact with user. The secret key a_{sk} is used to spend the authorization token owned by user who possesses the a_{pk}.

IDkey: is a vector of $idKey_{GID,i}$ as $IDkey = (idKey_{GID,i})_{i \in N}$

Where $idKey_{GID,i}$ is a private key delivered by a certifying authority $C\mathring{A}_i$ monitoring a set of attribute $\hat{U} \cap \tilde{A}_i$ and GID. Where GID corresponds to the notion of global identifiers introduced by chaise in [59] to link private keys together that were issued to the same user by different authorities.

The requester gets an IDkey by running an IDkeygen algorithm with the certifying authorities. Then it is recorded in the blockchain by its owners as a public value when minting an IDcoin.

IDcoin: For each IDkey, we associate an IDcoin. An IDcoin is a data object *idc*, to which we associate the following:

- An IDcoin commitment, denoted cm (idc): a string that appears on the ledger once *idc* is minted by the requester.
- An IDcoin serial number, denoted sn (idc): a unique string associated with the idc, used to prevent double spending.
- An IDcoin address, denoted a_{pk} (idc): an address public key, representing who owns idc.

Authorization token *TKN*: is a data structure constructed as follow:

$$TKN := (\text{SmartContract address} \,\|\, \text{cm}).$$

It is added in the TKNList by the SmartContract in a simultaneous way as the commitment of IDcoin appears in IDCMlist.

Commitments of minted IDcoins and serial numbers of spent IDcoins. For any given time T,

IDCMList$_T$ denotes the list of all IDcoin commitments appearing in Request Access RqTx and GetAccess GtATx recorded in the blockchain at time T.

TKNList$_T$ denotes the list of all authorization tokens appearing in GetAccess GtATx transaction recorded in the blockchain at time T.

SNListT$_T$ denotes the list of all serial numbers appearing in GetAccess transaction in the ledger at time T.

Merkle tree over commitments IDcoin: For any given time T, $IDCTree_T$ denotes a Merkle tree over ***IDCMList***$_T$ and rt_T its root. Moreover, the function $Path_T$ (idcm) gives the authentication path from a IDcoin commitment *idcm* appearing in ***IDCMList***$_T$ to the root of $Tree_T$. For convenience, we assume that the ledger at time T L_T also stores rt_{T*} for all $T^* < T$ (i.e., it stores all past Merkle tree roots).

Merkle tree over authorization token: For any given time T, $AUTree_T$ denotes a Merkle tree over ***TKNList***$_T$ and $rt^*{}_T$ its root. Moreover, the function $Path^*{}_T$ (idcm) gives the authentication path from a authorization token ***TKN*** appearing in ***TKNList***$_T$ to the root of $TKNTree_T$. For convenience, we assume that the ledger at time T L_T also stores $rt^*{}_{T*}$ for all $T^* < T$ (i.e., it stores all past Merkle tree roots).

SmartContract: is a program running in the blockchain that stores two values: (1) a Cipher text CT. (2) its corresponding challenge ***CH*** which is a random number linked to CT as: $CT := Encrypt(PP, CH, (M_i, \rho_j, PK_i)_{i \in I})$. The SmartContract embedded three functions:

(i) Decrypt(PP,IDkey,CT) → M
(ii) A boolean function IsEqual(M, *CH*)
(iii) AddToTKNList(TKN)

New transactions: We use three new types of transactions in our framework that are:

(1) GrantAccess transaction: it serves to deploy the SmartContract in the blockchain:

$$GrATx := (SmartContract\ SC)$$

(2) RequestAccess transaction: triggers the SmartContract and mint IDcoin:

$$RqTx := (IDkey;\ k;\ s;\ cm)$$

(3) GetAccess transaction: pour the RequestAccess IDcoin and holds the authorization token:

$$GtATx := (rt;\ sn;\ cm;\ \pi)$$

Blockchain: Our protocol is applied on top of a public permissionless ledger such as Ethereum. At any given time T, all users have access to the ledger at time T, which is a sequence of transactions. The ledger is append-only.

5.4.2 Algorithms

We will next describe functions that are needed in our system. The protocols that implement these functions will be described in the later section. Most, if not all, of these functions will be performed with respect to a given blockchain B, or need to make calls to the public parameters in our system. These are implicit inputs of the protocols.

A (PPDAC) scheme Π is a tuple of polynomial-time algorithms: (SystemSetup, *Authority Setup, IDkeyGen, encrypt, decrypt,* CreateAddress, GrantAccess, *RequestAccess, GetAccess, ValidateTransaction, executeSmart Contract, AllowAccess*)

System Setup: $(\mathbf{1^k}) \rightarrow PP$. Taking as input a security parameter $\mathbf{1^k}$, the global setup algorithm outputs the public parameter PP.

The algorithm Setup is executed by a trusted party. The resulting public parameters PP are published and made available to all parties (e.g., by embedding them into the protocol's implementation). The setup is done *only once*. Afterward, no trusted party is needed, and no global secrets or trapdoors are kept.

Authority Setup: $\boldsymbol{KG(1^k) \rightarrow (SK_i, PK_i)}$. Each authority in the system runs the authority setup algorithm with the security parameter 1^k, as input to produce its own secret key and public key pair $(SK_i; PK_i)$.

$\boldsymbol{Encrypt(PP, CH, (M_i, \rho_j, PK_i)_{i \in I}) \rightarrow CT}$. It takes as inputs: the public parameters PP, a random plaintext string CH, a set of access structures $(Mi, \rho i)_{i \in I}$ and a set of public keys $(PKi)_{i \in I}$. It outputs the cipher-text CT associated with this set of access structures.

$\boldsymbol{ExecuteSmartContract\ (RqT_x, SmartContract\ SC) \rightarrow True\ or \perp}$. It takes as input the RequestAccess transaction. It returns true or false. This algorithm recalls tree sub-algorithms that are: $\boldsymbol{(Decrypt, IsEqual, AddtoTKNList)}$ defined as follow:

$\boldsymbol{Decrypt(PP, IDkey, CT) \rightarrow M}$. Taking as input the public parameter PP, the cipher-text $\boldsymbol{CT}$ and *IDkey* from the RequestAccess transaction triggering the SmartContract. Where a Request Access transaction $RqTx$ is as follow: $RqTx := (IDkey, k, s, cm)$

IsEqual(M, *CH*). A Boolean function that compares the resulting message M and the recorded challenge *CH* as follow: If the Requester's attribute, associated to *IDkey*, satisfies the set of access structures $(Mi, \rho i) i \in I$, then M will be equal to CH and the function returns true, else the function returns false $\perp$.

AddtoTKNList(idcm) $\rightarrow$ ***(rt, path(idcm))***. It takes as input the IDcoin commitment and returns its authentication path in the Merkle tree having as root *rt*.

CreateAddress(PP) $\longrightarrow$ $(\boldsymbol{a_{pk}}; \boldsymbol{a_{sk}})$. The algorithm generates a new pseudonym address keypair: $(a_{pk}; a_{sk})$. Each user generates at least one address key pair in order to receive authorization tokens. The public key a_{pk} is published, while the secret key a_{sk} is used to redeem tokens sent to a_{pk}. A user may generate any number of address key pairs; doing so does not require any interaction

$$\boldsymbol{GrantAccess(CH, (PKi) \in I, (Mi, \rho i) i \in I) \rightarrow GrATx := (SmartContract\ SC)}$$

The algorithm GrantAccess gives as output a GrantAccess Transaction holding hidden RO access control policies, in form of SmartContract and managing access control for each device.

It takes as inputs: implicitly the public parameters PP, the plaintext CH, a set of access structures$(Mi, \rho i) i \in I$, and set of public keys$(PKi) \in I$ for relevant authorities.

$$\boldsymbol{IDkeyGen\left(GID; PP; \hat{U} \cap \tilde{A}; SK_i\right) \rightarrow idKey_{GID,i}\ and\ IDkey = \left(idKey_{GID,i}\right)_{i \in N}}$$

A requester and an authority engage in the ***IDkeyGen*** algorithm that takes as input the public parameter *PP*, the secret keySK_i, the requester's global identifier GID_U and a set of attributes $\hat{U} \cap \tilde{A}$. It produces an *IDkey* for *Rq*.

$$\boldsymbol{RequestAccess\left(PP, GID_U, a_{pk}\right) \rightarrow (IDcoin, RqTx := (IDkey; k; s; cm))}.$$

The algorithm RequestAccess takes as inputs the public parameters PP, the requester's global identifier GID_U, and the public address of the requester $\boldsymbol{a_{pk}}$ in which he wants to receive an authorization token. The algorithm outputs an IDcoin and a RequestAccess transaction ***RqTx***

$$\boldsymbol{GetAccess}(PP, rt, idcoin, path(idcm), a_{sk}) \rightarrow GtATx := (rt; sn; cm; \pi)$$

The algorithm GetAccess takes as inputs an IDcoin with its corresponding secret key a_{sk} (*required to redeem idcoin*). To ensure that IDcoin have not been previously minted, the GetAccess algorithm also takes as input the Merkle root *rt* with the authentication paths path form *Idcm*. It outputs a GetAccess transaction *GtATx*

$$\textbf{ValidateTransaction}(\textbf{GrantAccess}\ \boldsymbol{GrATx}, \textbf{RequestAccess}\ \boldsymbol{RqTx}, \textbf{GetAccess transaction}\ \boldsymbol{GtATx}) \rightarrow (\boldsymbol{true}, \boldsymbol{false})$$

The algorithm ValidateTransaction checks the validity of a transaction: All transactions must be verified before being considered well-formed. In practice, transactions can be verified by nodes in the distributed system maintaining the ledger, as well as by users who rely on these transactions.

$$\textbf{AllowAccess}(\textbf{GrantAccess Transaction}\ \boldsymbol{GrATx}, \boldsymbol{SNlist}) \rightarrow (\boldsymbol{true}, \boldsymbol{false})$$

The algorithm AllowAccess takes as input GrantAccess transaction and SNList from the current ledger, if the IDCoin spent by this GrantAccess transaction appears in SNlist then access is allowed, else no.

5.5 PPDAC construction

Phase 1: Setting up

System Setup: $(\mathbf{1}^{\boldsymbol{k}}) \rightarrow PP$.Taking as input a security parameter $\mathbf{1}^{\boldsymbol{k}}$, the global setup algorithm outputs the public parameter *PP*.

Authority setup: Suppose that there are *N* *Certifying* Authorities $\{C\mathring{A}_1, C\mathring{A}_2, \ldots, C\mathring{A}_N\}$, and each authority $C\mathring{A}_i$ monitors a set of attributes $\tilde{A}_i$. Each user *U* has a unique global identifier GID_U and holds a set of attributes $\hat{U}$. Each authority in the system runs the authority setup algorithm with the security parameter 1^k as input to produce its own secret key and public key pair $(SK_i; PK_i)$ as follow: $KG(1^k) \rightarrow (SK_i, PK_i)$.

NB: It is worth to note that this phase is conducted outside the blockchain.

Phase 2: The GrantAccess phase consists in defining access control policies over devices and hiding them inside SmartContracts: The resource owner is responsible for defining access policy and obfuscating the policy.

For each device, the RO acts as follow:

1. He generates a random number that we call a challenges: *CH*
2. He defines an Access structure $(Mi, \rho i) i \in I$
3. He employs the distributed DCP-ABE scheme to encrypt challenge *CH* under the corresponding access structure Mi, ρi and gets cipher texts *CT* through the following function: $Encrypt(PP, CH, (M_i, \rho_j, PK_i)_{i\epsilon I}) \rightarrow CT$.

   It takes as inputs: the public parameters PP, the plaintext message CH, a set of access structures $(Mi, \rho i) i \in I$ and a set of public keys $(PKi) \in I$.It outputs the cipher-text CT associated with this set of access structures.

4. He defines a SmartContract that:
   a. stores two values: (1) a Cipher text CT, (2) and its corresponding CH
   b. embeds three functions:
      (1) $Decrypt(PP, IDkey, CT) \rightarrow M$.
      Taking as input the public parameter PP, the cipher-text and *IDkey* from the RequestAccess transaction triggering the SmartContract. Where a Request Access transaction $RqTx$ is as follow: $RqTx := (IDkey, k, s, cm)$, the details of this transaction will be elaborated in the next phase.
      (2) A Boolean function that compares the resulting message M and the recorded challenge CH as follow: IsEqual(M, CH)
      If the Requester's attribute, associated to *IDkey*, satisfies the set of access structures $(Mi, \rho i)i \in I$, then M will be equal to CH, else the function returns false $\perp$.
      (3) If the output of compare function is true, then the SmartContract adds *idcm* associated to *IDkey* in $RqTx$ transaction to Authorization token list TKNList.
5. Then the RO deploys the defined SmartContract in the blockchain through a GrantAccess transaction: GrATx:= (SmartContract SC)

Finally, to prevent SmartContract and transaction malleability we use a digital signature. This digital signature plays the role of "MACS" to tie a SmartContract to the RO and his devices. The Ro samples a key pair(pk_{sig}, sk_{sig}), and computes $\sigma = sig_{sk_{sig}}(SC)$. Then, sets

$$GrATx := \left(SmartContract\ SC, pk_{sig}, \sigma \right)$$

Phase 3: The RequestAccess phase consists in successfully executing the SmartContract to get an access token

This phase is divided into two sub-phases as follow:

Requester (RQ)—Authority Interaction. In this sub-phase, the Rq interacts with the certifying authorities $(C\mathring{A}_i)_{i \in I}$ and runs the *IDkey* extraction algorithms as follow:

$$IDkey\text{Gen}\left(GID; PP; \hat{U} \cap \tilde{A}; SK_i \right) \rightarrow idKey_{GID,i} \text{ and } IDkey = \left(idKey_{GID,i} \right)_{i \in N}$$

Taking as input the public parameter PP, the secret keySK_i, the requester's global identifier GID_U and a set of attributes $\hat{U} \cap \tilde{A}$, the *IDkey*Gen generation algorithm outputs an *IDkey* for *Rq*. We assume that the communication between the RQ and the authorities is secured and private. This phase is running outside the blockchain network.

Requester-SmartContract Interaction. After getting the address of the SmartContract managing access to the target device, the requester triggers the SmartContract with a Request Access transaction $RqTx$ to correctly decrypt the challenge and get an authorization token. As mentioned, in the beginning of this section, in our setting, we do not use private keys as in traditional CP-ABE schemes, but instead we are using identity keys: **IDKey** imbedded inside an **IDcoin**. In order to release the intractability property of **IDKey**, we make use of zk-SNARKs (described above) and a commitment scheme as follow:

Let ***COMM*** denotes a statistically-hiding non-interactive commitment scheme (i.e., given randomness r and message m, the commitment is $\boldsymbol{c} := \boldsymbol{COMM_r(m)}$ subsequently, c is opened by revealing r and m, and one can verify that $COMM_r(m)$equals c).

The requester mint **IDcoins** holding the **IDKey** to the ledger through a RequestAccess $RqTx$ transaction. Unlike traditional CP-ABE schemes, where private keys are used to decrypt the cipher text and they are kept secret to their owner in order to be used only by him, in our setting we provide the same property of restricting the use of the **IDKey** to his owner even if it is public. To do so, we derive the **IDcoins** associated **to IDkey** in the following way:

We use three pseudorandom functions (derived from a single one). For a seed x these are denoted $PRF_x^{address}$. We assume that $PRF_x^{address}$ is moreover collision-resistant.

To provide targets for IDcoins, we use addresses: each user u generates an address key pair $(a_{pk};a_{sk})$. The IDcoins tied to IDkey contains the value a_{pk} and can be spent only with knowledge of a_{sk}. In this way, the usage of IDkey is controlled by his owner who possesses the a_{sk}. A key pair $(a_{pk};\ a_{sk})$ is sampled by selecting a random seed a_{sk} and setting $a_{pk}:=PRF_{ask}^{address}(0)$. A user can generate and use any number of address key pairs.

To “tie” an IDcoin to a specified IDkey, the requester Rq first samples ρ which is a secret value and computes $\rho' = PRF_{ask}^{IDkey}(\rho)$ then determines the ID coin’s serial number sn as $sn := PRF_{ask}^{sn}(\rho')$.

Then, the requester commits to the tuple $(a_{pk}:IDkey;\rho')$ in two phases: (a) u computes $\boldsymbol{k} := \boldsymbol{COMM_r(a_{pk} \| \rho')}$ for a random r; and then (b) the requester computes $\boldsymbol{cm} := \boldsymbol{COMM_s(IDkey \| k)}$ for a random s. The minting results in an IDcoin idc:= $(a_{pk};\ IDkey;\ \rho;$ r; s; cm) and a RequestAccess transaction $RqTx$:= ($IDkey$; k; s; cm).

Specifically, due to the interlinked commitment, anyone can verify that cm in RequestAccess transaction is a coin commitment of an IDcoin of $IDkey$ (by checking that $\boldsymbol{COMM_S(\ IDkey \| k)}$ equals cm) but cannot discern the owner (by learning the address key a_{pk}) or serial number (derived from ρ') because these are hidden in k. In this way the anonymity and untraceability of the **IDkey** is achieved.

The RequestAccess transaction is accepted by the ledger only if the requester presents the correct ***IDkey*** that matches the hidden access control policy by decrypting the cipher text $Decrypt(PP, \textbf{IDkey}, CT) \rightarrow M$ and in one hand.in the other hand, if the commitment is correct.

At this phase, the IDcoin commitment and respectively a corresponding authorization token signed with the SmartContract are recorded in the blockchain in two lists: an IDcoin commitment list IDcList and respectively an authorization token list **TKNList**.

The authorization token is an IDcoin commitment signed with the address of a SmartContract that added the IDcoin to the ledger. Subsequently, letting IDcoinList denotes the list of all IDcoin commitments on the ledger and respectively the **TKNList** denotes all the authorization token on the ledger. The requester may spend the authorization token by posting a GetAccess transaction that contains:

(i) An ID coin's serial number sn, the address of the SmartContract.
(ii) A zk-SNARK proof π of the NP statement "I know r such that $\boldsymbol{COMM_r(sn)}$ appears in the list IDcoinList of IDcoin commitments."
(iii) An IDcoin commitment idcm with the address of the SmartContract.

TKN: = (SmartContract address || idcm) appears in **TKNList**

Assuming that *sn* does not already appear on the ledger (as part of a past spend transaction), the requester can redeem the authorization token. (If *sn* does already appear on the ledger, this is considered double spending, and the transaction is discarded.) Furthermore, the end device will be able to verify that this authorization token was delivered by his corresponding SmartContract by checking the SmartContract address. The Requester anonymity is achieved because the proof π is zero knowledge, while *sn* is revealed, no information about r is, and finding which of the numerous commitments in **TKNList** corresponds to a particular GetAccess transaction is equivalent to inverting $\mathbf{f(x)} := \boldsymbol{COMM_x(sn)}$, which is assumed to be infeasible. Thus, the origin of the transactions is anonymous (Fig. 15).

Phase 4: Getting access and spending the authorization token

To get access to the protected device, the requester spends the authorization token using a GetAccess transaction as follow:

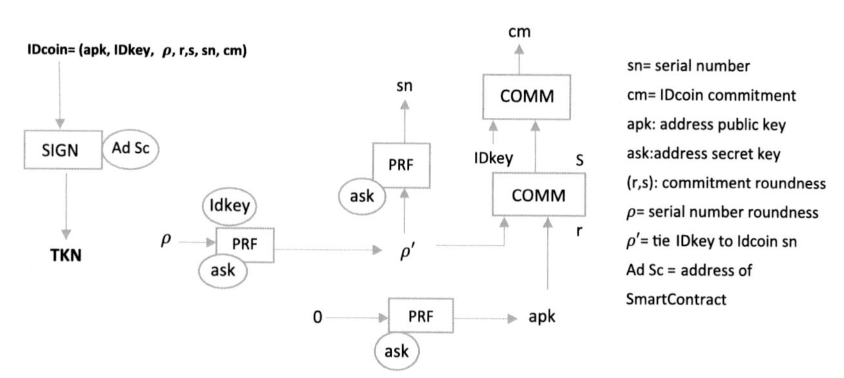

Fig. 15 Generation of IDcoins and authorization tokens.

Suppose that a requester, with address key pair $(a_{pk}; a_{sk})$, wishes to consume his IDcoin:$=$ $(a_{pk}; IDkey; \rho; \rho', \mathrm{r}; \mathrm{s}; \mathrm{cm})$, the requester produces a zk-SNARK proof π for the following NP statement, which we call GetAccess:

"*Given the Merkle-tree root rt, serial number sn, and IDcoin commitments cm, I know IDcoin and address secret key* a_{sk} *such that: i) the IDcoin is well formed, for IDcoin it holds that:* $\boldsymbol{k} := \boldsymbol{COMM_r}(\boldsymbol{a_{pk}} \| \boldsymbol{\rho'})$ and $\boldsymbol{cm} := \boldsymbol{COMM_s}(\boldsymbol{IDkey} \| \boldsymbol{k})$. ii) *The address secret key matches the public key:* $a_{pk} = PRF_{ask}^{address}(0)$. *iii) the IDcoin is correctly tied to Ideky:* $\rho' = PRF_{ask}^{IDkey}(\rho)$. *iv) The serial number is computed correctly:* $sn := PRF_{ask}^{sn}(\rho')$. v) *The coin commitment cm appears as a leaf of a Merkle-tree with root rt.*"

A resulting GetAccess transaction gets *GtATx* the form $GtATx := (rt; sn; cm; \pi)$ is appended to the ledger and it is rejected if the serial number sn appears in a previous transaction.

In this way, if the requester does not know the address secret key $\boldsymbol{a_{sk}}$ that is associated with the public key $\boldsymbol{a_{pk}}$. Then, the requester cannot spend the authorization token neither uses the IDkey because he cannot provide a new $\boldsymbol{a_{sk}}$ as part of the witness of a subsequent GetAccess transaction.

Furthermore, to prevent transaction malleability, and tie each IDkey to its owner, we use digital signature. The Requester samples a key pair(pk_{sig}, sk_{sig}), set $m = \left(\vec{u}, \pi_{grant}\right)$, Compute $\sigma = sig_{sk_{sig}}(m)$. Then, Set $GtATx := (rt; sn; cm; IDkey, \pi, pk_{sig}, \sigma)$.

Phase 5: The AllowAccess phase consists in checking the validity of the token by end-device and decide either allow or deny access

When the requester presents the authorization token to the end device, this later verifies that the authorization token that exists in the **TKNList** was added by the end device's SmartContract. If yes the access is allowed else no.

Compressing the authorization token list for lightweight verification: In the above NP statement, IDCMList is specified explicitly as a list of IDcoin commitment. This straightforward representation severely limits scalability because the time and space complexity of most protocol algorithms (e.g., the proof verification algorithm) grows linearly with IDCMList. Furthermore, authorization token corresponding to already spent IDcoins cannot be dropped from IDCMList to reduce costs, since they cannot be identified (due to the same zero-knowledge property that provides anonymity). As in [60], we rely on a collision-resistant hash function CRH to avoid an explicit representation of IDCMList. We maintain an efficiently updatable append-only CRH-based Merkle tree Tree(IDCMList) over the (growing) list IDCMList. Letting rt denote the root of Tree(IDCMList), it is well-known that updating rt to account for insertion of new leaves can be done with time and space proportional to the tree depth. Hence, the time and space complexity is reduced from linear in the size of IDCMList to logarithmic. With this in mind, we modify the NP statement to the following one: "I know r such that $\boldsymbol{COMM_r(sn)}$ appears as a leaf in a CRH-based Merkle tree whose root is rt." Compared with the naive data structure for IDCMList, this modification increases exponentially the size of IDCMList which a given zk-SNARK implementation can support. We do the same things with the authorization token List TKNList (see Fig. 16).

This ends the outline of the construction. We conclude by noting that, due to the zk-SNARK, our construction requires a one-time trusted setup

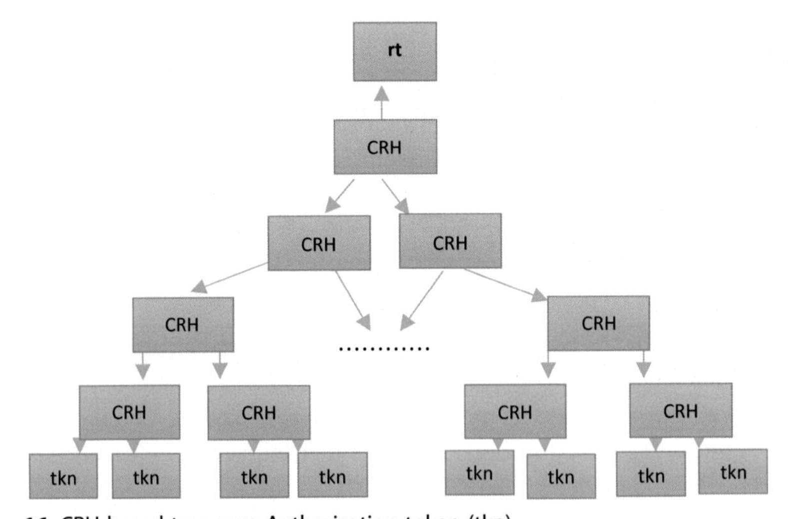

Fig. 16 CRH-based tree over Authorization token (tkn).

System Setup: Taking as input a security parameter $\mathbf{1}^k$ this algorithm outputs the public parameters **PP**
Given a group generator $\mathcal{G}$, a field $\mathbb{F}$ (publicly-verifiable preprocessing)
Suppose that there are N Certifying Authorities $\{C\AA_1, C\AA_2, \ldots, C\AA_N\}$, and each authority $C\AA_i$ monitors a set of attributes $\tilde{A}_i$

- Construct a bilinear group $\mathcal{G}(\mathbf{1}^k) \longrightarrow (\boldsymbol{e}, \boldsymbol{p}, \mathbb{G}, \mathbb{G}_T)$
  Let g, g_1, g_2 generators of group $\mathbb{G}$ and C an $\mathbb{F}$-arithmetic circuit
- Construct $C_{GrantAccess}$ for GrantAccess at security parameter $\mathbf{1}^k$
- Compute $keyGen(1^k, C_{GrantAccess}) \rightarrow (pk_{GrantAccess}, vk_{GrantAccess})$
- Compute $\mathcal{G}_{sig}((\mathbf{1}^k) \longrightarrow param_{sig}$
- **Set PP** $:=(\boldsymbol{g}, \boldsymbol{g_1}, \boldsymbol{g_2}, \boldsymbol{e}, \boldsymbol{p}, \mathbb{G}, \mathbb{G}_T, \boldsymbol{pk_{GrantAccess}}, \boldsymbol{vk_{GrantAccess}}, \boldsymbol{param_{sig}})$
- **Return te public parameter PP**

Authority Setup. Each certifying authority $C\AA_i$

- Chooses $(a_i, y_i, z_i, \delta_i) \leftarrow \mathbb{Z}_p$
- Compute $Q_i = e(g, g)^{a_i}$, $U_i = g^{y_i}$, $V_i = g_2{}^{y_i}$, $\Psi_i^1 = g^{\delta_i}$, $\Psi_i^2 = g_1{}^{\delta_i}$ where $i \in \{1,, N\}$
- Select $w_{i,j} \overset{\$}{\leftarrow} \mathbb{Z}_p$ for each attribute $a_{ij} \in \tilde{A}_i$ where $i \in \{1,, N\}$
- Compute $W_{i,j} = g^{w_{i,j}}$
- Compute $\Phi_{i,j} = g_1{}^{w_{i,j}} \, g^{\frac{1}{\delta_i + a_{ij}}}$
- Set $PK_i = \{Q_i, U_i, V_i, \Psi_i^1 \, \Psi_i^2, (\Phi_{i,j}, W_{i,j})_{a_{ij} \in \tilde{A}_i}\}$
- Set $SK_i = \{a_i, y_i, z_i, \delta_i, (w_{i,j})_{a_{ij} \in \tilde{A}_i}\}$
- Return $(SK_i; PK_i)$

CreateAddress. Takes as inputs the public parameters PP and returns a keypair address

- Randomly sample a_{sk} as seed of $PRF^{address}$
- Compute $a_{pk} := PRF^{address}_{ask}(0)$
- Return $(a_{pk}; a_{sk})$

Encryption. *For each* $j \in I$, *select an access structure* $(Mj, \rho j)$, *and a vector* $\vec{\boldsymbol{v}}_j = (\boldsymbol{s_j}, \boldsymbol{v_{j,2}}, \ldots, \boldsymbol{v_{n_j}})$ ***Where*** $(\boldsymbol{s_j}, \boldsymbol{v_{j,2}}, \ldots, \boldsymbol{v_{n_j}}) \overset{\$}{\leftarrow} \mathbb{Z}_p$ and $\mathcal{M}_j$ a $m_j \times n_j$ matrix

- *Compute* $\xi_{j,i} = \mathcal{M}_j^i \times \vec{\boldsymbol{v}}_j$ where $\mathcal{M}_j^i$ is the ith row in matrix $\mathcal{M}_j$
- *Select* $(\omega_{j,1}, \omega_{j,2}, \ldots, \omega_{j,m_j}) \overset{\$}{\leftarrow} \mathbb{Z}_p$
- *Compute* $E_0 = CH \prod_{j \in I} e(g, g)^{a_j s_j}$, $\{F_j = g^{s_j}, K_j = g_2{}^{s_j}, D_j = V_j^{s_j}\}_{j \in I}$
- *Compute* $(A_{j,1} = g^{y_j} \xi_{j,1} W_{\rho j(1)}^{-\omega_{j,1}}, B_{j,1} = g^{\omega_{j,1}}, \ldots, A_{j,m_j} = g^{y_j} \xi_{j,m_j} W_{\rho j(m_j)}^{-\omega_{j,m_j}}, B_{j,m_j} = g^{\omega_{j,m_j}})$
- *Set the cipher text* $CT = \{E_0, ((F_j, K_j)(A_{j,1}, B_{j,1}, \ldots, A_{j,m_j}, B_{j,m_j}))_{j \in I}\}$

IDkeyGen.to generate an identity token IDKey for a user with $GID = \eta$, and set of attributes $\acute{U} \cap \tilde{A}$ the certifying authority $C\AA_i$, acts as follow:

- *Select* $(\alpha_{U,i}, \beta_{U,i}) \overset{\$}{\leftarrow} \mathbb{Z}_p$
- *Compute* $P_i = g^{a_i} g^{y_i \beta_{U,i}} \boldsymbol{g_2}^{\alpha_{U,i}} \boldsymbol{g_2}^{\frac{z_i + \eta}{\alpha_{U,i}}}$, $L_i = g^{\beta_{U,i}}$, $R_i = g^{\alpha_{U,i}}$, $R'_i = \boldsymbol{g_1}^{\alpha_{U,i}}$, $G_i = g^{-\alpha_{U,i}}$, $G'_i = \boldsymbol{g_1}^{-\alpha_{U,i}}$
- *Compute* $(\Omega_x = w_x^{\alpha_{U,i}})_{a_x \in \acute{U} \cap \tilde{A}}$
- *Set* $IDkey_U^i = \{P_i, L_i, R_i, R'_i, G_i, G'_i \; (\Omega_x)_{a_x \in \acute{U} \cap \tilde{A}}\}$
- *Return Set* $IDkey_U^i$

Decryption. This algorithm takes as inputs *Set* $IDkey_U^i$ *and cipher-text CT and returns plain Text CH as follow:*

$$\frac{\boldsymbol{E_0} \prod_{j \in I} \boldsymbol{e}(\boldsymbol{R_i}, \boldsymbol{F_j}) . \boldsymbol{e}(\boldsymbol{G_i}, \boldsymbol{D_j}) . \boldsymbol{e}(\boldsymbol{G_i}, \boldsymbol{K_j}) \prod_{j \in I} \prod_{i=1}^{m_j} (\boldsymbol{e}(\boldsymbol{A_{j,i}}, \boldsymbol{L_i}) . \boldsymbol{e}(\boldsymbol{B_{j,i}}, (\boldsymbol{\Omega_{\rho j(i)}}))^{\boldsymbol{\theta_{i,j}}}}{\prod_{j \in I} \boldsymbol{e}(\boldsymbol{P_i}, \boldsymbol{F_j})} \rightarrow \boldsymbol{CH}$$

Where $\{\theta_{i,j} \in \mathbb{Z}_p\}_{i=1}^{m_j}$ ***are constants such that*** $\sum_{i=1}^{m_j} \theta_{i,j} \, \xi_{j,i} = s_j$ ***if*** $\{\xi_{j,i}\}_{i=1}^{m_j}$ ***are valid shares of*** $\boldsymbol{s_j}$ ***according to the access structure*** $(\boldsymbol{Mi}, \boldsymbol{\rho i})$

Fig. 17 PPDAC Privacy-preserving distributed access control (part1/2).

of public parameters. The trust affects soundness of the proofs, though anonymity continues to hold even if the setup is corrupted by a malicious party.

Algorithms Constructions: We describe the construction of the PPDAP scheme = (SystemSetup; AuthoritySetup, CreateAddress, GrantAccess, RequestAccess, GetAccess, VerifyTransaction; AllowAccess) in Figs. 17 and 18.

GrantAccess. It takes as inputs a set of access structure (Mi, ρi)i ∈ I and the address of the device to be protected a_{pk}. It outputs a SmartContract SC and a GrantAccess transaction GrATx

For each device the Ro acts locally as follow:

a) randomly Samples a string CH∈ $\mathbb{G}_T$,
b) Genertae $(pk_{sig}, sk_{sig}) = \mathcal{KS}i\mathcal{G}(\boldsymbol{param_{sig}})$
c) runs the encryption algorithm as follow: encrypt (CH, (Mi, ρi)i ∈ I)
d) the encryption algorithm returns $CT= \{E_0, ((F_j, K_j)(A_{j,1}, B_{j,1}, \ldots, A_{j,m_j}, B_{j,m_j}))_{j\in I}\}$ *where*

$$E_0 = CH \prod_{j\in I} e(g,g)^{\alpha_j s_j}$$

Then, The RO creates the Smart Contract called by an execute SmartContract algorithm SC as follow:

ExecuteSmartContract: it takes as input the challenge *CH*, the Ciphertext *CT* and RequestAccess *RqTx* transaction, it outputs a bit b, equals 1 if the Requester fulfills te defined access control policy

1. Parse the RequestAccess transaction *RqTx* as $(IDkey, k, s, idcm)$
2. Parse Idkey as $\{P_i, L_i, R_i, R'_i, G_i, G'_i \ (\Omega_x)_{a_x \in \hat{U} \cap \hat{\lambda}}\}$
3. Parse *CT* as:

$$(C0, (Xj\ , Yj\ , Ej\ , (Cj, 1, Dj, 1), \cdots, (Cj, \ell j\ , Dj, \ell j\ ))j\in I)$$

4. Runs the decryption algorithm as : $Decrypt(PP, IDkey,\ CT) \to M.$

As: $M= \frac{E_0 \prod_{j\in I} e(R_i,F_j)\cdot e(G_i,D_j)\cdot e(G_i,K_j)\prod_{j\in I}\prod_{i=1}^{m_j}(e(A_{j,i},L_i)\cdot e(B_{j,i},\Omega_{\rho(i)}))^{\theta_{i,j}}}{\prod_{j\in I} e(P_i,F_j)}$

5. Run isEqual (CH, M) : if CH== M , b=1, else b=0;
6. If b= 1, add cm to IDCMlist,
7. Return b

e) Set Compute $\sigma = sig_{sk_{sig}}(SC)$
f) Set GrATx := (SmartContract SC, pk_{sig}, σ)
g) Return GrATx

RequestAccess: It takes as input $IDkey$ and a requester addressa_{pk}. It outputs an $IDcoin_{IDkey}$ and a RequestAccess transaction $RqTx$

1 Randomly sample a secret value ρ
2 Randomly sample two COMM trapdoors r; s.
3 Compute $\rho' = PRF_{ask}^{IDkey}(\rho)$
4 Compute $IDcoin_{IDkey}$ serial number $sn := PRF_{ask}^{sn}(\rho')$.
5 Compute $k := COMM_r(a_{pk} \parallel \rho')$
6 Compute $idcm := COMM_s(IDkey \parallel k)$
7 Set $IDcoin_{IDkey}$ idc := (a_{pk}; $IDkey$; ρ; ρ',r; s; cm)
8 Set RequestAccess Transaction *RqTx*:= (*IDkey*; k; s; cm)
9 Return idc and *RqTx*

GetAccess: takes as inputs: PP, IDcoin *idc*, public address a_{pk}, path of cm (idc) in root *rt*. It outputs: a GetAccess Transaction *GtATx*

1 Parse idc := (a_{pk}; $IDkey$; ρ; ρ',r; s; cm)
2 Compute $\rho' = PRF_{ask}^{IDkey}(\rho)$
3 Compute $sn := PRF_{ask}^{sn}(\rho')$
4 Genertae $(pk_{sig}, sk_{sig}) = \mathcal{KS}i\mathcal{G}(\boldsymbol{param_{sig}})$
5 Set $\vec{u}$ =(rt, sn, cm, $IDkey$)
6 Set $\vec{v}$ =(path, idc, a_{pk})
7 Compute $\pi_{grant} = \boldsymbol{Prove}(pk_{grant}, \vec{u}, \vec{v})$
8 Set $m = (\vec{u}, \pi_{grant})$
9 Compute $\sigma = sig_{sk_{sig}}(m)$
10 Set $GtATx := (rt;\ sn;\ cm\ ; IDkey, \pi, pk_{sig}, \sigma)$
11 Return *GtATx*

AllowAccess: takes as input a GetAccess Transaction *GtATx* And SNlist from the current ledger L.
It Outputs a bit b=1 if access is allowed

1 Parse *GtATx*: $(rt;\ sn;\ cm\ ; IDkey, \pi, pk_{sig}, \sigma)$
2 If sn appears in SNlist , b=1 (access is allowed)
3 Else b= 0. (access is denied)

ValidateTransaction: takes as inputs te PP,*GtATx* , *RqTx* and a GrATx transactions, the current blockchain. It outputs a bit b = 1 if the transaction is valid

- Given a *RqTx* transaction
  1. Parse *RqTx as* : *IDkey*; k; s; cm
  2. Set $cm^* := COMM_s(IDkey \parallel k)$
  3. If $cm^* = cm$ b= 1 else b=0
  4. return b
- Given a *GtATx transaction*
  1. Parse *GtATx* : $(rt;\ sn;\ cm\ ; IDkey, \pi_{grant}), pk_{sig}, \sigma)$
  2. If sn appears in the blockchain set b=0;
  3. If *rt does not appear in the ledger set b=0;*
  4. Set $\vec{u}$ =(rt, sn, cm, $IDkey$)
  5. Set $\vec{v}$ =(path, idc, a_{pk})
  6. Set $m = (\vec{u}, \pi_{grant})$
  7. Compute $check_{pk_{sig}}(m, \sigma)$
  8. Compute a= $\boldsymbol{Verify}(pk_{grant}, \vec{u}, \vec{v})$
  9. Return $a \wedge b$
- Given GrATx
  1. Parse GrATx:= (SmartContract SC, pk_{sig}, σ)
  2. Compute b= $check_{pk_{sig}}(m, \sigma)$

Fig. 18 PPDAC Privacy-preserving distributed access control (part2/2).

6. Conclusion

In this chapter we have presented our approach to tackle privacy and security issues in IoT. More specifically, this approach consisted in providing a distributed access control framework that respects the following principles:

edge intelligence, *security through transparency*, *privacy by design* and *user-driven policy*. Then, we have introduced FairAccess as a novel, stronger and transparent access control solution that met the four defined principles. FairAccess leverages the consistency offered by blockchain-based cryptocurrencies to solve all IoT-related access control challenges. While FairAccess leverages the consistency offered by blockchain-based cryptocurrencies to solve the problem of centralized and decentralized access control in IoT and provide a promising user-driven and transparent access control tool, it fails to guarantee a strong anonymity for users due to the public nature of the blockchain and its inherent traceability problem. To overcome this issue, a fully anonymous privacy-preserving Distributed access control scheme (PPDAC) was introduced in this chapter. PPDAC was integrated over FairAccess to maintain Blockchain transparency features while ensuring strong privacy guarantees for users. In this direction, attribute-based access control systems have been used, where access control decisions are based on the attributes (rather than the identity). A policy-hiding access control scheme that protects both sensitive attributes and sensitive policies has been developed using a white box version of distributed multi-authority cipher-text policy attribute based encryption DMCP-ABE. In addition, to face the traceability problem of authorization tokens a sufficient non-interactive zero proof of knowledge (zk-SNARK) protocol has been introduced. Finally, the formal definition and a construction of the proposed scheme were provided.

To conclude we summarize the answers to the research questions formulated in Section 3.

1. *How to solve access control challenges in IoT by leveraging the blockchain technology?*
   The vision of our decentralized access control framework is a system of autonomous organizations hinged around one or many Resource Owners in possession of one or many resources identified with addresses and interacting between each other through transactions (requesting, granting, delegating and revoking access) under the control of their RO. The blockchain is a ledger keeping track and ensures the validity of access transaction among interacting organization. Each manages its own access policy, under only the control of his Resource Owner. Actually, in FairAccess, we use Bitcoin-like addresses to identify all interacting entities and SmartContract (a.k.a. chain code) to express fine-grained and contextual access control policies enveloped inside transactions. We opt for authorization tokens distributed by the blockchain. FairAccess provides several useful mechanisms using the blockchain. In fact, in FairAccess, the blockchain is considered as a

database or a policy retrieval point, where all access control policies are stored in form of transactions; it serves also as logging databases that ensure auditing functions. Furthermore, it prevents forgery of token through transactions integrity checks and detects token reuse through the double spending detection mechanism.

2. *How to provide a lightweight access control framework for constrained IoT devices despite the intensive computation power required by the consensus protocol adopted by the public blockchain?*
We answer to this question by proposing a hierarchical architecture where devices in the IoT support different degrees of FairAccess functionality, depending on their performance and storage capabilities.
3. *How to enable a privacy preserving access control that hides user's access control policy and overcome the problem of traceability and profiling commonly known in the public blockchain?*
We overcome this issue by introducing PPDAC a fully anonymous privacy-preserving Distributed access control scheme to be integrated over FairAccess to get a strong privacy guarantees access control scheme that preserves the Resource owner's access control policies and the requester's sensitive attributes and preserves in a strong way the anonymity of the requester and the Ro. The proposed scheme aims to maintain Blockchain transparency features while ensuring strong privacy guarantees for users. We use attribute-based access control systems, where access control decisions are based on the attributes (rather than the identity) of the requester: Access is granted if Alice's attributes in her certificates satisfy Bob's access policy. We develop a policy-hiding access control scheme that protects both sensitive attributes and sensitive policies. That is, nodes in the public blockchain can decide whether Alice's certified attribute values satisfy Bob's policy, without learning any other information about Alice's attribute values nor Bob's policy. Our construction for PPDAC uses a novel technique that combines CP-ABE, szNARK protocols and SmartContract.

References

[1] A. Benessia, Â.G. Pereira, The dream of the Internet of things, in: Science, Philosophy and Sustainability: The End of the Cartesian Dream 5, vol. 78, 2015. Retrieved from, https://books.google.com/books?hl=fr&lr=&id=_K7ABgAAQBAJ&oi=fnd&pg=PA78&dq=decentralizing++internet+of+things+ibm+watson+&ots=O98rKriJQ1&sig=Kz5bB4EJ80hSktroqnbo_ScvSZM.

[2] J. Gubbi, R. Buyya, S. Marusic, M. Palaniswami, Internet of things (IoT): a vision, architectural elements, and future directions, Future Gener. Comput. Syst. 29 (7) (2013) 1645–1660. https://doi.org/10.1016/j.future.2013.01.010.

[3] D. Miorandi, S. Sicari, F. De Pellegrini, et al., Internet of things: vision, applications and research challenges, Ad Hoc Netw. 10 (7) (2012) 1497–1516. Retrieved from, http://www.sciencedirect.com/science/article/pii/S1570870512000674.

[4] E. Borgia, The internet of things vision: key features, applications and open issues, Comput. Commun. 54 (2014) 1–31. https://doi.org/10.1016/j.comcom.2014.09.008.

[5] F. Mattern, C. Floerkemeier, From the Internet of Computers to the Internet of Things, Springer Berlin Heidelberg, 2010, pp. 242–259. https://doi.org/10.1007/978-3-642-17226-7_15.

[6] D. Guinard, A Web of Things Application Architecture—Integrating the Real-World Into the Web, PhD Thesis, ETH Zurich (19891), 220. https://doi.org/10.3929/ethz-a-006713673, 2011.

[7] F. Bonomi, R. Milito, J. Zhu, S. Addepalli, Fog computing and its role in the internet of things, in: Proceedings of the First Edition of the MCC Workshop on Mobile Cloud Computing—MCC'12, ACM Press, New York, 2012, p. 13. https://doi.org/10.1145/2342509.2342513.

[8] C. Jardak, J.W. Walewski, Enabling Things to Talk, https://doi.org/10.1007/978-3-642-40403-0, 2013.

[9] J. Holler, V. Tsiatsis, C. Mulligan, S. Avesand, From Machine-to-Machine to the Internet of Things: Introduction to a New Age of Intelligence, Academic Press, 2014. Retrieved from, https://books.google.com/books?hl=fr&lr=&id=wtfEAgAAQBAJ&oi=fnd&pg=PP1&dq=From+Machine-To-Machine+to+the+Internet+of+Things:+Introduction+to+a+New+Age+of+Intelligence+&ots=mICCLT9hgD&sig=UNQUKzbr7Co6cDF9jn8P9FL_01Q.

[10] O. Vermesan, P. Friess (Eds.), Internet of Things: Converging Technologies for Smart Environments and Integrated Ecosystems, River Publishers, 2013.

[11] J. Armitage, Paul Virilio: From Modernism to Hypermodernism and Beyond, Retrieved from, https://books.google.com/books?hl=fr&lr=&id=-WL75BiQBRYC&oi=fnd&pg=PP1&dq=PAUL++virilio+Politics+of+the+Very+Worst+&ots=0AmM0dctMV&sig=QpI5pdafuGVLP4hr88N3kSb9i_A, 2000.

[12] J.S. Kumar, D.R. Patel, A survey on Internet of things: Security and privacy issues, Int. J. Comput. Appl. 90 (11) (2014), MLA.

[13] C. Decker, J. Seidel, R. Wattenhofer, Bitcoin meets strong consistency, in: Proceedings of the 17th International Conference on Distributed Computing and Networking, 13 (2016), ACM.

[14] E. Vasilomanolakis, J. Daubert, M. Luthra, V. Gazis, A. Wiesmaier, P. Kikiras, On the security and privacy of Internet of things architectures and systems, in: Secure Internet of Things (SIoT), International Workshop on IEEE.ISO, 690, 2015, pp. 49–57.

[15] J.H. Ziegeldorf, O.G. Morchon, K. Wehrle, Privacy in the internet of things: threats and challenges, Secur. Commun. Netw. 7 (12) (2014) 2728–2742. https://doi.org/10.1002/sec.795.

[16] M. Langheinrich, Privacy by Design—Principles of Privacy-Aware Ubiquitous Systems, Springer Berlin Heidelberg, 2001, pp. 273–291. https://doi.org/10.1007/3-540-45427-6_23.

[17] J.H. Ziegeldorf, O.G. Morchon, K. Wehrle, Privacy in the internet of things: threats and challenges, Secur. Commun. Netw. 7 (12) (2014) 2728–2742. https://doi.org/10.1002/sec.795.

[18] F. Berman, V.G. Cerf, Social and ethical behavior in the internet of things, Commun. ACM 60 (2) (2017) 6–7. https://doi.org/10.1145/3036698.

[19] E.P. Goodman, The atomic age of data: Policies for the Internet of things, The Report Explores Some of These Policy Questions in the Context of the "Smart City" Use Case, ISO 690, 2015.

[20] C. Diaz, S. Gürses, C. Troncoso, Engineering privacy by design, Comput. Priv. Data Prot. (2011). Retrieved from, https://software.imdea.org/~carmela.troncoso/papers/Gurses-CPDP11.pdf.

[21] S. Gurses, J.M. del Alamo, Privacy engineering: shaping an emerging field of research and practice, IEEE Secur. Priv. 14 (2) (2016) 40–46. https://doi.org/10.1109/MSP.2016.37.
[22] S. Spiekermann, L.F. Cranor, Engineering privacy, IEEE Trans. Softw. Eng. 35 (1) (2009) 67–82. Retrieved from, http://ieeexplore.ieee.org/abstract/document/4657365/.
[23] J.A. Sánchez Alcón, L. López, J.F. Martínez, et al., Trust and privacy solutions based on holistic service requirements, Sensors (Basel) 16 (1) (2015) 16. Retrieved from, http://www.mdpi.com/1424-8220/16/1/16/htm.
[24] A. Cavoukian, Privacy by design in law, policy and practice, in: A White Paper for Regulators, Decision-Makers and Policy-Makers, 2011.
[25] A. Ouaddah, A. Abou Elkalam, A. Ait Ouahman, FairAccess: a new blockchain-based access control framework for the internet of things, Secur. Commun. Netw. 9 (18) (2016) 5943–5964. https://doi.org/10.1002/sec.1748.
[26] A. Ouaddah, A.A. Elkalam, A.A. Ouahman, Towards a Novel Privacy-Preserving Access Control Model Based on Blockchain Technology in IoT, Springer, Cham, 2017, pp. 523–533. https://doi.org/10.1007/978-3-319-46568-5_53.
[27] A. Ouaddah, H. Mousannif, A. Abou Elkalam, A. Ait Ouahman, Access control in the internet of things: big challenges and new opportunities, Comput. Netw. 112 (2017) 237–262. https://doi.org/10.1016/j.comnet.2016.11.007.
[28] R.S. Sandhu, Role-based access control, Adv. Comput. 46 (1998) 237–286. https://doi.org/10.1016/S0065-2458(08)60206-5.
[29] E. Yuan, J. Tong, Attributed based access control (ABAC) for Web services, in: IEEE International Conference on Web Services (ICWS'05), IEEE, 2005. https://doi.org/10.1109/ICWS.2005.25.
[30] L. Seitz, G. Selander, E. Wahlstroem, S. Erdtman, H. Tschofenig, Authorization for the Internet of Things Using OAuth 2.0, Retrieved from, https://tools.ietf.org/html/draft-ietf-ace-oauth-authz-01, 2016.
[31] S. Cirani, M. Picone, P. Gonizzi, L. Veltri, G. Ferrari, IoT-OAS: an oauth-based authorization service architecture for secure services in IoT scenarios, IEEE Sensors J. 15 (2) (2015) 1224–1234. https://doi.org/10.1109/JSEN.2014.2361406.
[32] R. Roman, J. Zhou, J. Lopez, On the features and challenges of security and privacy in distributed internet of things, Comput. Netw. 57 (10) (2013) 2266–2279. https://doi.org/10.1016/j.comnet.2012.12.018.
[33] S. Cirani, L. Davoli, G. Ferrari, R. Léone, A scalable and self-configuring architecture for service discovery in the internet of things, IEEE Internet Things J. 1 (2014) 508–521. Retrieved from, http://ieeexplore.ieee.org/abstract/document/6899579/.
[34] O. Said, M. Masud, Towards internet of things: survey and future vision, Int. J. Comput. Netw. 5 (1) (2013) 1–17. Retrieved from, http://citeseerx.ist.psu.edu/viewdoc/download?doi=10.1.1.741.3655&rep=rep1&type=pdf.
[35] C.-W. Tsai, C.-F. Lai, A.V. Vasilakos, Future internet of things: open issues and challenges, Wirel. Netw. 20 (8) (2014) 2201–2217. https://doi.org/10.1007/s11276-014-0731-0.
[36] S. Panikkar, S. Nair, P. Brody, V. Pureswaran, ADEPT: An IoT Practitioner Perspective, Retrieved from, http://ibm.biz/devicedemocracy, 2015.
[37] M. Antonakakis, T. April, M. Bailey, et al., Understanding the mirai botnet, in: USENIX Security Symposium, 2017, 1092–1110.
[38] S. Nakamoto, Bitcoin: A Peer-to-Peer Electronic Cash System, 2008, pp. 1–9.
[39] S. Underwood, Blockchain beyond bitcoin, Commun. ACM 59 (11) (2016) 15–17.
[40] M. Swan, Blockchain: Blueprint for a New Economy, Retrieved from, https://books.google.com/books?hl=fr&lr=&id=RHJmBgAAQBAJ&oi=fnd&pg=PR3&dq=BLOCKCHAIN+APPLICATION+BEYOND+FINANCE&ots=XPyHHY-Rg4&sig=xZ1yJTxrIgXGqd22L30op9po8-s, 2015.

[41] M. Ulieru, Blockchain 2.0 and beyond: adhocracies, in: Banking Beyond Banks and Money, 2016 Retrieved from, http://link.springer.com/chapter/10.1007/978-3-319-42448-4_15.
[42] J.L. Carter, M.N. Wegman, Universal classes of hash functions, J. Comput. Syst. Sci. 18 (2) (1979) 143–154. Retrieved from, http://www.sciencedirect.com/science/article/pii/0022000079900448.
[43] R.C. Merkle, A digital signature based on a conventional encryption function, in: Conference on the Theory and Application of Cryptographic Techniques, Springer, Berlin, Heidelberg, 1987 369–378, Retrieved from, http://link.springer.com/chapter/10.1007/3-540-48184-2_32.
[44] L. Lamport, R. Shostak, M. Pease, The byzantine generals problem, ACM Trans. Program. Lang. Syst. 4 (3) (1982) 382–401. Retrieved from, http://dl.acm.org/citation.cfm?id=357176.
[45] F. Tschorsch, B. Scheuermann, Bitcoin and beyond: a technical survey on decentralized digital currencies, IEEE Commun. Surv. Tutor. 18 (3) (2016) 2084–2123. Retrieved from, http://ieeexplore.ieee.org/abstract/document/7423672/.
[46] W. Wang, D.T. Hoang, Z. Xiong, et al., A Survey on consensus mechanisms and mining management in blockchain networks, 2018, arXiv preprint arXiv:1805.02707.
[47] V. Buterin, A next-generation smart contract and decentralized application platform, in: Ethereum, 2014, pp. 1–36. Retrieved from, http://buyxpr.com/build/pdfs/EthereumWhitePaper.pdf.
[48] K. Christidis, M. Devetsikiotis, Blockchains and smart contracts for the internet of things, IEEE Access 4 (2016) 2292–2303. Retrieved from, http://ieeexplore.ieee.org/abstract/document/7467408/.
[49] J. Caldwell, IBM Internet of Things Point of View and Strategy, 2015.
[50] M. Conoscenti, A. Vetrò, J.D. Martin, Blockchain for the Internet of Things: A Systematic Literature Review, Retrieved from, http://porto.polito.it/id/eprint/2650266, 2016.
[51] A. Ouaddah, B. Bellaj, FairAccess 2.0: a smart contract-based authorization framework for enabling granular access control in IoT, Int. J. Inf. Comput. Secur. (IJICS) (2018).
[52] F. Reid, M. Harrigan, An analysis of anonymity in the bitcoin system, in: Security and Privacy in Social Networks, Springer New York, New York, 2013, pp. 197–223. https://doi.org/10.1007/978-1-4614-4139-7_10.
[53] D. Ron, A. Shamir, Quantitative Analysis of the Full Bitcoin Transaction Graph, Springer, Berlin Heidelberg, 2013, pp. 6–24. https://doi.org/10.1007/978-3-642-39884-1_2.
[54] J. Bethencourt, A. Sahai, B. Waters, Ciphertext-policy attribute-based encryption, in: 2007 IEEE Symposium on Security and Privacy (SP'07), 2007, IEEE, pp. 321–334. https://doi.org/10.1109/SP.2007.11.
[55] L. Cheung, C. Newport, Provably secure ciphertext policy ABE, in: Proceedings of the 14th ACM Conference on Computer and Communications Security, ACM, 2007, pp. 456–465. Retrieved from, http://dl.acm.org/citation.cfm?id=1315302.
[56] N. Bitansky, A. Chiesa, Y. Ishai, O. Paneth, R. Ostrovsky, Succinct Non-interactive Arguments via Linear Interactive Proofs, Springer, Berlin, Heidelberg, 2013, pp. 315–333. https://doi.org/10.1007/978-3-642-36594-2_18.
[57] A. Beimel, עמוס בימל, Secure schemes for secret sharing and key distribution, Technion-Israel Institute of Technology, Faculty of Computer Science, 1996.
[58] J. Han, W. Susilo, Y. Mu, J. Zhou, M. Ho Allen Au, Improving privacy and security in decentralized ciphertext-policy attribute-based encryption, IEEE Trans. Inf. Forensics Secur. 10 (3) (2015) 665–678.

[59] M. Chase, Multi-authority attribute based encryption, in: Theory of Cryptography, Springer Berlin Heidelberg, Berlin, Heidelberg, 2007, pp. 515–534.
[60] T. Sander, A. Ta-Shma, Auditable, anonymous electronic cash, in: Advances in Cryptology—CRYPTO' 99, Springer, Berlin, Heidelberg, 1999, pp. 555–572. http://link.springer.com/10.1007/3-540-48405-1_35. (cit. on p. 122).

About the author

Dr Aafaf Ouaddah holds a PhD in Computer Science on her work on security and privacy in IoT through the blockchain technology from the Cadi Ayyad University, Marrakesh (Morocco) in 2017. She received the Engineer degree in Networking and Information Technology in 2013, from the National Institute of Posts and Telecommunication (INPT) graduate school engineering. Since 2017, she is a Chief Scientist at Mchain enterprise. Her research interests include the blockchain and new distributed ledgers as well as security and privacy in distributed systems, IoT, and Fog computing. She has published more than 10 research papers in various conferences, workshops and International Journals of repute including IEEE, Springer and Elsevier.

CHAPTER NINE

Blockchain with IOT: Applications and use cases for a new paradigm of supply chain driving efficiency and cost

Arnab Banerjee
Infosys Ltd., Bengaluru, India

Contents

Abstract

This chapter discusses the use cases of Blockchain with IOT along with other systems/practices to achieve a new paradigm of supply chain. The chapter identifies immediate challenges in various realms of supply chain ranging from product or order tracking, traceability and recall, counterfeiting, agri supply chain, automotive supply chain, digital homes, offices and warehouses to manufacturing and distribution supply chain. Elaborating the key features and capabilities of IOT and Blockchain it explains with example how the combination of these two powerful technologies along with peripherals

Advances in Computers, Volume 115
ISSN 0065-2458
https://doi.org/10.1016/bs.adcom.2019.07.007

systems and backbone ERPs can solve the listed business problems. It explores the process flows, technical architecture and business impacts with the IOT and Blockchain combination. The chapter ends with listing of the economic value and cost savings, the IOT Blockchain combination brings to supply chain and the opportunities lying in future.

1. Introduction

Supply chain in today's globalized economy is complex. A product sold in retail or online is designed by teams in more than one location, manufactured at some place, stored somewhere and delivered everywhere. Products pass through various countries, warehouses, weather conditions, handling methods, and storage situations before being delivered to end consumers. The physical realms of product flow have a twin with the digital aspect of the information flow. As more technological advancements are made this digital information is becoming more and more important. The timely availability of these information in a secure, transparent and structured way is key to faster and more accurate decision making and along with it the ability to save cost and judicious use of resources.

Clive Humby while giving a talk at ANA Senior marketer's summit in Kellogg Business School in 2006 compared data to crude oil. He indicated that like crude oil data can be broken down and analyzed to get great value. Companies have to learn to deal with this massive amounts of data to get insights and make actionable decisions [1]. More recently in 2017 Ajay Banga, CEO of Mastercards quoted that he believes that "Data is the new Oil" [1a]. A product from its concept or raw material to it featuring on the shelf or web portal generates enormous amount of data. Channelizing these data to analyze for future improvement, decision making and cost reduction is the direction and intent of every supply chain leader. Numerous components of a finished products, plethora of sourcing options, varied geographical and climatic conditions, storage situations, real time tracking need and physical handling makes it essential to track and monitor the physical, climatic and warehousing conditions. All these are only possible when the information is secure, structured, real time and easily available.

Today every organization is connected with systems and is intertwined. As we move ahead in the digital journey there are systems within an organization like ERP, MRP, WMS, CRM which improves the organizations processes driving efficiency reducing cost and helping it function better.

But it is the ability to interact and communicate across organizations seamlessly that can really drive benefits for decision making and other aspects. Focus is slowly moving from inter-organization data interactions to data interactions with physical machines and other realms. Today most of the inter-organization communications are driven by EDI, B2B messaging and manual reports and there is no mechanism of communicating with machines and other physical realms like pressure, temperature, etc. The inter-organization communications in these technologies are neither reliable or real time nor secure. These are mostly batch processed, needs manual interventions and comes with its own plethora of limitations. This chapter looks at the challenges faced by different type of supply chain with these existing technologies and how the modern and mature technology of blockchain with the internet of things (IOT) capability will solve the problems, drive efficiency and reduce cost.

2. Supply chain challenges

As per the Mckinsey & Company report in most of the industries today, supply chain is an independent organization with a Chief Supply Chain Officer (CSO) leading them [2]. The reason for this paradigm shift is the realization of the impact of supply chain on the overall bottom lines of balance sheet. As per O'Byrne [3], supply chain today has the maximum potential to save cost and drive efficiency in any kind of industry be it manufacturing, healthcare or distribution. Also the various asks of customers is like ability to track products real time, ability to trace back products or components to its origin (provenance), seamless product recalls, improve manufacturing process to reduce cost, drive product features based on customer feedback, more environment friendly products, sourcing for the best quality, sourcing at the best cost, provide a better customer service, etc., all falls within the spheres of supply chain.

Let us look at some of the pertinent supply chain challenges from various realms of supply chain and industries. These challenges are faced today with the current technology landscape where ERP, WMS, CRM are mostly implemented.

2.1 Product/order tracking

There are multiple stakeholders involved to make products available to customer as per demand. For example, a company manufacturing routers receives a demand from its customers. The demand is transferred to a contract manufacturer to be shipped to a warehouse. The product is shipped by contract manufacturer through a logistics service provider. The product reaches the warehouse of the company which is then shipped to customer. Due to multiple stakeholders in this demand/supply transaction the real time information sharing is a challenge. What these companies do is to use reports and EDI/B2B messaging which provides information of product reaching or leaving a certain location. What happens in between is an information void. Similar information disconnects is there for pharmaceutical and food industry also. Most of the current/existing solutions for tracing uses a warehouse management solution along with RFI tags [4]. This is information vid and supply chain challenge still needs a solution.

2.2 Product traceability/recall/anti-counterfeiting

To address the problem of counterfeiting, contamination of products and medicines, degradation of products and medicines, the US Food and Drug administrator (FDA) has issued Drug Supply chain and Security Act (DSCSA). The act mandates that all manufacturer, distributor and third party logistics provider develop an electronic and interoperable system to track medicines. This calls for an almost real time and highly secure data interchange which can be analyzed any time. Similarly, US FDA mandates the manufacturer to recall products ASAP from customers if US FDA has already asked for it (before it damages customer). Manufactures today have very limited access or information to end customer. Counterfeiting of products is another supply chain challenge as its extremely difficult to figure out where the counterfeit products are injected into the supply chain. All these three scenarios ask for a secure real time capability to trace back products reliably and quickly. These challenges are constantly evolving with ecommerce and omni channel supply chains and there are very few credible solutions available in market to solve such problems.

2.3 Agri supply chain

The Agri supply chain is complex and sensitive. It starts with the farmers and end with the retailer or consumption of the products at the hands of consumer. It is particularly complex for four reasons, first at the very beginning there is limited use of technology or sophistication as it lies with farmers, second most of the times the products undergo multiple transformations unless it is directly consumed, third it has multiple stake holders or value chain members with varied levels of sophistication, and finally it is particularly long and spans across time and geography. For a given end product uniquely identifying land of harvest, knowing the farmer who grew it and the environmental impacts of crop grown, farming practices for the crop, crop yield, fertilizers used, type of soil preparation, are of particular interest to many governmental agencies, retailers and even consumers [4]. With the lengthy supply chain and final consumer being oblivious to the upstream complexity it is not only intricate but also multifaceted.

2.4 Digital automotive

There are two key trends of digital transformation in automotive industry. First is autonomous cars and second is connected cars. Of the two it is the connected cars that is fast evolving, greatly transforming and can potentially disrupt the automotive supply chain. The concept of connected cars, i.e., cars connected to internet and sending data to cloud transmitting car function and other allied information can open up various options for car owners and other members of the automotive value chain. These include but not limited to predictive maintenance, car infotainment, safety, security, recall, drive information, advanced navigation, customer insights and insurance information. The new cars sold that will be connected to the Internet are predicted to rise to 98% by 2020 [5]. Today the connected cars are restricted to the OEMs partnering with technology companies. Example Renault-Nissan-Mitsubishi are slated to roll out Microsoft and Android connected vehicles by 2021 [6]. But the actual supply chain benefits lie when the automotive partners like dealers, service companies, component manufacturers and insurance companies are connected. They have access to the same data source and then can take decisions or reach out to owners helping them save money, time and giving them better experience. These are sensitive information and are complex to capture, connect and share with limited to no solution existing today.

2.5 Digital homes and offices

The concept of smart homes and office are evolving and so are their usage. Most of the usage of connected homes and offices are around a smart device which can or cannot be controlled from a mobile app and which controls various equipment like thermostats, televisions, washing machine, refrigerator or lights. But with advent of more applications and services, it is becoming more clear that the benefit of technology can only be achieved when these devices and their applications are connected to services.

Smart homes today help efficient usage of energy, improves quality of life and maximizes security. As the market is currently immature and evolving, the real benefit of smart homes or offices will come when the hardware manufacturer, consumer electronics, retail outlets, software and ecosystem developers, utility service providers and home security solutions are all connected. The connects has to move from devices to devices and service to service. It is a case where individually the products are being developed and the need is to have a holistic look and bring players/partners together. Lack of a mature technology is resisting the development in this direction.

2.6 Transparency in distribution industry

One of the most challenging aspects in distribution industry is that of trust. Trust between distributor and supplier for availability of products, for price of a product in a specific country/geography or for a segment of a customer or for the shipment of the goods should be irrefutable. Apart from these another area of trust unique to distribution industry is ship and debit claims. Today the major challenge is the ship and debit claim is the manual reconciliation done at both the supplier and distributors end. Volume and complexity in claims leads to millions of dollars stuck in claims for distributors. The claim always suffers from trust deficit for products shipped and received, onhand, price discrepancy, agreements value, etc. Point of Sale reporting is another piece of the zigsaw of distribution industry which drives quite a few monetary decisions like commissions, volume rebates, etc., and suffers because of trust and transparency deficit. Most of these problems are solved today through manual interventions, reports, discussion and emails. This is primarily because there is limited to no inter-company technology platform. EDI, B2B messaging doesn't helps solve these problems. So every distributor needs a solution for this problem and in need of a mature technology to bridge this gap.

2.7 Manufacturing systems

Technology has always transformed core manufacturing processes. This is evident from the way over the years manufacturing has evolved with global supply chains and localization of products. It is the real time integration among partners which has forced companies and products managers to look for ways to share data for analysis. This has given rise to consideration of data security, data transparency, data access and real time decision making based on accurate and untampered data. This has given rise to the needs of digital twins which helps simulate real world impacts/actions to improve the physical world with a simulation in cyber world [7]. Real time decision making in multi-partner eco system and digital twins for manufacturing are only possible when there is a safe, transparent and secure data sharing and availability mechanism. Today such real time data sharing and digital twins are not so prevalent due to lack of mature technology.

3. Current generation solutions

Based on the diverse type of industry, process flows, products, types of supply chain strategy it is very evident from the discussion above that to serve customer and to have an effective decision making there is a need for smooth inter-organization data sharing. Today these are managed in some way or the other and the way those are managed will pave way to understand how it can be solved with better use of technology. Order tracking and product traceability are currently managed through EDI/messaging system and also through reports and gated scans at various check points. This indicates that the need of the future system is to provide the data like these scans real time from anywhere. Similarly, EDI/messaging system hints that the information has to reach the correct and exact destination so that informed decisions can be taken. Agri supply chain struggles with lack of adoption of technology at grass root level. This may be due to ease of use, lack of technology provisioning and cost. Due to this the intended transparency of crops, land use, environmental impacts are missing from most of the products. The capture of grass root data and sharing of these data transparently and securely across multiple agencies are a challenge today which restricts the agri supply chain and its information being available at the hands of end consumer. Same is the challenge for digital automotive, digital homes, distribution and manufacturing that are challenged by real time capturing and sharing of data securely across multiple agencies/stakeholder and analyzing these in the digital twins. All these indicate that the need of the hour from technology is to provide a means and mechanism of capturing data safely and transmitting it to multiple stakeholders securely and maintaining the transparency. This chapter further explores the technology of Internet of Things and its combination with Blockchain as a credible and mature solution mechanism to the discussed problems.

4. IOT with blockchain: A powerful combination

Based on the discussion above it is clear there are three aspects of the requirement. Characteristically these can be separated as, first is the ability to capture data real time from the source, uninterrupted so that the data is transformed from physical to cyber world seamlessly. It has to be cost effective and it has to be completely human, system, operations and process agnostic. Second, the data captured has to be secured to the best extent possible. The data should be impeccable, trustworthy and untampered. Third, there has to be a seamless mechanism to share this data almost real time across enterprises/partners without any process dependency and minimal additional cost. The need for all the three requirements is forcing enterprises to look beyond the existing mechanism of ERP systems, point to point interfaces, EDI, messaging systems and reporting. This is where the newer technology and its mechanism, architecture and maturity comes to fore. For the discussed challenges in supply chain and the way things are handled today it is the use of Distributed Ledger Technology (DLT) based Blockchain system along with the Internet of Things (IOT) that can solve most of the problems. The technical capability and architecture of data capture, storage, security and distribution greatly supports this combination. This combination can also provide many unforeseen benefits like seamless synchronization, real time analytics and mobility aspects to data.

4.1 Blockchain

Blockchain system is a network of members connected together through a distributed ledger technology. Blockchain network of systems records transactions as blocks and are Immutable and Secure (to a great extent) and chronologically maintains the data blocks across transactions. These blocks of data are shared among a number of computers/nodes rather than being stored on a central server. The blockchain can also be described as an incorruptible digital ledger of transactions that are programmed to record. Blockchain can tokenize a physical product into a virtual/digital token. One of the powerful features of Blockchain is on the controlled data access to certain members only based on smart contracts. In blockchain the transaction data can be verified by rules (smart contracts) unlike a single centralized authority. The biggest benefit of Blockchain is the ability to seamlessly and securely share data with partner organizations.

Blockchain as being used today offers three types of architectures to help companies integrate among themselves. These are:

4.1.1 Open blockchain or permission-less blockchain architecture

In this architecture there is no one partner or node owning or reading/writing/auditing the blockchain. Here, all the partners together owns it and everyone is equal in the network. Bitcoin, Litecoin network are example of open blockchain architecture. The open blockchain network generally do not work if the network is for a specific purpose or by a specific owning organization. This blockchain architecture works on the consensus mechanism. The consensus mechanisms are drawn through the Proof of Work. The proof of work drives the need for authenticators or miners. These play a crucial role as they are the gatekeepers who onboard members, verify transactions and calculate credits. This architecture requires significant computational power to maintain a large-scale distributed ledger and standardize calculations among numerous participants and authenticators. Due to this challenge, open blockchain architecture is not so readily acceptable. Though Bitcoin network is one of the best and earliest known blockchain system there are not many like it.

4.1.2 Permissioned or private blockchain network

This architecture is more suitable if it's for a specific purpose. In this architecture there is a particular organization or a single defined owner of the network. The owner of the network defines the business specific rules and/or smart contracts for other members. Smart contracts are rules defined to engagement among members. These standardized and well-defined criteria are the cornerstone of the permissioned blockchain. In the permissioned network participant entry is decided by existing members, a regulatory authority or a consortium. Some examples of private blockchain networks are Oracle Blockchain Cloud Services developed on Hyperledger Fabric.

4.1.3 Consortium blockchain network

Consensus network is like a semi private blockchain architecture which is owned and controlled by a preselected group of members. These group of members decide on future memberships, rules, etc. The right to read the blockchain may be public, or restricted to the participants. For example, 1 might imagine a consortium of 15 financial institutions, each of which operates a node and of which 10 must sign every block in order for the block to be valid. A consortium blockchain provides many of the same benefits affiliated with private blockchain like efficiency and transaction privacy. R3 is an example of consortium blockchain.

The blockchain network can be connected to multiple applications. The capability of connecting blockchain network to Enterprise Resource Planning (ERP) system has been the most credible of them all [8,9]. This has opened up the opportunity to connect and transmit the transacting system data securely to partners for seamless data sharing, decision making and transparency. As rightly stated by Litan et al. [10] Blockchain networks have emerged as a promising innovation to affirm the integrity of data shared among constituents in multiparty process collaboration. This has opened up chances of real time collaboration and analytics which was not possible few years back. Most of the organizations were trying to work it out through the EDI or reports mechanism which was slow and human intensive.

For the supply chain challenges discussed in previous section the Blockchain system of networks is definitely a credible option to solve such problems. But only Blockchain system as a standalone solution may not be enough. The basic limitation is the capture of data and availability of the same to different systems. Blockchain system will provide a secure and seamless measure to share data. But the capture of data itself poses a challenge specially when we consider the source if data is not always a transacting system like ERP. It could be physical information and data coming from any source and anywhere. So offline capability, system and device independence becomes a key shortcoming. This brings in the aspect of Internet of Things to seamlessly capture data and have an architecture to share it as needed.

4.2 IOT: Internet of things

Internet of Things (IOT) is a concept to connect or capture any kind of information from device or machine to transmit from its source to any destination independent of platform using internet. The IOT can also be described as a giant network of connected devices communicating and sharing information among themselves. IOT particularly is of interest in the supply chain domain as it provides the ability to capture and share data from remote location or from a process automatically which is otherwise not possible. It opens up the possibility of virtually connecting every stake holder of supply chain either directly or indirectly. This can help mitigate problems which were considered unsolvable because of non-availability of information within supply chain framework.

The data collection and sharing by IOT devices are based on technologies and standards like radio frequency identification (RFID) and wireless communication standards. RFID technology provides an easy and cost effective mechanism of physical data capture for movement of goods in factories, warehouses or retail stores. Conversion of these movements of goods information into a technology platform enables visibility of data and its availability for various analysis. RFID tags and tag readers are required to capture this information. Tags are required on the source from which the data needs to be captured, while the tag readers read the tags and captures the information. These capture any kind of movement real time. For Wireless communication standards IOT systems generally use Wireless Sensor Networks (WSNs) which forms an important part of the overall architecture that collects data from the physical world or machines and convey it to central controllers which will then point and transmit it to a database to store the information and use it for further analysis [7]. There are various other technologies available today for IOT devices like ZigBee, Z-Wave, WiMax, LPWAN, etc. Internet of things connects the information between three types of things. Although there is no clear definition of things but these can be broadly classified as:

1. Connected People—This involves connecting the people socially, with society and community for a better living.
2. Connecting Products to people—This involves connecting devices individually or as a group to people. This can be like connecting home devices to people for a better living or domestic use products with people ranging from washing machine to refrigerators.
3. Connecting Assets with machines—These are kind of industrial applications where boilers, furnaces, turbines and machines are connected using sensors to other machines or computers for monitoring and decision making. These also enable digital twins' technology. These can be extended to trucks, ships, places and warehouses for connections and monitoring.

An IOT system needs capability to connect to any type of heterogeneous systems. It should be able to install or connect to any operating system with high reliability and in real time. It should have the ability to start and stop the flow of data and needs to be a configurable system. It should be dynamic in nature in the sense it should be able to sleep or wake up and get connected or disconnected as per design. An IOT system needs to have the ability to detect physical world phenomenon and its changes and convert to a digital signal. The last and most important aspect is the scale of data handling. The devices should have the ability to handle the enormity of data [11,12].

This certainly opens up lot of opportunities to explore and solve in supply chain but also brings with it the possible and credible issue of data security. Another issue is the storage of these IOT data securely. IOT is all about capture and transmitting of data from origin to its recipients but IOT doesn't solves or provides any mechanism to store data of its own.

IOT as a technology serves two purpose first it provides the capability to capture data and an architecture to transmit it to its destination. But two key aspects of IOT which cannot be ignored in today's digital world is its inability to store data and the security of the captured data. As per research and advisory firm Gartner estimates the number of installed IOT devices will reach 20.4 billion by 2020 [13]. Business Insider predicts by 2020, 24 billion IOT devices will be installed with majority of them being used by business or government [14]. IBM quotes IDC and predicts there will be 30 billion IOT devices connected to the internet by 2020 [15]. This sheer volume indicates there has to be a credible, scalable and secure mechanism to capture, store and transmit data across networks/partner organizations. IoT systems are dependent on a centralized architecture. Information is sent from the IOT system to internet where the data is stored and processed. As the number of IOT devices are going to be in billions a centralized system will limit scalability as well as expose numerous weakness compromising security. For supply chain applications it is of utmost importance to have a scalable and secure mechanism as sensitive data are going to cross organization boundaries for decision making.

4.3 Blockchain and IOT combinations: Solution to supply chain challenges

Looking at the needs of the supply chain and business needs/challenges like security of data for capture and access, seamless and device independent data capture, independence from physical location of supply chain/trading entity, real time and reliable availability of data, and cost effectivity indicates the kind of technology needed. Apart from it the data capture and analysis will need to be scalable and reliable for actual benefit. The supply chain problems discussed so far can be solved or minimized with the adoption of Blockchain and IOT solutions together. Blockchain and IOT together seamlessly brings the physical world data into computing environment and stores it in a distributed ledger in a multi-partner scenario thereby bridging the trust gap. Refer to Fig. 1 for a high level architecture. The main reason why the combination of Blockchain and IOT will solve the supply chain challenges are:

4.3.1 Scalability

Multiple IOT devices can feed data to a gateway leading the data to a blockchain network which works on distributed ledger technology. There will be multiple gateways feeding data into the blockchain network.

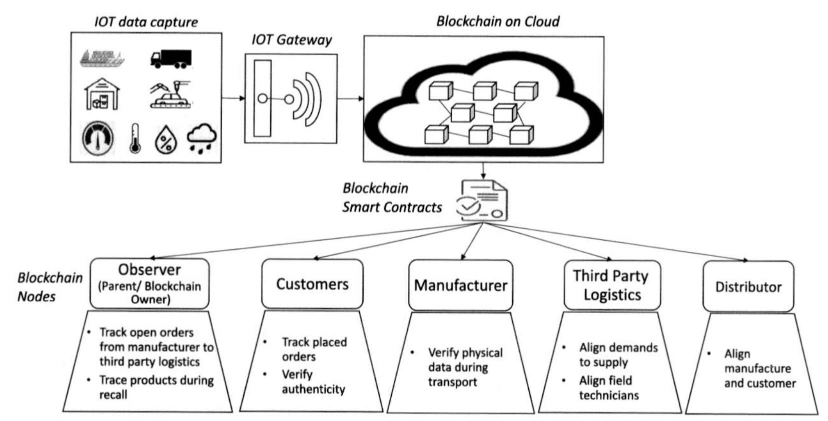

Fig. 1 High level IOT Blockchain architecture.

The IOT gateway is being designed for such high volume. Blockchain network today are enterprise grade system handling large volume of data. Thus it's a scalable approach for a large scale data driven supply chain organizations to handle volume of data.

4.3.2 Reliability

Gartner expects 200 billion internet-connected things by 2020 [16]. Many of these internet-connected things will leverage, exploit, and use IOT devices. IOT devices are being designed and tuned for high velocity, high volume, low power, low latency data transfers, short and long range data transfers. Researches are being carried out to reduce data loss, data lock in, data corruption and connectivity issues. More and more advanced chip sets and architectures are being designed to improve the standards for the same. Technologies like ZigBee, WiFi, WiMax, Z-Wave, LPWAN are being developed for greater reliability.

4.3.3 Security

The data security threat in this IOT-Blockchain system lies with the IOT device and its network. Once the data is in Blockchain network the data is tamper proof and immutable and thus secure. Various types of devices and its integration with gateway using WiFi and other wireless communications do pose threat to the IOT data. Data security breaches are more likely to happen in IOT network and the security and safety of it is same as network security. IOT security is being researched and worked upon to make this technology combination successful [17].

4.3.4 Cost effectivity/cost of ownership

Many of the Blockchain system is available as a service, costed at per unit of transaction data. For example, Oracle Blockchain Platform Cloud Service is available on per transaction basis [18]. Blockchain cost ownership is thus data driven. With Blockchain available on cloud platform and no upfront cost of infrastructure the solution becomes affordable and many companies are trying it out on trial basis. Also companies are offering free trial versions to ease out the adoption. IOT does needs investments in sensors and companies such as Oracle are providing connectors with such sensors out of the box. Overall the cost of ownership is there and is mainly driven by number of sensors and the volume of data. But with the benefits it can bring to the business with no upfront cost and diminishing cost of ownership, the investments are very affordable.

4.3.5 Transparency and accessibility

With the data stored in blockchain network all blockchain services (including enterprise grade solutions like Hyperledger based Oracle Blockchain services) provides mobile application based access to the data with all security and accessibility. With Blockchain system as the cardinal data storage system being fed from IOT system the smart contracts in blockchain makes data accessible to partners seamlessly and transparently.

Fig.1 depicts a high level Blockchain IOT conceptual architecture for supply chain solutions. It can be seen that it brings in all the supply chain partners together into the blockchain system (common platform) with data being captured from various systems through the IOT mechanism. The data captured can be anything from physical or environmental data to product specific data. This figure depicts a cloud based blockchain architecture. The genesis of the architecture lies in capture of data through any kind of mechanism like sensors or RFIDs and transmitting it to the gateway. The gateway will submit the data to the blockchain. In this architecture the data captured by IOT is directly fed into blockchain with no intermediate data storage. This architecture has no IOT platform as such to store data and then transmitting it to blockchain system. Thus this architecture eliminates the need for infrastructure for captured data in IOT platform. On the contrary because of a lack of IOT platform, data cannot be encrypted before transferring over the network with only possibility of basic authentication. Having an IOT platform can help with advanced authentication and encryption before transfer over the network. The architecture as depicted in Fig. 1 will have the least cost of ownership and should be the first step by any company in this direction.

This architecture of the combination of Blockchain and IOT can help solve various supply chain problems. Let us go through the listed supply chain problems and its solutions thereof to be offered by this technology and architecture.

Product/Order Tracking and Product Traceability/Recall/Anti-Counterfeiting—These are two of the supply chain challenges identified. The solution to both these problems is very similar. The main challenge in product tracking and traceability in supply chain are multiple stakeholders, real time tracking of physical goods movements information across stakeholders and ability to trace back ownership at any given point in time in the life cycle of the product. This information is not only needed and worthy for stakeholders within the supply chain but also for end customers who want real time information of its demand. The information voids in these areas are transparency and losing customer satisfaction. The combination of Blockchain and IOT can help track the information of physical goods movement in the supply network and make it available to any stakeholder. This can provide the information reliably, real time and in a controlled way which are otherwise not possible. The core of the solution lies in having sensors at various key supply chain cardinal points which can capture the deemed information and transmit it to a cloud based gateway system which will ultimately transmit it to a blockchain network. The data once available in blockchain network can then be securely shared across partners using smart contracts thus bringing in control and authenticity. Smart contracts can decide the level of access and information to partners.

Fig. 2 shows a supply chain partner coordination mechanism with the IOT Blockchain combination. The mechanism will not only help establish a track and trace of the orders but also it will help establish product traceability which can be used for authentication and product recall. The mechanism will also provide real time monitoring of the orders.

Benefits and usage of this mechanism to various partners are listed below:

Parent Company—This is the company which receives customer orders to be fulfilled. It can have mechanism to share order details in the blockchain network which can in turn help contract manufacturer (and suppliers) with transparency in demand (components). Based on IOT data available in blockchain from other partners it can derive following:

- Real time order tracking based on blockchain data (Component tracking from supplier, services tracing, logistics tracking, manufacturing tracking).
- Traceability of product from manufacture to customer.

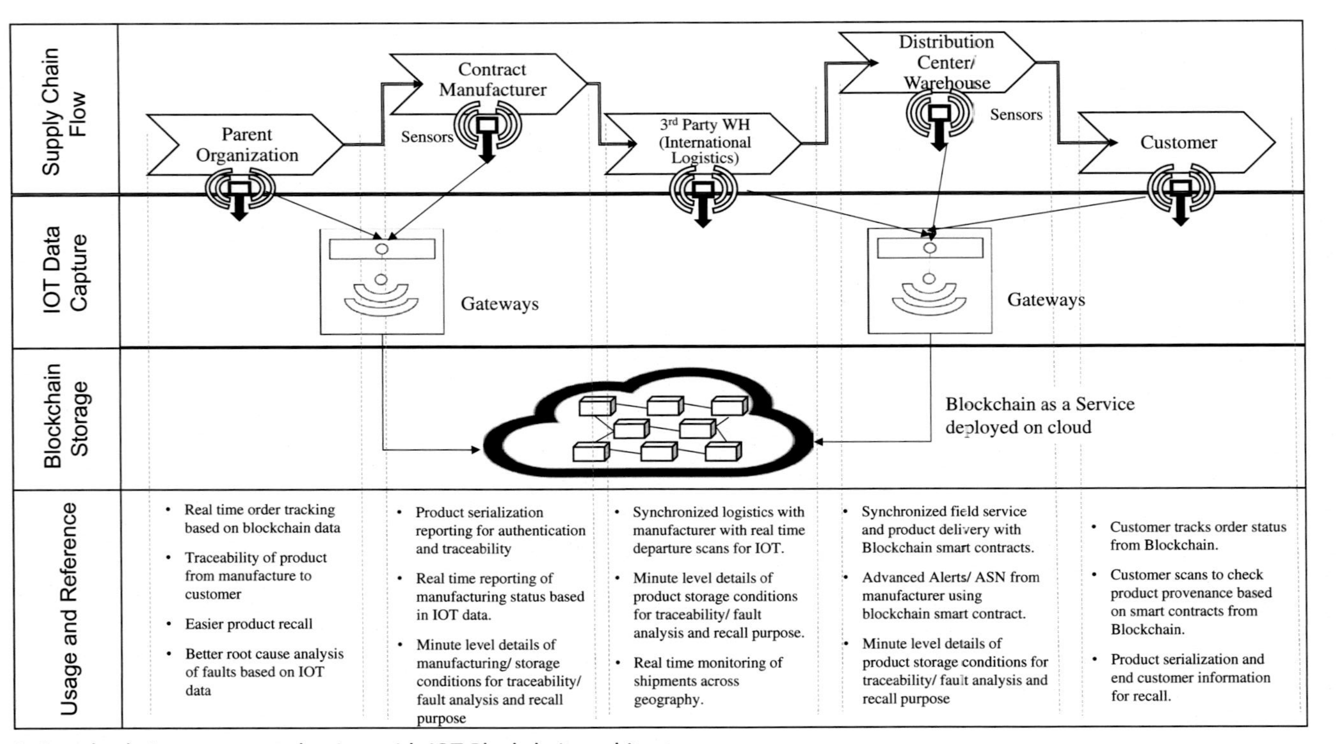

Fig. 2 Supply chain partner mechanism with IOT Blockchain architecture.

- Easier product recall.
- Better root cause analysis of faults based on data available in blockchain and captured by IOT.

Contract Manufacturer—Today the manufacturing hubs of the world are in the low cost economies. Real time IOT data available in blockchain (specially sub-assembly or contract manufacturers) can help all supply chain partners with better visibility and informed decisions. Some of the benefits of the data availability to and from manufacturer are:

- Product serialization reporting for authentication and traceability.
- Real time reporting of manufacturing status based on IOT data.
- Minute level details of manufacturing/storage conditions for traceability/fault analysis and recall purpose.
- Confirmation on arrival of sub-assembly from feeder plants to main plant.
- Suppliers material shipment schedules and confirmations.
- Break down information/potential supply chain disruptons.

Third Party Warehouses—As this is an in transit storage real time data of arrivals and departures, this info helps all supply chain partners. Also the physical storage condition data captured by IOT is of immense importance as these are not easily monitored and beyond the control of most of the organizations. These provides following benefits:

- Synchronized logistics with manufacturer with real time departure scans with IOT.
- Minute level details of product storage conditions for traceability/fault analysis and recall purpose.
- Real time monitoring of shipments across geography.

Distribution Centers/Warehouses—As this is the last storage point before supply to customers or retails the real time data of arrivals and departures helps inform customers of upcoming delivery, exact schedules of arrival, etc. Benefits are listed below:

- Synchronized field service and product delivery with Blockchain smart contracts.
- Advanced Alerts/ASN from manufacturer using blockchain smart contract.
- Minute level details of product storage conditions for traceability/fault analysis and recall purpose.
- Synchronized warehouse staffing as per schedules there by reducing unnecessary staffing.

End Customer—Customers today have very limited opportunity to verify the authenticity of a product and to know its provenance. Similarly end customer's information is almost off limits for most of the manufacturers which makes it very difficult in case of a recall. With the IOT, data can be captured when product is delivered to customer and storing it in a blockchain can make this information available to multiple stakeholders:

- Parent company's ability to know the end customer product is delivered to.
- Customer scans to check product provenance based on smart contracts from Blockchain.
- Product serialization and end customer information for recall.
- Possibility of product feedback from end customer to manufacturer.

Fig. 2 depicts a use case where an order from customer placed on a company is fulfilled with manufacturing at a contracts manufacturer and how the IOT blockchain mechanism can help track and trace in such multi-partner scenario spread across geography.

4.3.6 Agri supply chain

The challenges in the agri supply chain are mainly in the last leg farm level connectivity, capture of various farm and farmer information and availability of such information to other partners for traceability, authenticity and provenance purpose. The farm and farmer level electronic data capture is particularly challenging due to lack of infrastructure, lack of farmer motivation to adopt to digital methods and training the farmer. The solution to the problem is twofolds. One to incentivize farmers to digitize and second bring in the technology like IOT and blockchain together making it affordable and adaptable to revolutionize the grass root agri supply chain. For the problem of incentivizing farmers following are the considerations:

- Incentive program to encourage digitization of farm products from the farmer. It means farmers have to convert farm and produce level information to a digital format and share. These could be through some mobile or web application.
- Farmers are provided with incentive for digital data of land, type of harvest, location of land and not just farm data and made easy to digitize through some mobile or web application. Even better if photos of the same can be loaded. Specifically, applicable for big farm products like palm fruit, etc.

- Every data entered into the system by farmer is incentivized so that digitization is preferred by them. The incentives provided should be in monetary values.
- Transfer of ownerships needs to be registered when products/produce are sold by farmer to agents, mills or third party. These also need to be captured in digital format and farmers are to be incentivized for digital transactions.

Fig. 3 how IOT blockchain combination can help revitalize the agri supply chain. This architecture will make possible real time provenance, traceability by bringing in information of land, date of harvest, products harvested, type of soil, farm practices adopted, farmer details, pesticide use that can eventually add value to final product at the hands of the customer. Incentives and payments are a critical part of this solution and will go hand in hand with the supply chain solution. Incentives will sensitize and motivate farmers to adopt digital methods.

Following are the various agri supply chain partners and their roles in this digital solution.

Let us look at the various partners and their usage of IOT blockchain system in Fig. 3:

Farmer—As per the depiction farmers are to be provided with mobile and web applications to digitize the farm products and all other information. IOT will play a key role in capturing this information from the fields and transmitting it to the Blockchain. This solution is also possible without IOT based data capture and in that mechanism farmer may need to enter those data manually. The farmer data has to be captured through mobile or web application separately and is not through IOT system.

The IOT data capture of harvested products can automatically assign lot/serial numbers there by helping individual tracking of products. The data captured through IOT or manually entered can be transmitted to blockchain that can be available for any reporting and analysis till end consumer. As farmers sell the products to agents and others, the transfer of ownership is registered there by enabling traceability and establishing provenance. There can be many other members of the agri value chain like agents, etc. and those are not depicted in Fig. 3. Those can also be made a part of the blockchain network.

Processing Mills—Most of the farm products be it palm fruit, rice, coco, coffee or any other fruit like pomegranate, oranges, apple are cleaned, extracted, packaged in processing mills. IOT sensors in those premises and

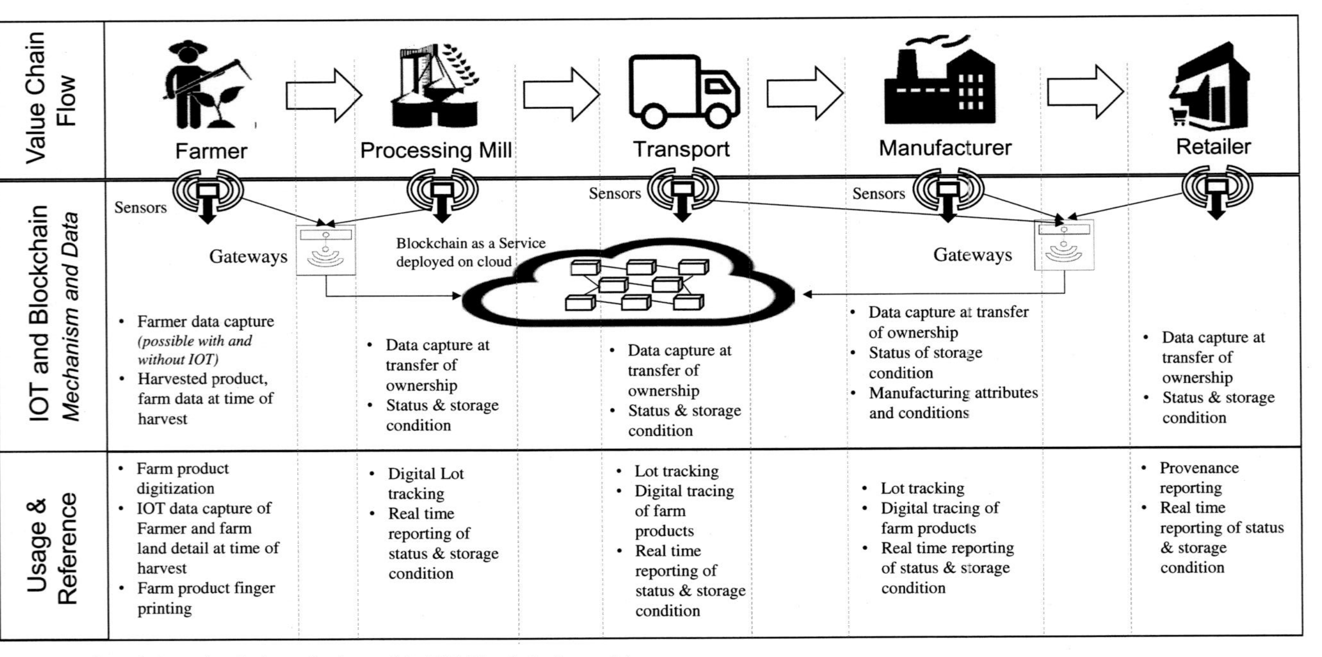

Fig. 3 Agricultural Supply chain solution with IOT Blockchain architecture.

storage locations can greatly enhance the traceability and help end customers with information of products storage criteria like the physical conditions of pressure, temperature, humidity, etc. The arrival/ departure scans (possible through RFID) can pin point the products arrival/departure time and duration at any location. Having this information in blockchain can greatly improve the traceability of a harvest to any partner in case of problem or establishing provenance. This information plays a key role in certification of oils like RSPO certified palm oils (RSPO stands for Roundtable on Sustainable Palm Oil).

Transport—Safe transport of any perishable edible products like chocolates, fruits, juices are an integral part of safe and fresh product at the hands of consumer. Today consumers have very limited to no information on the transport of these products. IOT data capture at the time of transport or exit scan of warehouse can greatly help bridge this important gap. The sensors in the trucks/ocean freight, warehouse can capture the physical storage conditions and make it available in blockchain network. Based on smart contracts various partners can access this information and make informed decision for the product. Even this information can be rendered to end customer if required for authenticity and transparency.

Manufacturer—There are numerous agri products which are used in manufacturing of edible and non-edible products. This can range from consumer edible products like cookies, chocolates, tin foods, baby products to consumer products like skin care products, soaps, detergents to industrial products like grease and oils. Tracking the transfer of ownership to manufacturer is an important aspect in provenance. IOT sensors can help track this when the transaction happens. Storing them in blockchain will help establish provenance and traceability to any stakeholder as per the smart contracts to access data. During manufacturing and after, the status of various environmental/physical and product related attributes can help determine the quality of final product, its freshness and other characteristics at the hands of consumer. Aberrations during manufacturing (in cases of baby products) can be really dangerous and today this information is not available. With the IOT blockchain mechanism this can be made available and analyzed and help establish the root cause analysis if anything goes wrong or any problems detected during random sample tests.

Retailer—For a retailer emphasis on quality and freshness of food is of utmost importance. As per Banerjee and Venkatesh [4] there are approx. 48 million people in the United States become sick every year due to food borne diseases, so the focus on quality can hardly be ignored. With issues like

E. coli infection happening in fast food restaurants, the retailers and end consumer needs to get a very clear visibility on manufacturing process, harvesting process, storage and other aspect. All these are possible when the data are available in blockchain right from farm to fork which can be traced and analyzed. In other words, the IOT Blockchain architecture can help to get the information from source/origin all the way to final consumption. The storage condition in retail stores plays a critical role in maintaining the freshness and sterility of the product. These can be checked and controlled with the use of IOT sensors to track and Blockchain network to store the information for analysis and prevention purposes.

The entire solution as depicted in Fig. 3 entails that the agri supply chain with IOT and blockchain can enable a farm to fork traceability/provenance and this can only happen when farmers are incentivized to digitize.

4.3.7 Digital automotive supply chain

Today the technology in automotive industry is focusing more on enhancing the car's ability to connect with the outside world and in-car experience than vehicle's internal performance [19]. These are the cars connected to internet passing data of cars status and working condition using sensors and data is stored in cloud. Tesla is a very good example of a connected car where car itself operates like a computer. We have reached an era where cars rolling out of factory are connected to internet and it is maturing. Today almost all OEMs have connected car offering ranging from mid-tier to premium segments. The information or statistics of car and its operations are available in cloud mostly to the OEM and are used for analysis of operating data by OEMs, provide a superior car owning experience to owners, generate newer revenue streams, advanced navigation and/or predictive vehicle centric alerts. But the next level of digital automotive which is lurking and not yet developed is when the OEM-dealers-service providers-component manufacturer and insurance companies are all inter-connected with a single source of truth. It is very important for car OEMs to understand that they are not just competing among themselves but they are even competing with Uber, Google and many other companies who are developing various services and features using disruptive technologies. These ranges from autonomous driven vehicles, used car buying process (Carvana, fair, vroom), car financing, security system, automated and online service to name a few. This is where the IOT Blockchain combination solution can help achieve this unsurmountable problem of seamless service, security, predictive alerts and experience to car owners.

Fig. 4 depicts a futuristic automotive supply chain where every supply chain partners are connected to each other with the IOT being the data capture mechanism and blockchain being the central repository cutting across systems and locations. It brings to focus key members of the supply chain who are going to benefit from the IOT blockchain architecture.

Fig. 4 shows the automotive partners who can be brought together in the single blockchain platform. It is to be noted that all the partners will be a node in the blockchain network. IOT sensors installed in the cars will capture the data and will be transmitted to the centralized blockchain network. Each and every vehicle will have a unique vehicle identification number which will be tokenized into blockchain. OEMs while getting the VIN (Vehicle Identification Number) will register the manufactured vehicle into the Blockchain. The moment car changes ownership from OEM to dealer the blockchain network will register it. IOT sensors in the car will transmit data via the IOT gateway to the blockchain network. These data can be accessed by dealers, service companies, insurance companies and component manufacturers. This will help with predictive analytics for maintenance or failures. Upon transfer of ownership from OEM to dealers to owner it will be establish traceability. This architecture can enable remote diagnostics capability for any problem experienced by owner saving time and effort. Component manufacturers will get advanced alerts from service technicians for parts required, it will help establish provenance of parts and keep track of parts changed in the vehicles and reducing supply chain leadtime. Insurance companies will have a transparent visibility on status of the vehicle history, information about its previous owners and also usage of vehicle. It will also get to know the previous vehicle owners driving history which can help them access premiums. Vehicle specific information will not be lost with change of ownership as it happens today because the vehicle will be tracked through unique VIN number in the blockchain. This will immensely help insurance companies and will enable them to have better insight on vehicle performance, driving history, maintenance rendered on the vehicle there by calibrating insurance accordingly.

4.3.8 Digital homes and offices

Smart homes are not just about the devices but it is about the services also. Devices combined with services will make consumers' lives more comfortable, more secure and more efficient. These include and not limited to home security, energy control, lighting control, health, transportation, environment and information dissipation that will be improving the efficiency of

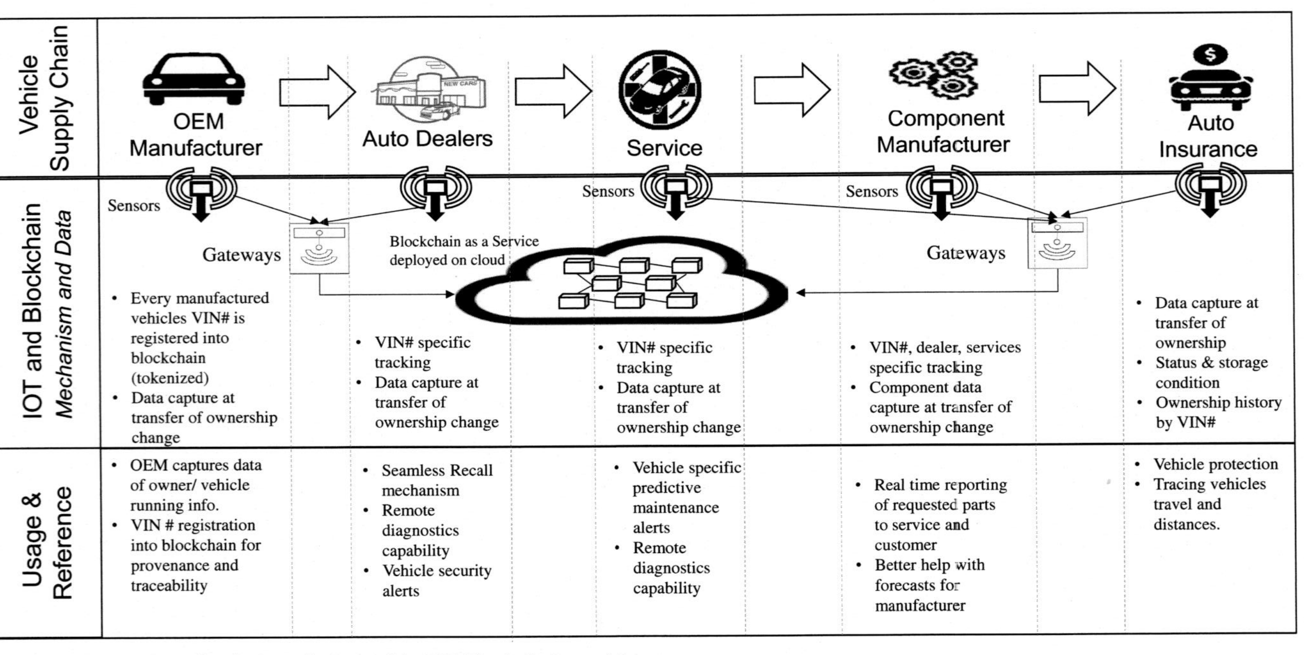

Fig. 4 Automotive Supply chain solution with IOT Blockchain architecture.

daily life of consumers. Today the limitation is around the interoperability of these devices which is primarily due to lack of common protocols, accessibility issues and single repository of information. There are some key smart home players (in specific areas) which are already there in market and those needs to be combined for a better customer experience. Fig. 5 depicts these key areas and how they can interact with each other. These products already exist and have matured to an extent in the market and can help improve and take the user experience to the next level.

In Fig. 5 one of the first use of IOT and blockchain is in home security. Currently these systems do not operate over a blockchain system and thus cannot give partial access to someone or a pin pointed access to a neighbor. To achieve this the use of IOT and Blockchain for a stringent and flexible authentication is required. Blockchain system will hold the smart contracts to enable or deny login to a particular user. The smart contracts (in Blockchain) along with IOT can also enable partial access to house or office. Comcast in the United States is already experimenting with the use of IOT and Blockchain for security systems [20] for partial and temporary accesses in home security. But what is to be noted is these are still in silos of the company (provider of service).

Another major use case of IOT and Blockchain coming together for consumers is in home/office/warehouse lighting, temperature and alarms controls. These ranges from continuous monitoring of temperature and lighting to alarm setting as per the need of the consumer. The ability to remotely manage temperature, alarms, smoke detectors, lighting devices, refrigerators, washing machines, dishwashers, etc. are the key goals of the devices today. These devices are working independently and in silos. These systems are focused on the purpose of automation and information dissipation to consumers today. The need of the hour is to bring them together so that they can lead the next generation of service. These can be to connect these devices to utility company for bill settlement or root cause for electrical faults. There can be capabilities developed to connect the retail stores with washing machine or dishwashers for placing orders on detergents or washing pods seamlessly with/without user intervention. An alarm in one system can be a helpful information for other system also like a drop in temperature can be an indicator for security breach or a fire alarm can be a good information to utility service provider. Similarly, if the security information and other aspects like detectors and controls are transparent to insurance companies it will help them provide better coverage and ultimately save cost to consumers.

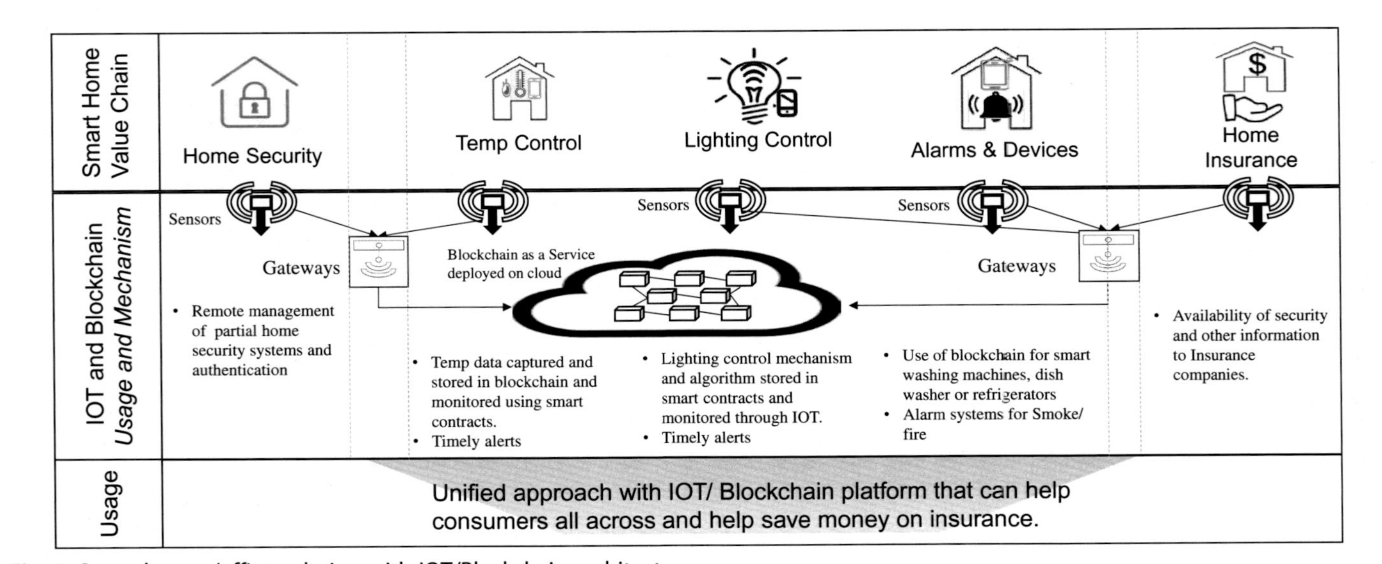

Fig. 5 Smart homes/office solution with IOT/Blockchain architecture.

The IOT Blockchain integration can be expanded to the entire spectrum of devices and services from lighting, temperature, washing machine, refrigerator, dish washer, alarms and controls for homes, offices and warehouses. The real benefits will start to surface when with one system or application will be able to control everything smart at home or office. This application will be based on an integrated IOT and blockchain platform system rather than an individual application. IOT in these cases will play a critical role as it is the sensor which will pick up aberrations to trigger alarms or other actions. The data being available in blockchain will help harmonize and cross reference it across systems (bringing in transparency).

4.3.9 Manufacturing systems and distribution industry

Manufacturing is globalized today. It has multiple partners spread across geographies. It is very important that these manufacturing systems are synchronized. Due to multiplicity of the partners it is almost impossible to bring all of them in the same information platform unless it is driven by blockchain. IOT system for data capture and Blockchain network for data repository is an all-powerful combination that can revolutionize the manufacturing systems. Fig. 6 depicts a high level architecture of how multiple ERP systems from multiple partners can be brought together using IOT/Blockchain combination. This figure helps us understand the distinct role of IOT to capture data from individual ERP and transmitting them to blockchain network. The figure also depicts the data can be captured directly from ERP system and interfaced to blockchain (without IOT). The data captured from IOT systems can be like those of shipping, logistics, production output, manufacturing lines

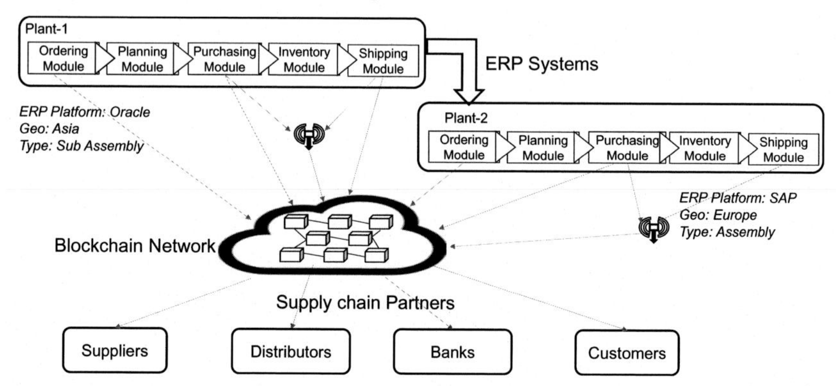

Fig. 6 A representation of multi system synchronization using IOT and Blockchain.

problem, material quality issues, etc. The data captured (without IOT) from the ERP system could be like orders, supply, master data and those can be directly fed to blockchain. Blockchain system makes the data, partner and platform independent and make its availability seamless for any analysis or decision making. For example, as shown in Fig. 6 the supply information from ERP like the on hand, work in process data, manufacturing problems from sub-assembly plant (plant 1) is a critical piece of information to multiple partners if it's a feeder plant to multiple assembly plants (like plant 2). This is where the information captured through IOT and available in blockchain will need to be individual plant system independent and available for every supply chain partner. Smart contracts and controlled access of data to partners can be enabled in blockchain network. These all will be possible from this IOT Blockchain solution.

Similarly, for heavy manufacturing industries having furnaces, turbines and boilers it is the digital twins which is the need of the hours. Digital twins are virtual replicas of physical devices to run simulations. Digital twin is a computer program, that takes real-world data about a physical object or system as inputs and produces outputs predications or simulations of how that physical object or system will be affected by those inputs. Digital twins exist only because of IOT systems. IOT sensors are placed on the physical device and they transmit the data which is used for analysis. These data can be the status of the system or action of some inputs. If these data are transmitted to the blockchain then it will help make it available to partners as per their access and will democratize the data for analysis from different partners as per their expertise. Most of the major manufacturing companies like GE, Chevron, Schneider Electric have already invested in digital twins and are benefitting in millions from it [21].

In distribution industry the supply chain transaction can be fed into blockchain which can help make this information available to all supply chain partners. These include but not limited to manufacturers and suppliers which can help POS reporting and Ship and Debit claims. Fig. 7 depicts a simplified view of the distribution industry supply chain transaction which can be fed into blockchain. This simple diagram represents how POS report and Ship and Debit claim can be extracted for the supplier/manufacturer from blockchain instead of individual systems (ERP or other transaction system). This will enable the following benefits:

4.3.9.1 Improve operational efficiency

By streamlining ship and debit claim process, point of sales reporting through the blockchain there will be no need for separate inventory

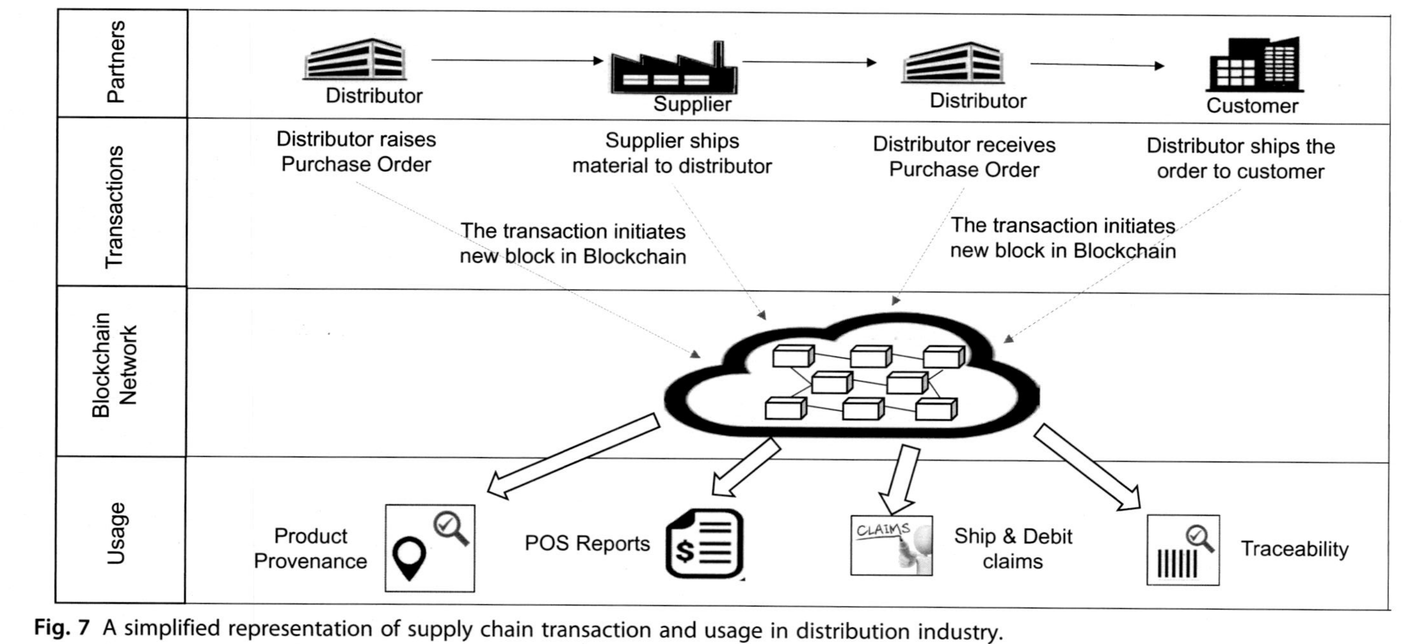

Fig. 7 A simplified representation of supply chain transaction and usage in distribution industry.

reporting and reconciliation. It will improve on claim process as it will reduce manual intervention for both supplier and distributor. This will reduce manual reporting and reconciliation thus freeing up resources for other purposes thus improving efficiency.

4.3.9.2 Reduce claim rejection and accelerate cash flow

With data extracted directly from the blockchain system there will be untampered data for claim which will be accurate and completely acceptable to all partners. Also the transparency by virtue of blockchain system in stock traceability with sales and purchase order details can provide correct cost to customer there by improving the claim reduction and accelerating cash flow into system.

In the distribution business the IOT can play a role in the warehousing and logistics area where it can help trace the temperature or physical condition while in storage or transport. It can track for humidity, pressure, temperature or shocks for sensitive products. It has been discussed in detail in the solution for product traceability.

The solution discussed in this section provides a unified view that Blockchain with IOT is a very effective supply chain solution. Blockchain is slow in its adoption and growth, but its application is broad and comprehensive. With IOT it will become a strong and significant architecture that will enable many tools and reports to help managers improve the supply chain operations. This combination will democratize data, breaking barriers of individual systems. The solution will ultimately add value to product, streamline supply chain and empower customer.

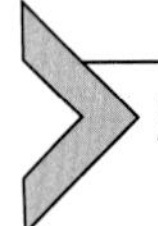

5. The economic value of IOT-blockchain combination for supply chain solutions

Gartner predicts that blockchain will add $176 billion in business value by 2025, and $3.1 trillion by 2030 [22]. Columbus [23] shares the forecast from various researches and states that the global IoT market will grow from $157B in 2016 to $457B by 2020. Columbus [23] also states that Discrete Manufacturing, Transportation and Logistics, and Utilities will lead all industries in IOT spending by 2020, averaging $40B each. Bain & Company predicts business to business IOT segments to generate more than $300B annually by 2020. Boston Consulting Group predicts IOT market to reach $267B by 2020. The combination of IOT and Blockchain would definitely help to achieve these numbers and may be more than what is being predicted.

The IOT-Blockchain combination can drive value in different ways for any supply chain solution. Obviously there is an amount of investment associated with its establishment, process definition and implementation. But once it is implemented it can help reduce cost by getting rid of peripheral systems, reducing human dependency, eliminate redundant processes, simplifying warehousing processes and automating supply chain processes which are reports and human driven. It can also help bolster brand value with ability to track and trace, establishing provenance and smoother recalls. It can also boost revenue with unlocking of opportunities which were not possible due to systems working in silos. With synchronized products and services across segments and the adoption of a single platform for supply chain with backbone of blockchain and supported by IOT the revenues are going to grow better that what it is today.

The economic value can immediately start with process refinement leading to cost reduction. With time as the processes mature with the technology newer revenue generation models will emerge. The salient feature of this supply chain solution is trust and time. It is of utmost importance to highlight that this entire discussion is invalid without the trust among supply chain partners. Also the technology of IOT and blockchain are maturing and it will need time to mature, develop and evolve further before it totally assimilates and augments the supply chain processes.

Digital revolution is characterized by the convergence of different technologies for the betterment of customer and business processes. The future of IOT Blockchain combination will also converge, blurring the lines of digital and physical world. It will use Artificial Intelligence and Machine Learning algorithms and use data from Blockchain for decision making. As a Blockchain IOT platform it will be able to bring in more partners and their data for decision making which will make world a better place to live!

6. Conclusion

IoT and blockchain as separate technologies are restricted in their applications in any field. In supply chain and logistics, it can derive significant benefit with the combined architecture of IOT, blockchain and other supply chain systems like ERP. The architecture presented is generic and can help any type of supply chain orchestration and few specifics are highlighted and discussed. The chapter individually discusses the problem faced today in agri supply chain, auto supply chain, product or order tracking, traceability and recall, counterfeiting, digital homes, manufacturing and distribution supply chain and how these challenges can be solved with the IOT

Blockchain combination. Ranging from generic supply chain flows to specific agri or automotive supply chain the benefits are similar in eliminating the intermediaries, bringing in transparency and establishing provenance or tracking it to the detailed level. Blockchain and IOT as a technology are maturing and growing together. They are codependent on one another and stands maximum changes of development with both being together. IOT desperately needs the blockchain's features as it brings in security, immutability and smart contracts while the blockchain is in need of IOT to channelize the feed of data to convert every aspect as a big-time opportunity for supply chain to become more effective. IOT Blockchain combination will revolutionize the supply chain in the way the partners interact today. The combination will add value to products, invoke trust among partners, reduce supply chain cost, improve process efficiency, avoid information voids and empower customers.

References

[1] M. Palmer, Data is the new oil, in: ANA MarketingMaestros, 2006. 3 November. http://ana.blogs.com/maestros/2006/11/data_is_the_new.html; (a) D. Reid, 'Mastercard's boss just told a Saudi audience that 'data is the new oil', CNBC, 2017. https://www.cnbc.com/2017/10/24/mastercard-boss-just-said-data-is-the-new-oil.html. Accessed 20 July 2019.

[2] K. Alicke, J. Rachor, Supply Chain 4.0—The Next-Generation Digital Supply Chain, McKinsey & Company, 2016. https://www.mckinsey.com/business-functions/operations/our-insights/supply-chain-40--the-next-generation-digital-supply-chain. Accessed 17 May 2019.

[3] R. O'Byrne, 7 ways everyone can cut supply chain costs, in: CSCMP's Supply Chain Quarterly, 2011. https://www.supplychainquarterly.com/topics/Strategy/scq201102seven/. Accessed 18 April 2019.

[4] A. Banerjee, M. Venkatesh, Product Tracking and Tracing With IOT and Blockchain, Infosys Ltd, 2019. https://www.infosys.com/Oracle/insights/Documents/product-tracking-tracing.pdf.

[5] T. Leeson, How Auto Makers Benefit From the Connected Supply Chain, https://blogs.opentext.com/how-auto-makers-benefit-from-the-connected-supply-chain/, 2018.

[6] M.J. Foley, The First Cars Using Microsoft Connected Vehicle Platform at Scale Are Coming, ZDNet, 2019. https://www.zdnet.com/article/the-first-cars-using-microsoft-connected-vehicle-platform-at-scale-are-coming/. Accessed 19 May 2019.

[7] R.Y. Zhong, X. Xu, E. Klotz, S.T. Newman, Intelligent manufacturing in the context of industry 4.0: a review, Engineering 3 (2017) 616–663.

[8] A. Banerjee, Infosys, https://www.infosys.com/Oracle/white-papers/Documents/integrating-blockchain-erp.pdf, 2018.

[9] A. Banerjee, Chapter three—blockchain technology: supply chain insights from ERP, Adv. Comput. 111 (2018) 69–98, 2018.

[10] A. Litan, B. Lheureux, A. Pradhan, Integrating Blockchain With IoT Strengthens Trust in Multiparty Processes, Gartner Research, 2019.

[11] C. Elena-Lenz, Internet of Things: Six Key Characteristics, Frog, 2015. https://designmind.frogdesign.com/2014/08/internet-things-six-key-characteristics/. Accessed 23 May 2019.

[12] K.K. Patel, S.M. Patel, Internet of things-IOT: definition, characteristics, architecture, enabling technologies, application & future challenges, Int. J. Eng. Sci. Comput. 6 (5) (2016) 6122–6131.
[13] R.V. Meulen, Gartner Says 8.4 Billion Connected "Things" Will be in Use in 2017, Up 31 Percent From 2016, Gartner Research, 2017. https://www.gartner.com/en/newsroom/press-releases/2017-02-07-gartner-says-8-billion-connected-things-will-be-in-use-in-2017-up-31-percent-from-2016.
[14] Business Insider Intelligence, There will be 24 billion IoT devices installed on earth by 2020, in: Business Insider, 2016. https://www.businessinsider.com/there-will-be-34-billion-iot-devices-installed-on-earth-by-2020-2016-5?IR=T. Accessed 20 May 2019.
[15] K. Lewis, Digital Transformation Is a Journey, An Enterprise Scale IoT Platform: Watson, Internet of Things Blog, https://www.ibm.com/blogs/internet-of-things/enterprise-scale-iot-platform-watson/, 2017. Accessed 20 May 2019.
[16] M. Hung, Leading the IOT, Gartner, 2017. https://www.gartner.com/imagesrv/books/iot/iotEbook_digital.pdf. Accessed 14 June 2019.
[17] A. Pal, The Internet of Things (IoT)—Threats and Countermeasures, CSO, 2019. https://www.cso.com.au/article/575407/internet-things-iot-threats-countermeasures/. Accessed 23 June 2019.
[18] Oracle, Blockhain Platform Pricing Options, https://cloud.oracle.com/blockchain/pricing, 2018. Accessed 23 April 2019.
[19] Mckinsey & Company, What's Driving the Connected Car, Mckinsey & Company—Automotive & Assembly, 2014. https://www.mckinsey.com/industries/automotive-and-assembly/our-insights/whats-driving-the-connected-car. Accessed 11 June 2019.
[20] N. Davis, Blockchain for the Connected Home: Combining Security and Flexibility, Comcast Labs, 2018. http://labs.comcast.com/blockchain-for-the-connected-home-combining-security-and-flexibility. Accessed 11 June 2019.
[21] A. Eshkenazi, Real Benefits From Digital Twins, APICS Supply Chain Management Now, 2018. http://www.apics.org/sites/apics-blog/think-supply-chain-landing-page/thinking-supply-chain/2018/09/21/real-benefits-from-digital-twins. Accessed 21 May 2019.
[22] B. Granetto, R. Kandaswamy, J. Lovelock, M. Reynolds, Forecast: Blockchain Business Value, Worldwide, 2017–2030, Gartner Research, 2017. https://www.gartner.com/en/documents/3627117. Accessed 26 May 2019.
[23] L. Columbus, 2018 Roundup of Internet of Things Forecasts and Market Estimates, Forbes, 2018. https://www.forbes.com/sites/louiscolumbus/2017/12/10/2017-roundup-of-internet-of-things-forecasts/#671f2fe51480. Accessed 11 June 2019.

About the author

Arnab Banerjee, Ph. D.
Infosys Limited, India

Principal Consultant, Enterprise Application Services

E-Mail: Arnab_banerjee08@infosys.com

Linkedin: https://www.linkedin.com/in/arnab-banerjee-a44697b/

Arnab Banerjee is a Principal Consultant with Enterprise Applications Services of Infosys Ltd. He consults in the area of ERP, IOT and Blockchain technology. His consulting experience of more than 17 years' spans across North America, Europe and Asia. In his blockchain consulting role he helps implement Oracle Blockchain cloud services. It spans from identifying the use case, solution architecture to implementation. For IOT applications consulting role he helps implement enterprise grade industrial IOT solution architecture integrated with ERP systems. He is an ERP implementation specialist in the supply chain domain. His research interests include information technology, IOT and Blockchain applications integration with ERP and other satellite systems. He has extensive publications in the area of reverse supply chain, lean/agile/leagile initiatives, theory of constraints, supply chain transformations and humanitarian logistics. Over the years he has published more than 35 research papers in various peer reviewed international journals, international research conferences, case studies, book chapters, white papers and point of view articles. He holds a PhD in Supply Chain Management, a Master's in Industrial Engineering and a graduate degree in Mechanical Engineering. He is a certified Six Sigma black belt champion. He is listed in Who's Who of the World 2015 edition.

CHAPTER TEN

Integration of IoT with blockchain and homomorphic encryption: Challenging issues and opportunities

Rakesh Shrestha, Shiho Kim
Yonsei Institute of Convergence Technology, Yonsei University, Seoul, South Korea

Contents

Abstract

The advancement of new technology has taken a huge leap in the last few decades. It is bringing a drastic change in each step of human life that are capable of performing intelligent tasks. The internet of things and blockchain are disruptive technologies that have received a huge attention from industry, academic and financial technologies. There is a risk of privacy leakage of sensitive information in the centralized IoT system

Advances in Computers, Volume 115
ISSN 0065-2458
https://doi.org/10.1016/bs.adcom.2019.06.002

because the centralized servers can access the plain text data from the IoT devices. There is an extensive interest in applying the blockchain in the IoT system to provide IoT data privacy and decentralized access model. However, the previous blockchain-based IoT systems have issues related to privacy leakage of sensitive information to the servers as the servers can access the plaintext data from the IoT devices. So, we present the potential of integration of blockchain based-IoT with homomorphic encryption that can secure the IoT data with high privacy in a decentralized mode. In addition, we provide comparison of the recent technologies toward the securing and preserving the privacy of the IoT data using blockchain and homomorphic encryption technology. We have also highlighted the research challenges and possible future research directions in the integrated blockchain-based IoT with homomorphic encryption.

1. Introduction

With the rapid development of internet and technology, we are in the fourth industrial revolution or Industry 4.0, where everything will be connected with each other in the cyber world. The Industry 4.0 will bring a huge revolution in the convergence of technologies such as big data, AI, robotics, internet of things, 5G, cloud computing, blockchain, and cryptocurrency. The Internet of Things (IoT) has a huge potential to provide different types of exhilarating services across many areas from industry, business, Intelligent Transportation System (ITS), social media, healthcare and smart cities. The IoT is a network of resource-constrained heterogeneous devices with very low computing power, low memory and limited battery life, connected with each other using the Internet. IoT will have a deep impact on internet of vehicles, internet of drones, internet of military things and so on. Since, every sector has their own smart things connected over the internet, we can call it as internet of everything. There is no doubt regarding the advantages of IoT ecosystem. Despite the advantages of the IoT, there are some issues that need to be solved carefully. The IoT devices generate a huge amount of data that might lead to security and privacy issues, but most of the IoT device manufactures cannot provide complete data security. IoT applications deal with mass user private information that leads to serious threats related to security and privacy of the user data. Currently, IoT works on a centralized system and the data among the IoT devices are exchanged in a client-server access model. The issue with this model is that the central server might act maliciously and can hijack and make fraudulent use of the user's private data. One possible solution to this issue is to introduce a decentralized IoT system, which can securely store the user's private data and give IoT data users the full authority over their own data [1].

The decentralized IoT data access model assures that user's private data is not manipulated by the centralized servers and provides efficient coordination of external computing resources of the IoT devices for better performance. In order to achieve this objective, peer to peer scheme such as blockchain plays an important role. There is a rapid increase in research from industry and academic fields to utilize the security concept of blockchain in IoT systems.

The blockchain (BC) is a peer to peer decentralized and distributed computing paradigm providing privacy and security that underpins the Bitcoin cryptocurrency [2]. Blockchain can operate in a trustless network environment without the need for centralized servers and having trust on peer nodes. Blockchain consists mutual database secured by consensus-based verification and proof of work mechanism. The entire ledger or database is distributed over all the peer nodes in the blockchain networks based on consensus algorithms. The blockchain can provide security and anonymity of the user's data because the same copy of the ledger is stored in each peer node based on secure cryptography. The blockchain is being used in different fields beyond the scope of cryptocurrencies because of the special features of blockchain such as data immutability, publicly verifiability, transparency, auditability, security and privacy. Some of them are intelligent transportation system [3], supply chain management, Industry [4], agriculture [5], security, smart energy [6] and many more. Furthermore, smart contract features can be used to set policies, control user data and monitor access rights. It is a programmable document that is executed autonomously when the set of pre-defined conditions are met. The use of robust cryptography and self-executing smart contracts makes blockchain a strong candidate for IoT privacy and security. The blockchain can provide IoT data security and privacy in a decentralized manner. However, the previous blockchain-based IoT systems still have issues related to information leakage to the servers because the servers can access the plaintext data from the IoT devices. If we can integrate the blockchain-based IoT with the homomorphic encryption, then the homomorphic encryption has high potential in securing the IoT data by providing privacy to the user's IoT data.

The Homomorphic Encryption (HE) is an encryption mechanism that can resolve the security and privacy issues. It allows the third party service providers such as cloud servers to perform certain type of operations on the user's encrypted data without decrypting the encrypted data, while maintaining the privacy of the users' data. There are basically three types of homomorphic encryption and they are Partially Homomorphic Encryption

(PHE), Somewhat Homomorphic Encryption (SHE) and Fully Homomorphic Encryption (FHE). In HE, mathematical operation on the plaintext during encryption is equivalent to another operation performed on the cipher text.

In this article, we discuss the integration of IoT, blockchain and homomorphic encryption, and its advantages in the IoT ecosystem. The integration of blockchain with IoT using homomorphic encryption provides a secured, decentralized IoT system where the IoT data are securely stored in the blockchain instead of centralized servers. In this way, all the peer IoT nodes can verify the data publicly. The integration of homomorphic encryption supports any degree of homomorphic multiplication and addition on encrypted data without decrypting the data first. So, the nodes performing homomorphic encryption cannot get any information from the user's encrypted data, thus, providing privacy and confidentiality. In addition, the scalability issue can be solved as more and more IoT devices with higher computation capability join the blockchain networks and provide their service for creating and verifying blocks along with performing homomorphic encryption.

The rest of the chapter is divided into the following sections. In Section 2, we present a brief background of IoT, blockchain and homomorphic encryption. In Section 3, we discuss the security and privacy issues related to different layers of IoT. In Section 4, we present in detail about the impact of IoT, blockchain and homomorphic encryption. Section 5 presents the collaborative security by integrating IoT, blockchain and homomorphic encryption focusing on security aspect of IoT devices. Section 6 is related to application and use cases of the integrated blockchain-based IoT with homomorphic encryption mechanism. In Section 7, we highlight the research challenges and discuss the possible future research directions in the blockchain-based IoT with homomorphic encryption scheme. Finally, in Section 8, we summarized the chapter.

2. Background

In this section, we will give a brief description and overview of the state-of-the-art of each technology.

2.1 Internet of things (IoT)

There are billions of embedded devices in the world, ranging from small devices to large devices. The small devices such as sensor nodes, RFIDs, wearables, health monitoring systems, smart phones, smart home appliances, and drones are resource-constrained devices with a very low computing power, low memory and limited battery life. On the other hand, there are large devices such as smart cars, smart devices, financial utilities and machineries used in industrial control systems (ICS) that have sufficient resources, powerful computing capability, high memory, and storage space. Hence, the interconnection of all the heterogeneous computing devices through internet to exchange data is called Internet of Things (IoT) as shown in Fig. 1. The elements of IoT ecosystem are intelligence and IP connectivity that does not require manual control or human interactions. According to International Data Corporation (IDC) [7], the IoT market will reach $1.2 trillion in 2022 connecting more than 30 billion IoT devices [8]. IoT devices make the life of the people much easier and simpler. There is a very high demand of IoT devices and its production is increasing rapidly along with the advancement in technology. Further, the IoT devices are improving and becoming more intelligent with the use of AI. The application of IoT devices will improve the quality of people's life as well as boost the world economy. IoT is a complex network of heterogeneous smart things that interact and communicate with each other and to the internet. The heterogeneous IoT devices connect to the internet through low power, limited bandwidth, and wireless communication mode such as Wifi (802.11 b/g/n/p), Bluetooth (BLE, 802.15.4), infrared, LoRa, ZigBee, LTE, 4G and 5G. Let us take an illustration where different devices connected with each other using internet and cloud computing. There is a platform for integration of vehicles with smart infrastructures that are enabled with IoT devices, which is known as Internet of Vehicle (IoV). The IoV industry incorporates VANET communication services along with intelligent traffic management and smart cities. The IoV brings vehicles, traffic management, drivers and other traffic related services together in a seamless manner through internet connectivity [9].

Despite the popularity of IoT, there is no specific standard reference model for IoT due to the use of heterogeneous devices [10]. Several researchers have proposed different layered IoT architectures. The authors in Ref. [11] presented a comparison between different layers of IoT architecture discussed in different literatures. Although different literatures have proposed different architecture layers, they almost agree in three key architecture layer and their task or purpose is similar. Considering the previous research works, IoT can be divided into three basic architectural layers [12] as shown in Fig. 2.

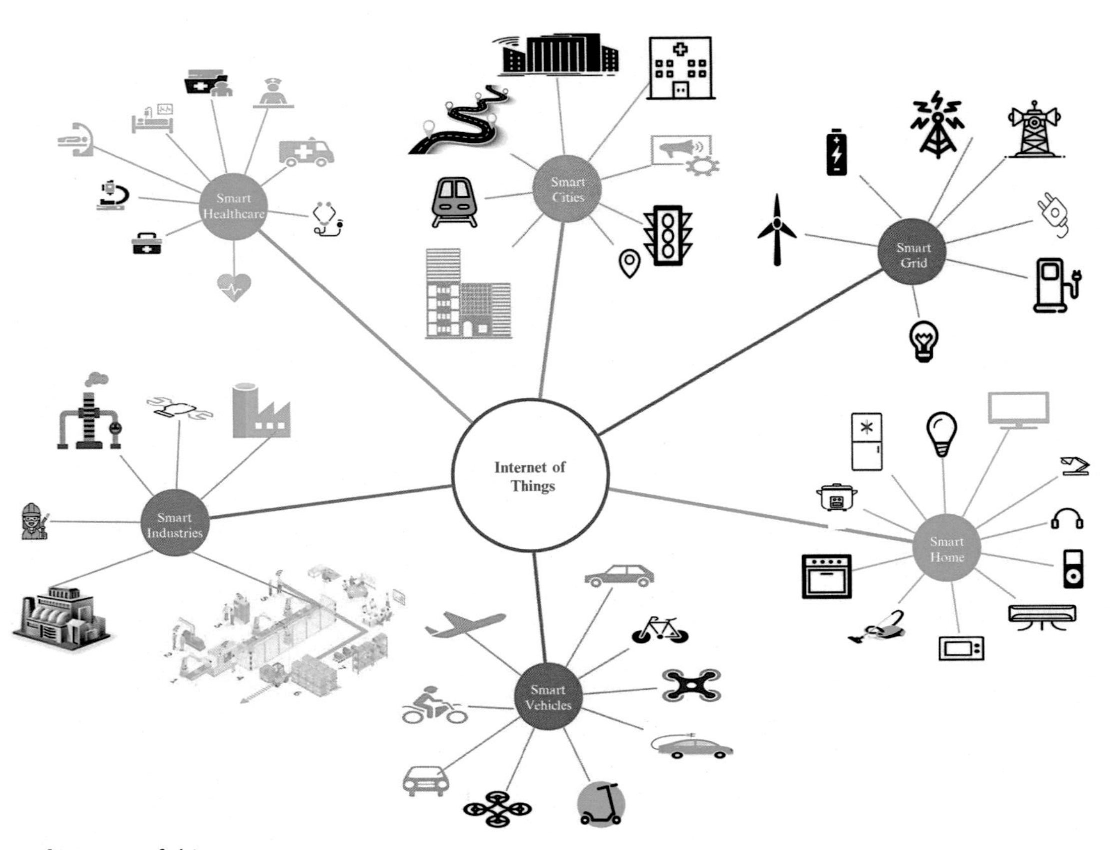

Fig. 1 Overview of internet of things.

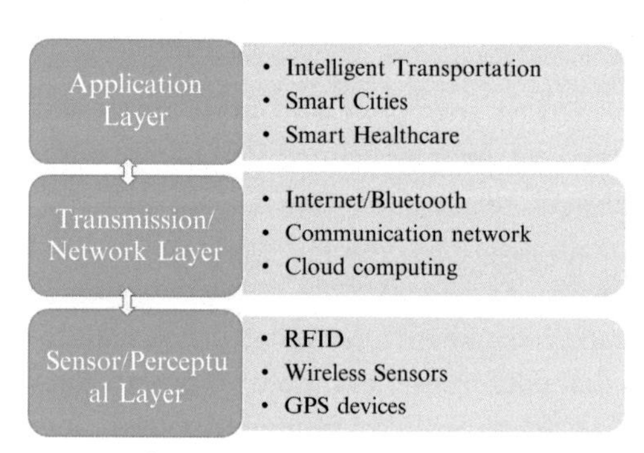

Fig. 2 Basic architecture of IoT.

The sensor or perceptual layer: It consists of physical devices such as RFID sensors that collects raw data (i.e., environmental conditions like temperature, pressure, humidity, etc.) from the real world, converts them to digital data and transfers the digital data to the upper layer.

Transmission or network layer: This layer is responsible for initial processing and reliable transmission of received information from other layers. It transmits data from source node to the destination node based on different communication medium such as internet, wireless communication, mobile communication, and low power communication.

Application layer: This layer provides different personalized smart applications and management services based on the user requirements such as intelligent transportation, smart home, and smart healthcare. This layer interacts directly with the end IoT devices. This layer is responsible for providing a control mechanism to access the data.

At the current state, IoT system is a complex integration of heterogeneous devices that is based on a so-called trusted centralized architecture for authentication. The IoT consists of several smart devices that are connected with the central server. The centralized system is managed by third party service providers in a client-server access model where the users' personal data and other information are stored on the servers. There are issues with the centralized system, i.e., there might be a single point of failure or the service providers might access the IoT data illegitimately. Moreover, the malicious devices might access the data on the server with false authentications or device spoofing. Hence, the centralized IoT architecture have issues such as data insecurity, privacy, scalability, flexibility, availability, and efficiency within the existing IoT networks. This is discussed in detail in Section 3.

2.2 Blockchain

With the sudden growth and interest in cryptocurrency, blockchain has attracted the interest of people from different areas such as industries, healthcare, government, financial, real state, academics and so on. The boost in the market capitalization of cryptocurrencies gives rise to various applications of blockchain. The blockchain is an evolving decentralized and distributed computing paradigm providing privacy and security in peer-to-peer networks that underpins the Bitcoin cryptocurrency [2]. In technical words, blockchain is a protected, mutual database and chronologically ordered chain of blocks secured by consensus-based verification and proof of work mechanism. One of the most important attribute of blockchain is that it can operate in a distributed and decentralized fashion without the need for third party service providers or centralized authority. Blockchain can operate in a trustless network environment without having trust on peer nodes while maintaining anonymity.

Some of the main advantages of blockchain are decentralization, anonymity, chronological order of data, distributed security, transparency and immutability [13]. The chaining of blocks is performed by adding the hash of the previous block to the current block, then the hash of the current block to the next block in a consecutive order [2,14] as shown in Fig. 3. There are

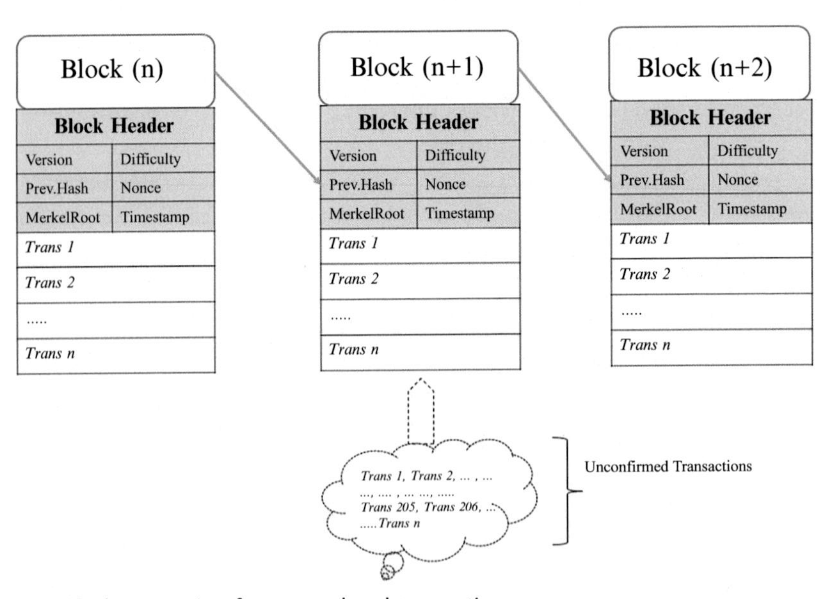

Fig. 3 Block generation from unordered transactions.

two types of blockchain. In the public blockchain, any node can join or participate in the blockchain network without permission and no one has control over the network. The information stored in the public blockchain can be seen and validated by the members of the nodes. In private blockchain, permission from the administrator is required to join the network and administrator has control over the nodes activities. Each blockchain consists of a first block known as genesis block, which is the root or origin of all the blocks. The block consists of a block header and a block body as shown in Fig. 3. The block header is composed of previous block's hash, nonce, timestamp and Merkle root [15] while the block body is composed of transaction lists and other variables. Each block is interlocked like a chain with the preceding block using the hash of the preceding block. The miner collects the transactions from the peer nodes and creates a new block, which is then sent to the blockchain network. The blockchain is verified by consensus of anonymous nodes in generation of blocks. The independent validation of each new block by every node on the network ensures that the miners cannot cheat. The consensus of all the participating mining nodes of the network can be used as ground truth for the next block. There are several blockchain consensus mechanisms for public and private blockchain [16]. The consecutive hashes of blocks guarantee that the transaction appears in a chronological order. The blockchain is considered secure, as the combined computational power of the malicious nodes does not dominate the computational power of the honest nodes [2,17]. The Proof of work (PoW) concept makes sure that the miner is not manipulating the network to make fake blocks. A PoW is a mathematical puzzle that is very hard to crack therefore securing the blockchain double spending attack. It is very difficult to determine a nonce in proof of work, but once determined it is easy to validate the new block [18]. Moreover, tampering and faking a block is very difficult and requires a significant computation power to change the successor blocks. Then, the preceding transaction cannot be changed without changing its block and all the following blocks, preventing the double spending attack.

Furthermore, the smart contract capabilities in blockchain provides additional security features that can be used to set policies, control user data and monitor access rights. The smart contract is a set of pre-defined contractual rules written in a program that is executed autonomously when the set of pre-defined conditions are met [19]. Even though the blockchain supports integrity and non-repudiation to some extent, confidentiality and privacy of the data or the devices are not preserved.

2.3 Homomorphic encryption

Due to privacy leakage of sensitive data, the conventional encryption systems are not completely secure from an intermediary service like cloud servers. The homomorphic encryption is a special kind of encryption mechanism that can resolve the security and privacy issues. Unlike the public key encryption, which has three security procedures, i.e., key generation, encryption and decryption; there are four procedures in HE scheme, including the evaluation algorithm as shown in Fig. 4. The HE allows the third party service providers to perform certain type of operations on the user's encrypted data without decrypting the encrypted data, while maintaining the privacy of the users' encrypted data. In homomorphic encryption, if the user wants to query some information on the cloud server, he first encrypts the data and stores the encrypted data in the cloud. Then, after sometime, the user sends query information to the cloud server. The cloud server runs a prediction algorithm on encrypted data using HE without knowing the contents of the encrypted data. Then, the cloud returns the encrypted prediction back to the user and the user decrypts the received encrypted data using the user's secret key, while preserving the privacy of his data as show in Fig. 4. In HE, mathematical operation on the plaintext during encryption is equivalent to another operation performed on the cipher text [20]. Let us consider a simple homomorphic operation on the plaintext with the corresponding cipher text operation.

$$\text{Let } E(m_1) = m_1^e \text{ and } E(m_2) = m_2^e,$$

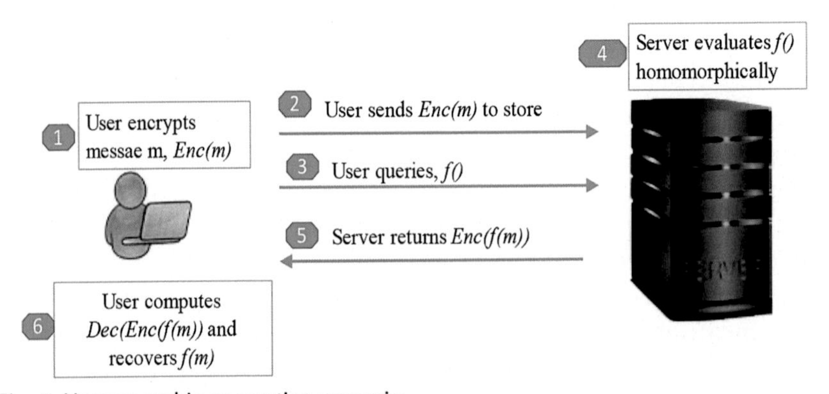

Fig. 4 Homomorphic encryption scenario.

Then,
Addition Homomorphism:

$$E(m_1) + E(m_2) = m_1^e + m_2^e = (m_1 + m_2)^e = E(m_1 + m_2)$$

Multiplication Homomorphism:

$$E(m_1) \times E(m_2) = m_1{}^e \times m_2{}^e = (m_1 \times m_2)^e = E(m_1 \times m_2)$$

The HE can be categorized into three groups based on the number of mathematical operations on the encrypted message. They are: Partially Homomorphic Encryption (PHE), Somewhat Homomorphic Encryption (SHE) and Fully Homomorphic Encryption (FHE) as shown in Fig. 5.

- Partially Homomorphic Encryption (PHE): In PHE scheme, only one type of mathematical operation is allowed on the encrypted message, i.e., either addition or multiplication operation, with unlimited number of times,
- Somewhat Homomorphic Encryption (SHE): In SHE, both addition and multiplication operation is allowed but with only a limited number of times.
- Fully Homomorphic Encryption (FHE): FHE allows a large number of different types of evaluation operations on the encrypted message with unlimited number of times.

Let P be the plaintext space, i.e., $P = \{0, 1\}$ which consists of input message tuple $(m_1, m_2, \ldots m_n)$. Let us represent the Boolean circuit by C and ordinary function notation as C $(m_1, m_2, \ldots m_n)$ to represent the evaluation of the circuit on the message tuple [21]. The general HE is described below:

- Gen$(1^\lambda, \alpha)$ is the key generation algorithm that generates output keys triplets, i.e., secret key-pair (*sk* and *pk*) along with evaluation key (*evk*), where λ is security parameter and α is auxiliary input, $(sk, pk, evk) \leftarrow \text{KeyGen}(\$)$

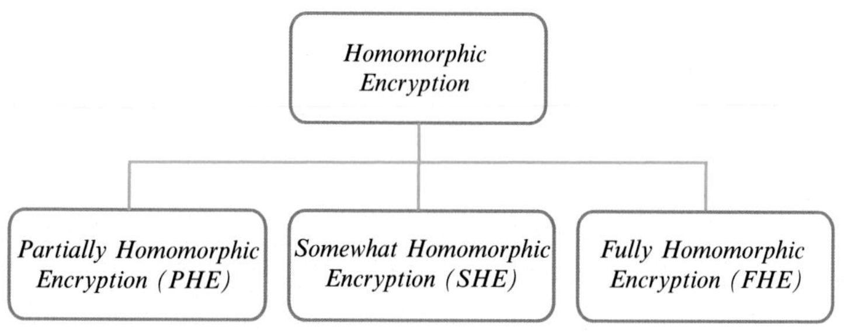

Fig. 5 Types of homomorphic encryption.

- Enc(pk,m) encrypts a message (m) with the public key (pk) and outputs a ciphertext ($c \in C$), $c \leftarrow Enc_{pk}(m)$
- Dec(sk, c) decrypts a ciphertexts with the secret key (*sk*) and recovers message (m) as the output, $m \leftarrow Dec_{sk}(c)$
- $Eval(evk, C, c_1, c_1, \ldots, c_n)$ produces evaluation output by taking *evk* key as input, a circuit $C\epsilon C$ and tuple of input ciphertexts, i.e., $c_1 \ldots c_n$ and previous evaluation results, $c^* \leftarrow Eval_{evk}(evk, C, c_1, c_2, \ldots c_n)$.

The HE has been applied in different fields because of its ability to compute on encrypted data, while providing security and privacy to the users. There is a wide application of HE in diverse fields such as healthcare, medical applications, financial sector, forensic applications, social networking advertisements, and smart vehicles, where the users' privacy can be maintained.

3. Issues in internet of things

Due to the enormous amount of heterogeneous IoT devices, they have different data formats and contents based on specific application and functionalities. Although, IoT community is working on global standardization of IoT, however, there are issues with compatibility, management and interoperability [22]. It is very hard to develop a standard security scheme that is suitable for all IoT devices. The security and privacy issues in IoT are discussed below.

3.1 Security and privacy issues

The existing IoT security mechanisms are not robust and they are vulnerable to several security and privacy issues. Privacy is a basic human right and it is a big issue in finance and health care system. So, while using IoT devices we should consider the privacy of the users' data strictly. It is difficult to implement the conventional security mechanism in resource constraint IoT devices. IoT devices' are vulnerable and are frequent target of attackers due to access control mechanism, software updates, weak applications and APIs [23]. Moreover, low memory, limited resources, inadequate computational power, and weak trust in third party cloud applications leads to security and privacy issues. The IoT devices deployed in remote and hostile environments are vulnerable to physical compromise and data theft. In addition, IoT devices have security issues while using controlled access in centralized cloud [24]. The centralized cloud is vulnerable to a single point of failure, data breach and unauthorized data sharing [25]. There is a privacy issue based on centralized system due to lack of protection against the

IoT devices fair data collection and use. Furthermore, there are no privacy protection models of IoT devices that meet the users' expectations.

There should be a proper security and privacy mechanism against potential attacks in every layer of the IoT architecture because each layer is dependent with other layers [26].

i. Sensor layer security: In this layer, the IoT devices should implement secure identification and lightweight encryption mechanism to prevent the malicious nodes from impersonating and joining the network. A secure key agreement is required for transmission of information between the legitimate IoT nodes.

ii. Transmission layer security: In this layer, a secure communication channel is required to prevent from different types of attacks such as DoS, and packet compatibility issues. This layer is prone to data privacy leakage and data confidentiality.

iii. Application layer security: This layer is prone to security and privacy leakage due to inappropriate access control mechanism that provide unauthenticated and unauthorized access to the system. Furthermore, this layer should be protected from software vulnerabilities, and system management failure due to misconfiguration and information protection, etc.

The IoT uses cloud services as a trusted third party due to resource constraints of IoT devices. The central system has control over the data stored in the cloud servers and they can manipulate the user data [27]. The security and privacy issues in IoT can be tackled by eliminating the centralized server based approach used for managing and maintaining the data. This can be done introducing the blockchain which is a distributed ledger-based technology.

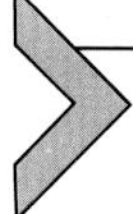

4. Impact of integration of IoT, blockchain, and homomorphic encryption

In this section, we focus on the state-of-the-art integration of IoT, blockchain and homomorphic encryption and analyze possible data security and privacy. The blockchain had gained lots of attraction in many fields including IoT ecosystem. The blockchain can enable a decentralized management system with secure and trustworthy transactions in IoT. IoT can leverage the advantages of blockchain to solve the key issues related to security and privacy. The convergence of IoT and blockchain can secure the sharing of information between different heterogeneous IoT devices. The blockchain platform provides data trustworthiness by guaranteeing the IoT data immutability and unforgeablility. The convergence of IoT with blockchain provides a synergic effect by leveraging the key benefits of blockchain technology. The blockchain helps the IoT system to operate in a decentralized and untrusted environment so that it is robust against different types of threats and attacks.

The blockchain has several advantages and among them decentralization, immutability, authenticity, data integrity, non-repudiation, reliability and availability are the key factors from security point of view. The blockchain smart contract provides additional features that can be used to set policies, monitor access rights and control user data. The smart contract executes autonomously when the set of pre-defined policies or conditions are met [19]. With blockchain technology, nodes achieve decentralization, when all the miners and full nodes maintains same copy of the blockchain state among all the distributed member nodes. Some of the properties of blockchain to secure IoT devices are discussed below.

Reliability: Unlike cloud services provided by trusted third party that stores data on the central server, where it has control over the data and it might manipulate the data easily. The blockchain maintains data reliability using Merkel tree and hash functions. If any malicious nodes attempt to tamper or modify the data by changing the hash values, the system will detect the malicious behavior and immediately reject the modified blocks.

Authenticity: In blockchain, authenticity is maintained by using cryptographic keys that are used by the nodes for signing and decryption. The blockchain will store these keys for encryption and verification. The cryptographic keys prevent blockchain from critical attacks that affects the IoT system such as Distributed Denial of Service (DDoS), man-in-the-middle, and replay attacks.

Non-repudiation: The non-repudiation is achieved in blockchain-based IoT by each IoT device signs the transaction and add them to the block in the blockchain as a result the sender cannot deny that the particular data is provided by him [28].

Decentralization: The IoT devices store data in the cloud that may not be available readily, when the services provided by the cloud servers are down due to cyber-attacks, software bugs or upgrade, power problems, etc. [25]. By using block chain in IoT, all the IoT data are stored in the decentralized and distributed blockchain and hence it is readily available when needed.

Availability: IoT ecosystem consists of a large number of small IoT devices. These IoT devices join and leave the network dynamically. Even when several nodes fail or leave the network, the data is always available in the blockchain because the same data is stored in multiple nodes.

Security and reliability: The data security and reliability is also a key requirement in IoT. The data are stored in public ledger as a block after the consensus of the honest miner nodes using PoW mechanism [14], this means that the data stored in the blockchain are highly reliable.

Despite the fact that the blockchain supports reliability and non-repudiation to some magnitude, privacy and confidentiality of IoT devices or data are not maintained properly. Because of the public nature of the bitcoin blockchain, the transaction information such as the sender and receiver public key and the amount of bitcoin transferred between them can be traced with some effort that might hamper the anonymity of bitcoin blockchain.

4.1 Integration of blockchain in IoT

There are several research conducted based on integration of blockchain in IoT. There is an increasing number of devices with integrated blockchain capabilities available on the market such as [29–31]. Some of the blockchain-based IoT in literatures are given below:

i. Dorri et al. [32] proposed a Lightweight Scalable Blockchain (LSB) for smart home IoT devices and its applications using overlay networks. The overlay networks consist of high resource devices that forms distinct clusters and the cluster head maintains the public blockchain. The clusters help to minimize the overheads leading toward privacy and security in IoT devices. The authors optimized the LSB algorithm, including the lightweight consensus algorithm, distributed trust method, throughput management as well as transaction traffic separated from the data flow. They present different types of attacks in IoT and blockchain and outline the specific defense mechanisms that ensures LSB is resilient to these types of attacks. Their simulation results show that their proposed scheme can overcome the scalability issues while reducing packet overhead and delay in blockchain. They further enhanced this scheme by including various core components and functions of the smart home tier comprising high resource devices that can act as miners [33].

ii. Dorri et al. [3] also proposed a distributed solution for smart vehicle security and privacy. The authors re-used the LSB blockchain concept to reduce blockchain overhead and provide decentralized security and privacy by using overlay networks for smart vehicles. The security of the vehicles is maintained by using symmetric encryption for all the transactions and the data are exchanged in a trustworthy manner. The privacy of the vehicles is maintained by locally storing the privacy sensitive data such as location information. Their qualitative analysis showed that their proposed architecture is resistant against common security attacks.

iii. The authors in Ref. [34] proposed a data storage and protection scheme for large scale IoT based on blockchain. The authors used the blockchain technology to store the IoT data in a distributed manner and used certificate less cryptography for transaction security. In the IoT system, a large group of devices act as miners that controls the data stored in the distributed blockchain thus eliminating the traditional centralized cloud based servers. Their certificate less cryptography provides effective authentication in IoT by reducing redundancy caused by PKI system. Moreover, they introduced the concept of the IoT data trading schemes.

iv. The authors in Ref. [35] proposed lightweight blockchain in connected vehicles for forensic applications. They used permissioned blockchain framework in connected vehicles for forensic analysis, such as traffic accidents, vehicle maintenance history, insurance, and vehicle diagnostic reports. This information is stored in the fragmented ledger. The authors integrate Vehicular Public Key Management (VPKI) scheme in the private blockchain for membership registration for providing privacy to the vehicles. The forensic framework based on fragmented ledger requires minimal storage and consumes low processing overhead; however, the fragmented ledger only stores hash values and does not ensure data correctness and data availability.

v. The authors in Ref. [36] proposed a scalable access management scheme for blockchain based IoT. A fully distributed easy-to-manage access control mechanism has been introduced for blockchain based IoT that eliminates the need of centralized access management system. Their scheme attempts to minimize the communication overhead between nodes in the blockchain by using a single smart contract. The lightweight manager nodes define the access control policy of the systems, which interacts with the smart contract and improves the scalability issues. The scheme is based on the proof of concept consensus mechanism that has been implemented in realistic IoT scenarios.

vi. The authors in Ref. [37] focus on preserving the privacy of the IoT blockchain ecosystem using Attribute-based Encryption (ABE) scheme. The ABE scheme provides access control as well as confidentiality using only single encryption. The end-to-end privacy and confidentiality of the information exchanged between the IoT devices are ensured by the ABE scheme. This scheme uses multiple distributed management nodes that overcomes the need for centralized access control servers. The numerical results show that their scheme can achieve privacy using ABS scheme with minimal computation overhead.

vii. The authors in Ref. [38] proposed IoTChain, a secured architecture for blockchain based IoT. The IoTChain provides end-to-end security solution by using authorization blockchain based on ACE framework [39] and security model based on OSCAR architecture [26] along with a self-healing group key scheme. The IoTChain uses smart contracts based on Ethereum for authorization requests. The ACE authorization servers act as miners and store the block in the blockchain after verification of the transactions. The authors also proposed a Proof-of-Possession (PoP) concept to bind the identity of the clients to access token.

viii. The authors in Ref. [40] proposed BCTrust, which is a blockchain based authentication mechanism for Wireless Sensor Network (WSN) devices that has resource constraints such as energy, storage, and computation. Similar to other schemes, the authors have used smart contracts based on Ethereum for transactions and only the group of trustworthy nodes have access to write on the blockchain. The BCTrust provides a decentralized authentication system with the global view of the blockchain network, thus the migration operation can be realized autonomously, transparently and securely. However, the authors have not discussed the scalability issues related to the network size as the number of nodes increases drastically.

ix. Similarly, Catenis Enterprise [41] introduced blockchain of things that aims to solve the enterprise Industrial IoT (IIoT) security problems that is built on blockchain messaging and digital asset infrastructure. It offers different services or products to the customers and provides professional services to help business with integration. It supports cross platform integration such as Ethereum, Hyperledger and other popular blockchains and allows secure message, smart properties, smart contract and digital asset transmissions. It integrates all heterogeneous IIoT systems that reduce the device development cost, while improving the developer productivity. It ensures that every message is automatically encrypted, and provides end-to-end encryption. The comparison of different types of blockchain used as a security mechanism in IoT is given in Table 1.

4.2 Integration of homomorphic encryption in IoT

i. The authors in Ref. [42] proposed HE for securing vehicles' location as a security mechanism, while preserving privacy in the opportunistic VANETs. They used HE to protect the location privacy by encrypting and verifying the location and distance information. The HE scheme encrypts and compares the distance information of the vehicles. The comparison result will not reveal any information to the third party. It provides security against the attacker vehicles while preserve the privacy of the vehicles.

ii. The authors in Ref. [43] proposed a FHE algorithm for preserving location and identity privacy in vehicle to Grid (V2G) networks. In V2G, the Electric Vehicles (EVs) share their information such as identity, electricity consumption behaviors, parking and charging locations to the power grid using two-way communication. The authors proposed a secure communication link between the EVs and the smart grid using FHE, when EVs are out of the communication coverage during running state. Besides security, HE ensures that only authenticated entity can retrieve and access the sensitive information. However, there is a drawback in this scheme during the initial deployment of V2G networks, due the limited number of EVs for message transmission to the smart grid in running state. However, during the initial deployment, there are limited numbers of EVs for message transmission that may hamper the smooth operation of the system.

iii. Similarly, the authors in Ref. [44] proposed a novel Randomized Authentication (RAU) protocol for preserving privacy based on Paillier's public key HE scheme that is suitable for both MANET and VANET environments. Their scheme preserves the user's privacy by allowing the individual users to self-generate a large number of

Table 1 Comparison of blockchain used as a security mechanism in IoT.

Reference	Scheme/scope	Blockchain type	Consensus mechanism	Features
[32]	Smart home IoT	Private blockchain	Time-based consensus	Improves scalability issues while reducing packet overhead and delay in blockchain
[3]	Smart vehicles security	Public BC managed by overlay nodes	–	Decentralized security and privacy by using overlay networks for smart vehicles and resistant against common security attacks
[34]	Data storage and protection	Private blockchain	–	Certificate less cryptography provides effective authentication in IoT and reduces redundancy
[35]	Connected vehicles for forensics applications	Permissioned blockchain	Proof of concept	Used fragmented ledger that consumes minimal storage and consumes low processing overhead
[36]	Fully distributed IoT access management	Private blockchain	–	Lightweight management nodes minimize communication overhead between nodes by using a single smart contract
[37]	Privacy of the IoT blockchain ecosystem using ABE scheme	–	–	Achieve privacy using ABS scheme with minimal computation overhead
[38]	IoTChain	Private blockchain	Proof-of-Possession (PoP)	Provides end-to-end security solution by using authorization blockchain
[40]	Authentication mechanism for WSN	Public blockchain	Proof of concept	Migration operation can be realized autonomously, transparently and securely
[41]	Enterprise industrial IoT	Permissioned blockchain	Proof of Authenticity	Enterprise security infrastructure software based on blockchain for IoT that delivers end to end secure messages to the target device

authenticated IDs by using zero-knowledge proof to achieve anonymous authentication. In case of disputes or malicious activities by the attacker nodes, the attackers can be traced and identified by using a pair of authentication servers that execute a collaboration protocol. It is lightweight and useful for small IoT devices; however, it provides only statistical security.

iv. The authors in Ref. [45] proposed a privacy-preserving route reporting scheme in VANETs for traffic management based on homomorphic encryption. The authors proposed Privacy-preserving Route reporting for Infrastructure-Based VANETs (PRIB), where each vehicle encrypts its route information using HE scheme and sends the cipher texts to RSU. In PRIB scheme, there are four stages, each stage works toward the security and privacy of the vehicles. However, there is computational overhead in PRIB scheme.

v. On the other hand, the authors in Ref. [46] improved the route reporting in PRIB scheme. The improvised route reporting is more practical than the original scheme [45]. Their improved scheme has five stages, including the batch verification stage. All other stages are similar to PRIB scheme. In this additional batch verification stage, the RSUs can verify multiple route reporting packets in a very short time using batch verification technique before sending the aggregated routes to the Traffic Management Center (TMC). Their experimental results show that their scheme has lower computation overhead and faster verification time than that of PRIB's scheme, which has additional pairing operations of n vehicles' route reporting packets.

vi. The authors in Ref. [47] used a novel type of homomorphic encryption known as Linearly Homomorphic Authenticated Encryption (LinHAE) for ground controllers in drones for safe autonomous flights. One of the issues with the drone system is that if the attacker steals the secret keys of ground controller then the whole system will be compromised. The LinHAE scheme supports linear operation between ciphertexts with fast encryption, evaluation and verification procedure for real-time controller and provides security against forgery and eavesdropping attacks. Even if the adversary obtains the ciphertext, he cannot have access to any information about the ciphertext. However, in this scheme, there is a possible obstacle with interplay between the dynamic propellers and the controllers. The comparison of different types of homomorphic encryption used as a security mechanism in IoT is give in Table 2.

Table 2 Comparison of homomorphic encryption used as a security mechanism in IoT.

References	HE scheme	Scope	Pros and Cons
[45]	PHE	Encrypts route information	*Pros*: Third party cannot infer any vehicle's route information *Cons*: Computational overhead as it requires $n+1$ pairing operations
[42]	SHE	Location privacy	*Pros*: Security against malicious vehicles and avoids privacy leakage *Cons*: High computation overhead, while calculating square of the distance between vehicles
[43]	FHE	Location and identity privacy in V2G	*Pros*: Protects the system from the attackers against tampering, intrusion and interception *Cons*: Limited EVs for message transmission during initial deployment
[44]	PHE	Pseudo identity	*Pros*: Lightweight computation and storage useful for IoT devices *Cons*: Not very efficient as it only statistically secure
[46]	PHE	Improved version of PRIB [45]	*Pros*: Lower computation overhead and faster verification time than PRIB
[47]	LinHAE	Secure ground controller	*Pros*: Provides security for drone controller against forgery and eavesdropping attacks *Cons*: Possible obstacle with interplay between dynamics and controller

4.3 Integration of homomorphic encryption in blockchain

A very few research has been done in integration of homomorphic encryption as a security in blockchain. Blockchain has gained a level of popularity in terms of secure distributed and decentralized ledger. However, blockchain is still not perfect in terms of security and privacy of the nodes. If homomorphic encryption can be used as an additional security features in blockchain, then it can enhance the security and privacy to the nodes.

The authors in Ref. [48] proposed Enigma, which is a peer-to-peer decentralized computation platform that enables users to store and share their data with guaranteed privacy using cryptographic schemes. Usually, the users store their private and sensitive data in less transparent centralized system but this leads to data leakage due to lack of privacy in the centralized servers. Enigma is a private decentralized platform used for recording public states in the blockchain and recording private states in off-chains. Enigma connects to the existing blockchain for executing public states such as transactions between nodes that can be audited. Enigma offloads private and resource intensive tasks such as high computation tasks to an off-chain network using private contracts, which is an advanced form of smart contracts but they handle private information. The off-chain network is a decentralized Distributed Hash Table (DHT) that is used to store users' encrypted private data. The blockchain stores access control protocols and references to the encrypted private data. Enigma uses an API to perform these tasks. The advantage of using off-chain network is to store large encrypted private data and perform huge computations tasks without leaking the data to any other nodes ensuring both privacy and correctness. In Enigma, each node has encrypted data and a distinct view of shares that guarantees privacy computation. A secure Multiparty Computation (MPC) based on homomorphic encryption is used so that multiplicative and additive homomorphism could be performed directly on the shares without interaction. Enigma uses three decentralized databases and they are public ledger, DHT and MPC. The public ledger is the blockchain ledger that stores the history of transactions in immutable manner. The off-chain data such as private data and shares are stored in the DHT and are accessible similar to the blockchain. The users encrypt the data locally before sending to the DHT and only the signing users can request the data back. The MPC is similar to DHT but underlying technology is different. In MPC, the secret shares are distributed to candidate computing nodes that stores their shares in their local view. Only the data owner can request for the data back by executing some function and can reconstruct the secret value by collecting shares from the different nodes locally. MPC guarantees the verifiable secret-sharing scheme using homomorphic encryption. In addition, Enigma implements security deposits and incentivizes fee to the good nodes that guarantees the correctness and fairness of the system.

The authors in Ref. [49] proposed Nebula blockchain that stores all the data transaction records securely in the immutable blockchain integrating homomorphic encryption. The genes are DNA sequence or molecule that encodes the blueprint of human beings. The total amount of DNA found in a cell is known as its genome. Several types of diseases will be recognized and new type of drugs will be developed if the genomic data is shared between the researchers. The authors proposed Nebula blockchain to eliminate centralized genomic data generation and intermediaries genomic companies that might cause data privacy leakage during genome sequencing. The data owner can obtain their private genomic data from Nebula sequencing facilities based on Nebula blockchain. The Nebula blockchain connects the data owner and data buyer directly without intermediaries so the personal genomic data can be protected by using blockchain. The Nebula blockchain is based on the blockstack platform [50] and Ethereum. The shared genome data is encrypted and analyzed based on homomorphic encryption and Intel Software Guard Extensions (SGX) so that the data buyer cannot see the genome data in plain text format. The genome data owners encrypt and share their private genome data with other secure compute nodes. Later, SGX combined with homomorphic encryption decrypts the encrypted data of the data owner that has been shared secretly. The bioinformatics computation require only addition operation so additive homomorphic encryption scheme is executed. Individually data owner encrypts using an additively homomorphic encryption the value 1 or 0, representing presence or absence of a genomic variant. Then, a computing node sums up all encrypted values outside the SGX enclave. The obtained encrypted sum is decrypted in the SGX enclave. Therefore, additively homomorphic encryption decreases the large number of decryptions to single decryption [49]. The data owner's identity is pseudo-anonymous by using secure cryptographic mechanism and all the genome data transactions are recorded in Nebula blockchain. The data owners or the genomic databanks stores their data at the edge of the Nebula network like off-chains while retaining ownership of the data. The genomic data purchased in Nebula is done through smart contracts that will speed up the data purchase by automating the procedure of signing contracts, making payments, and transferring genomic data.

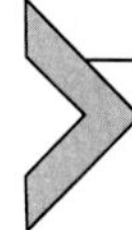

5. Collaborative security by integrating IoT, blockchain and homomorphic encryption

Recently, IoT has received enormous attraction from industry, business, intelligent transport, smart cities and academic sector. Similarly, blockchain has been the center of attraction in different fields such as finance, industry, academics and government sector. The IoT devices monitor the environment and collects the data from the environment using sensors and communicate with peer nodes to perform automated services and works. The IoT devices have limited storage and computation capabilities so the conventional IoT devices store all the data in a centralized server. As the user's data are entrusted with the centralized system, there is the issue of privacy and security. Even though, IoT devices encrypt their personal data and upload to the centralized servers; they are not able to perform homomorphic encryption on the encrypted data. The potential issues of IoT system have been discussed in Section 3. On the other hand, blockchain provides decentralized and distributed access model based on trustless networks. The member nodes of the network can interact with peer nodes without trusted intermediary. The use of robust cryptography and self-executing smart contracts makes blockchain a strong candidate for IoT privacy and security. Several research groups have a huge interest in integrating blockchain with IoT. The blockchain can provide IoT data security and privacy in a decentralized manner. In Section 4.1, we have discussed the integration of the blockchain in IoT and provided some of the previous research work in securing the privacy of the resource constrained IoT system. However, the previous blockchain-based IoT systems still have issues related to privacy leakage of sensitive information to the servers. In addition, several researchers have realized the IoT system based on homomorphic encryption as discussed in Section 4.2. In the previous works, the IoT devices collect all the environmental data using sensors and then encrypt and store the IoT data in the centralized servers. With the introduction of more and more IoT devices, there will be an enormous amount of IoT data streams at high speed. As a result, the centralized servers cannot handle such a large volume of data efficiently and there will be scalability issues. Moreover, there is always the risk of single point of failure while using the central servers.

If we can integrate the blockchain-based IoT with the homomorphic encryption, there will be a synergic effect due to the combination of these three technologies. We can carefully select and utilize the advantages of these technologies to build more secure IoT smart devices in the near future. We cannot neglect the ever-rising trends and evolving technologies such as blockchain and IoT without which we cannot imagine our future. Although blockchain-based IoT system attempts to provide some degree of security and privacy in IoT data, it still has some privacy issues and are not free from vulnerabilities. The integration of blockchain based-IoT with homomorphic encryption has a great potential in securing the IoT data with high privacy. Only few researches have been done on the integration of IoT, blockchain and homomorphic encryption. We have discussed them in the following paragraphs.

Zhou et al. [51] integrated the blockchain-based IoT with homomorphic encryption to overcome the privacy issues in IoT systems. They proposed a novel blockchain-based IoT system with secure data storage using homomorphic encryption called "Beekeeper." The proposed blockchain-based IoT system is established using the Threshold Multiparty Computing (TSMPC) protocol, where servers perform homomorphic computation on secret shares and generate responses. The servers store the encrypted data from the IoT devices and process users' encrypted data based on homomorphic computation without decrypting them. In this way, the servers cannot learn anything from the users' encrypted data. In addition, it provides immutability because it stores the transactions accepted by the validators in the blockchain and provides credibility due to the publicly verifiable data. The servers and the IoT devices communicate with each other based on the transactions of the blockchain. Since it is based on blockchain, incentives are provided to the nodes that processes and finalizes the data in the blocks that motivates the honest nodes to work toward creating the good blocks in the blockchain. The Beekeeper attempts to overcome scalability issues by attracting the external computing resources to join their network that improves the network performance. The authors implemented the BeeKeeper prototype system based on Ethereum private blockchain that contains a leader, servers and IoT devices. They performed the simulation of BeeKeeper system and evaluated the transaction performance generated between the devices.

However, the BeeKeeper has some limitations as discussed below [52]:

- In BeeKeeper, the servers can perform only one-degree homomorphic multiplication and any degree homomorphic additions on the encrypted data from the IoT devices.
- The sending of encrypted data to servers and request to the server is limited to only one IoT device.
- A pairing mechanism is used, which is a costly computing mechanism to verify the conformity of the response sent by servers.

To overcome the limitation of the BeeKeeper, Zhou et al. proposed BeeKeeper 2.0, which is a modified and improved version of the previous BeeKeeper [52]. The improvements in the BeeKeeper 2.0 are mentioned below:

- In BeeKeeper 2.0, the servers can perform any degree of homomorphic multiplication and addition on the encrypted IoT data, i.e., full homomorphic encryption.

- The service automation is added in the IoT devices, i.e., the devices can start the service protocol automatically.
- To verify the conformity of the response of the servers, the costly pairing mechanism has been removed.
- In the enhanced BeeKeeper 2.0, each IoT device can send encrypted data and request to the servers.
- The system performance as well as scalability is increased due to the addition of new blockchain nodes acting as servers.

We will explain about the BeeKeeper 2.0 mechanisms in more detail. The authors in Ref. [52] enhanced the BeeKeeper scheme and proposed a confidential BeeKeeper 2.0. They applied Decentralized Outsourcing Computation (DOC) scheme in the IoT system to achieve trusted blockchain-enabled IoT system. In the DOC scheme based on blockchain, the servers in the blockchain are capable of using fully homomorphic computations on the encrypted IoT data generated from the IoT devices. In other words, the servers in the blockchain networks are used as a decentralized outsourcing computation for fully homomorphic computation, i.e., they can perform any number of homomorphic addition as well as any degree of homomorphic multiplications provided by the IoT data owners. In addition, the servers cannot learn anything from the IoT users' encrypted data and if the servers act maliciously, the validators and IoT data owners in the network detect them. In this scheme, the validators check the validity of the BeeKeeper 2.0 transaction payloads such as signature verification key and commitments of secret data along with other usual verification procedures.

The working principle of BeeKeeper 2.0 is shown in Fig. 6. There are five basic transactions exchanged between the devices in the BeeKeeper 2.0. The description of transactions along with the verification algorithms are explained in Table 3.

- Step 1: The IoT device executes the *Setup* algorithm that generates two transactions TX_{VK} and TX_{CS}. Then, sends them to the validators of the blockchain networks.
- Step 2: The validators of the blockchain networks verify the TX_{VK} and TX_{CS} transactions executing *VerifyTx_*TX_{VK} and *VerifyTx_*TX_{CS} algorithms, respectively. The validators accept both the transactions TX_{VK} and TX_{CS} and stores them in the blockchain, if they are valid else the procedure return to Step 1. Moreover, each server verifies TX_{CS} core-share transaction by executing *CheckEnc_CS* algorithm, after the TX_{CS} is stored in the blockchain.

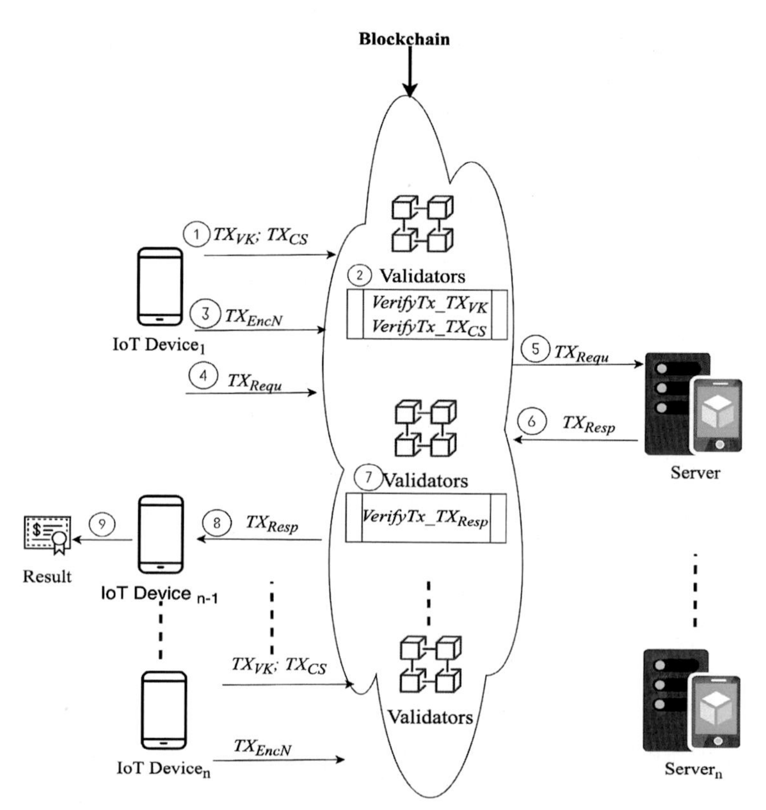

Fig. 6 Working principle of BeeKeeper 2.0. *Adapted from L. Zhou, L. Wang, T. Ai, Y. Sun, "BeeKeeper 2.0: confidential blockchain-enabled IoT system with fully homomorphic computation,* Sensors, *18 (11) (2018).*

- Step 3: The IoT device generates TX_{EncN} transaction by executing the *EncNum* algorithm and sends this transaction to the blockchain network.
- Step 4: Again, the IoT device generates TX_{Requ} transaction by executing the *Request* algorithm and sends this transaction to the blockchain network.
- Step 5: Each server executes *Respond* algorithm, after obtaining the TX_{Requ} transaction and
- Step 6: Server sends the TX_{Resp} transaction to the validators in the blockchain networks.
- Step 7: The network validators verify the TX_{Resp} transaction by executing the *VerifyTx_*TX_{Resp} algorithm. The validators accept TX_{Resp} transaction and stores the transaction in the blockchain, if it is valid else rejects it.

Table 3 Description of the transaction and verification notations.

Transaction/ algorithm	Description
TX_{VK}	Includes a Verification Key (VK)
TX_{CS}	Includes encrypted core-shares and commitments of core-shares (CS)
TX_{EncN}	Includes a set of encrypted numbers
TX_{Requ}	Includes a request
TX_{Resp}	Includes commitment and encrypted response
Setup	Algorithm that generates two transactions TX_{VK} and TX_{CS}
$VerifyTx\_TX_{VK}$	Algorithm to verify TX_{VK}
$VerifyTx\_TX_{CS}$	Algorithm to verify TX_{CS}
$VerifyTx\_TX_{Resp}$	Algorithm to verify TX_{Resp}
CheckEnc_CS	Algorithm to verify encrypted core-share in TX_{CS}
CheckEnc_Resp	Algorithm to verify encrypted responses

- Step 8: After the TX_{Resp} transaction is stored in the blockchain, the IoT device receives and verifies the encrypted response in the TX_{Resp} transaction by executing *CheckEnc_Resp* algorithm. The IoT device accepts the encrypted response if it is valid else rejects it.
- Step 9: The IoT device executes *Recover* algorithm to obtain the desired result after collecting t valid responses.

The BeeKeeper 2.0 provides the credibility of the system by storing the publicly verifiable information in the blockchain such as verification key, responses, encrypted numbers and commitments of core-shares and the validators in the network verify them. The authors used Hyperledger Fabric blockchain for testing the performance and evaluation of the BeeKeeper 2.0. Their results show that the servers can process the encrypted data at most 10 degree polynomials with the acceptable theoretical computation time between the request and recover stage, i.e., approximately 3.3 s.

Some of the advantages of integrated blockchain-based IoT with homomorphic encryption are discussed as follows:

- Decentralization: The integrated system provides decentralization where the data are stored in blockchain maintained by distributed peer nodes instead of centralized clouds. Any blockchain nodes can become a server of the IoT devices when desired by the peer nodes and owner of the IoT devices.

- Public ledger: As the data are stored in the blockchain, the IoT devices and the servers can verify the data publicly.
- Full homomorphic encryption. The servers in the blockchain network are able to compute any-degree homomorphic multiplications and any number of additions on encrypted data based on devices' request.
- Confidentiality: One most important advantage is that the servers are not able read any information from the encrypted data that provides confidentiality.
- Scalability: The scalability and the computation power improve with the additional blockchain nodes keen to act as servers.
- Lightweight. The resource constrained IoT devices do not require considerable memory and computation resources because (i) the blockchain stores the encrypted data (ii) the servers handle the computation on encrypted data to achieve desired results and (iii) the blockchain validators performs the verification tasks. So, the IoT devices are relieved from high processing and computation requirements and tasks.
- Fault-tolerant. The integrated system functions well provided that a threshold number of servers are honest and keen to serve IoT devices.

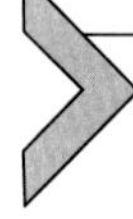

6. Use cases of blockchain-based IoT using homomorphic encryption

The authors in Ref. [53] proposed a conceptual security model based on the linearly homomorphic signature scheme in the ID-based crypto systems in the construction of IoT to overcome the disadvantages of public key certificates. There are several schemes, which stores IoT data in the cloud based on blockchain technique. They use cloud services for storage while providing outsourced computation service. However, it is difficult for the receiver to detect and obtain the correct and authenticated data from the centralized server. The centralized server might provide incorrect computation outcomes from user's data. This scheme [53] can overcome the shortcomings of using the central cloud computing services for data and computation authentication. In this scheme, the user generates the signature for all the data stored in the nodes and then these data and signatures will be uploaded in the network nodes. In addition, the data pointers as well as the access information will be calculated and kept on the blockchain. The data computation can be outsourced to the network nodes because all the data are signed with homomorphic signature, this means homomorphic signature can calculate any function. Thus, it reduces the computation and communication overhead. After the user receives the computation results along with the aggregated homomorphic signatures, they compare the computation results. The users accept the results if the signatures are valid else they will reject the results. By integrating blockchain-based IoT with homomorphic signature, (i) the IoT data and their respective signatures remain immutable because of blockchain, (ii) IoT data computation can be performed in the distributed network nodes and (iii) result verification can be done while keeping the authentication. This scheme can provide security from forgery and ID attack under random oracle model.

7. Challenges and future research direction

This section deals with the main challenges faced by integrating blockchain in IoT using homomorphic encryption. We have realized the huge potential of the blockchain integrated with IoT using homomorphic encryption. This integration can solve most of the privacy and reliability issues in IoT ecosystems. However, integrating blockchain in IoT cannot fully secure the IoT system. The concept of blockchain was initially designed for cryptocurrency, so the IoT devices should be equipped with cryptocurrency or transactions handling functionality. IoT includes billions of heterogeneous devices that generate trillions of transactions everyday as a result, the size of the blockchain increases and there will be scalability issues. For illustration, the data generated by millions of IoT sensor nodes deployed in smart cities will result in the bloating of the block size. The resource constrained IoT devices acting as miners or full nodes with limited storage capability cannot handle such a huge volume of data. Some research has been done in implementing fog computing as an alternative to overcome this issue [54]. However, the rapid increment in IoT devices from smartwatch to smart industrial equipment would produce large-scale data that hampers the security, scalability and real-time data delivery requirements, which needs to be resolved. In addition, when the blockchain size increases, the storage size, bandwidth along with the computation power for participating in consensus mechanism increases. We listed out some of the challenges faced by blockchain-based IoT using HE.

- Anonymity and Data privacy: Most of the IoT devices such as wearables used in health care systems are linked with the personal data of the user. The anonymity and data privacy of the users should be addressed efficiently to prevent data leakage. Some of the blockchain provides anonymity (e.g., Bitcoin); however, in case of IoT system, the anonymity should provide guarantee to hide the users' or smart devices' personal information such as identity, location, transactions, and private data.

Since, the IoT devices are very small and resource constrained, securing the device as well as the data stored in it against tampering, theft and hacking is a challenging issue, and requires expensive secure cryptographic mechanisms and hardware devices.

- Security: Many industries that produce IoT devices, want to launch the devices in the market without considering security issues seriously. These devices are vulnerable to hacks and attacks due to security breaches, lack of proper firmware updates and serious bugs. On the other hand, integrating blockchain with IoT and homomorphic encryption cannot guarantee data security because if the attacker breaches and corrupt the IoT data before committing the data in the blockchain, then the corrupted data will be in the blockchain forever. Similarly, the devices might not work properly due to sensors or actuator failure, environmental conditions, power outage, etc. These issues cannot be identified until the IoT devices are tested, as a result, they might produce corrupt output data. Hence, IoT device should be secure with sophisticated hardware and software to prevent from rising number of attacks in IoT system.
- Scalability and Storage capacity: As already discussed, billions of IoT devices produce trillions of transactions and most of the blockchains can process few hundreds of transactions per seconds. This challenge will escalate with the integration of the blockchain in IoT using homomorphic encryption. IoT ecosystem requires inherent storage space and it has to face the scalability issues. In case of IoT system, servers are used to store the IoT data. While blockchains are designed to store the encrypted data in public ledger and each ledger is stored in distributed peer nodes in a decentralized way. The blockchain is not designed to store such enormous IoT data so this results an impending bottle neck in integrating IoT [55]. Moreover, homomorphic encryption consumes huge amount of computation power that cannot be handled by the resource constrained IoT devices.

Nevertheless, there are ways to tackle these issues. The IoT devices generate tremendous amount of data but not all of these data are useful. Therefore, filtering, extracting, normalizing, hashing and compressing these data might reduce the storage space. Another way to manage blockchain size on IoT devices is to use new blockchain architecture such as sidechains or treechains [11]. The blockchain scalability issue can be improved by storing the IoT data on an off-chain network of private nodes in the form of Distributed Hash Table (DHT) [48], so that the blockchain does not have to store all the transaction data. The blockchain stores the reference to the data only, thus reducing the blocksize. Furthermore, the blockchain size can be reduced and homomorphic encryption can be utilized by using the concept of regional/local [18] and global blockchain, where resource constrained IoT devices and standard IoT devices are categorized to store data depending upon their battery, storage, processing and computation capabilities.

- Mining and Consensus: In blockchain-based IoT system, mining and consensus mechanism is an inherent characteristic. Simply using PoW based consensus in resource constrained IoT devices make it infeasible due to consumption of large power in solving the puzzle. Several lightweight consensus mechanisms that are suitable for blockchain-based IoT system have been proposed, but they are in early stage and have not been tested properly. Due to IoT devices limited resource, mining is still a challenging issue.

  However, recent IoT devices integrated with high computation power can withstand the mining and consensus mechanism. An appropriate consensus mechanism should be developed through academic and industrial research that can leverage the distributed nature and adapt the consensus mechanism of blockchain in IoT ecosystem.
- Real-time requirements: Some of the blockchain-based IoT devices such as IoV requires information disseminated in real time environment to prevent accidents. In case of blockchain-based IoT using homomorphic encryption, the IoT devices have to interact with blockchain and cryptographic functionalities that increases delay in information dissemination. The security benefits offered by the cryptography emanate at the cost of delay and requirements of higher computational power [55].
- Legal and regulatory issues: There are research going on the standard and proper legal systems for maintaining IoT device compatibility. It seems that there are not enough efforts contributed by IoT groups in developing protocols and standards. There are not enough laws and regulations for data protection, while sharing data between two heterogeneous devices. On the other hand, blockchain as a cryptocurrency has lots of controversy concerning legal regulations. Many countries have tighter laws and some countries have banned the cryptocurrency due to money laundering and unregulated transactions. The integration of blockchain

and IoT will face even bigger challenge regarding legal and regulatory issues for securing the user information.

Hence, proper standards, laws and regulations should be developed for using the IoT data based on the blockchain in a discriminatory manner. Moreover, there should be laws against the liability issues of the blockchain-based IoT devices. There should be awareness in policy plans and other regulatory co-ordinations along with the rapid growth of IoT devices [56]. In this way, most secure and trusted blockchain-based IoT network can be built.

Some of the future research directions of blockchain-based IoT with homomorphic encryption are as follows:

- We need to dig deeper regarding the use of public blockchain and HE in IoT because of the associated energy and computation cost due to the block mining on resource constrained IoT devices. We need to think of an efficient consensus and incentive mechanism to secure the blocks in IoT devices.
- How to combine blockchain with homomorphic encryption is also a part of research challenge.
- In the near future, there will be billions of IoT devices generating trillions of transactions every day. We should consider a special type of blockchain that takes no transaction fee or at least have negligible transaction fee.
- An efficient and lightweight blockchain-based IoT security system using HE should be designed to support resource constrained IoT devices.
- Need to consider different types of threat model for the integrated blockchain-based IoT with homomorphic encryption. The integration of technologies gives benefits, but it also adds new type of threats that might have a negative impact.
- An energy efficient consensus mechanism should be designed for the resource constrained IoT devices and management of the miner nodes.
- The IoT devices are installed a remote and hostile environments or in publicly accessible locations. The attacker or malicious people might take control of the devices physically or might tamper the devices hardware or software. In this case, the security and privacy of the data stored in the devices should be guaranteed by using blockchain and homomorphic encryption.
- A secure mechanism is required to detect the attacks or detect malicious devices in the network.

8. Summary

This section summarized the chapter. We presented an overview of IoT, blockchain and homomorphic encryption and then discussed the state-of-the-art of combination of those technologies. We dig into the security issues and challenges of the IoT system. We extensively surveyed the integration of IoT with blockchain and IoT with homomorphic encryption; however, they cannot provide full security for the IoT ecosystem. Then, we analyzed the convergence of IoT, blockchain and homomorphic encryption that provides some degree of security in the IoT system. We presented some application and use cases of blockchain-based IoT with homomorphic encryption. The collaborative security based on homomorphic encryption mechanism in blockchain-based IoT ensures that IoT data in the blockchain is encrypted, thus addressing the privacy issues related with public blockchains. The homomorphic mechanism over the IoT data stored in the blockchain smart contracts preserve privacy. We then present some of the challenges in the convergence of these technologies and then provide future directions. We believe that this work provides a useful guide for interested researchers and students in the field of convergence of IoT, blockchain and homomorphic encryption.

Glossary

Blockchain Security Blockchain provides security by recording and verifying the transactional data in a transparent ledger using consensus mechanism. However, there is still issue of privacy and trust in blockchain that needs to avoids potential cyberattacks. It is necessary to build a strong security mechanism into blockchain technology from the beginning through additional cryptographic technique.

IoT Data security While using IoT devices we should consider the privacy of the users' data strictly. It is difficult to implement the conventional security mechanism in resource constraint IoT devices. IoT devices' are vulnerable and are frequent target of attackers. In addition, the resource constrained features such as low memory, limited resources, inadequate computational power, and weak trust in third party cloud applications leads to IoT data security.

Homomorphic Encryption Homomorphic Encryption is an encryption mechanism that can resolve the security and privacy issues. It allows the third party service providers such as cloud servers to perform certain type of operations on the user's encrypted data without decrypting the encrypted data, while maintaining the privacy of the users' data.

Smart Contracts The smart contract is a self-executing program with predefined contractual policy that provides additional security features such as control user data and monitor access rights in blockchain. The smart contract is executed autonomously when the set of predefined conditions are met. If the mutual contract between the two parties is breached, the smart contract handles the issue by removing the ability for peer-to-peer nodes to conduct transactions without the need for a broker.

Collaborative Security In IoT system, a single security mechanism cannot effectively secure the IoT data. A collaborative security mechanism is required to detect and prevent a wide range of attacks in IoT system. A collaborative security mechanism has potential to protect the IoT networks and data from hardware, software or information threat from the malicious attackers.

References

[1] Y. Zhang, J. Wen, The IoT electric business model: using blockchain technology for the internet of things, Peer Peer Netw. Appl. 10 (4) (2017) 983–994.
[2] S. Nakamoto, Bitcoin: A Peer-to-Peer Electronic Cash System, White paper, 2008. www.bitcoin.org.
[3] A. Dorri, M. Steger, S. Kanhere, R. Jurdak, BlockChain: a distributed solution to automotive security and privacy, IEEE Commun. Mag. 55 (12) (2017) 119–125.
[4] Z. Li, J. Kang, R. Yu, D. Ye, Q. Deng, Y. Zhang, Consortium blockchain for secure energy trading in industrial internet of things, IEEE Trans. Ind. Inf. 14 (8) (2018) 3690–3700.
[5] F. Tian, An agri-food supply chain traceability system for China based on RFID & blockchain technology, in: 13th International Conference on Service Systems and Service Management (ICSSSM), 2016, pp. 1–6.
[6] J. Gao, et al., GridMonitoring: secured sovereign blockchain based monitoring on smart grid, IEEE Access 6 (2018) 9917–9925.
[7] IDC, IDC forecasts worldwide technology spending on the internet of things to reach $1.2 trillion in 2022, in: Worldwide Semiannual Internet of Things Spending Guide, IDC, 2018 [Online]. Available, https://www.idc.com/getdoc.jsp?containerId=prUS43994118. Accessed 28 April 2019.
[8] D. Lund, C. MacGillivray, V. Turner, M. Morales, Worldwide and Regional Internet of Things (IoT) 2014–2020 Forecast: A Virtuous Circle of Proven Value and Demand, IDC, 2014.
[9] T. T. Dandala, V. Krishnamurthy, and R. Alwan, "Internet of vehicles (IoV) for traffic management," in 2017 International Conference on Computer, Communication and Signal Processing (ICCCSP), 2017, pp. 1–4.
[10] A. Al-Fuqaha, M. Guizani, M. Mohammadi, M. Aledhari, M. Ayyash, Internet of things: a survey on enabling technologies, protocols, and applications, IEEE Commun. Surv. Tutorials 17 (4) (2015) 2347–2376.
[11] I. Makhdoom, M. Abolhasan, H. Abbas, W. Ni, Blockchain's adoption in IoT: the challenges, and a way forward, J. Netw. Comput. Appl. 125 (2019) 251–279.
[12] M. Khari, M. Kumar, S. Vij, P. Pandey, Vaishali, Internet of things: proposed security aspects for digitizing the world, in: 2016 3rd International Conference on Computing for Sustainable Global Development (INDIACom), 2016, pp. 2165–2170.
[13] X. Chen, Blockchain challenges and opportunities: a survey Zibin Zheng and Shaoan Xie Hong-Ning Dai Huaimin Wang, Int. J. Web Grid Serv. 14 (4) (2018) 352–375.
[14] A.M. Antonopoulos, Mastering Bitcoin, first ed., O'Reilly Media, Inc, USA, 2015.
[15] R.C. Merkle, A digital signature based on a conventional encryption function, in: Proceedings of CRYPTO'87, 1987, pp. 16–20.
[16] R. Shrestha, R. Bajracharya, S.Y. Nam, Blockchain-based message dissemination in VANET, in: 2018 IEEE 3rd International Conference on Computing, Communication and Security, 2018, pp. 161–166.

[17] N.Z. Aitzhan, D. Svetinovic, Security and privacy in decentralized energy trading through multi-signatures, blockchain and anonymous messaging streams, IEEE Trans. Dependable Secure Comput. 15 (5) (2018) 840–852.
[18] R. Shrestha, R. Bajracharya, A.P. Shrestha, S.Y. Nam, A new-type of blockchain for secure message exchange in VANET, Digit. Commun. Netw. (1) (2019) 1–15.
[19] V. Buterin, A Next-Generation Smart Contract and Decentralized Application Platform, *When Satoshi Nakamoto,* Ethereum, 2015, pp. 1–36.
[20] A. Acar, H. Aksu, A.S. Uluagac, M. Conti, A survey on homomorphic encryption schemes, ACM Comput. Surv. 51 (4) (2018) 1–35.
[21] F. Armknecht, et al., A guide to fully homomorphic encryption, in: IACR Cryptology ePrint Arch, 2015, pp. 1–35.
[22] A. Banafa, IoT Standardization and Implementation Challenges, IEEE Internet of Things, 2014. [Online]. Available, https://iot.ieee.org/newsletter/july-2016/iot-standardization-and-implementation-challenges.html. Accessed 9 April 2019.
[23] OWASP, Top 10 Application Security Risks—2017, https://www.owasp.org/index.php/Top_10-2017_Top_10, 2017.
[24] R. Shrestha, R. Bajracharya, S.Y. Nam, Challenges of future VANET and cloud- based approaches, Wirel. Commun. Mob. Comput. 2018 (2018) 1–15. no. Article ID 5603518.
[25] N. Kshetri, Can blockchain strengthen the internet of things? IT Prof. 19 (4) (2017) 68–72.
[26] F. Ayotunde Alaba, M. Othman, I.A.T. Hashem, F. Alotaibi, Internet of things security: a survey, J. Netw. Comput. Appl. 88 (2017) 10–28.
[27] E. Gaetani, L. Aniello, V. Baldoni, R. Lombardi, M. Federico, Andrea, Sassone, Blockchain-based database to ensure data integrity in cloud computing environments, in: Italian Conference on Cybersecurity, 2017, pp. 1–10.
[28] X. Liang, J. Zhao, S. Shetty, D. Li, Towards data assurance and resilience in IoT using blockchain, in: 2017 IEEE Military Communications Conference (MILCOM), 2017, pp. 261–266.
[29] Ethembedded, Ethembedded, [Online]. Available, http://ethembedded.com/, 2017. Accessed 9 April 2019.
[30] Raspnode, Raspnode.com, [Online]. Available, https://raspnode.com/, 2017. Accessed 9 April 2019.
[31] Ant Router, Ant Router r1-ltc the Wi_ Router That Mines Litecoin, [Online]. Available, https://antminerprofitability.com, 2017. Accessed 8 April 2019.
[32] A. Dorri, S.S. Kanhere, R. Jurdak, P. Gauravaram, LSB: A Lightweight Scalable BlockChain for IoT Security and Privacy, 2017, pp. 1–17, arXiv:1712.02969.
[33] A. Dorri, S.S. Kanhere, R. Jurdak, P. Gauravaram, Blockchain for IoT security and privacy: the case study of a smart home, in: 2017 IEEE International Conference on Pervasive Computing and Communications Workshops (PerCom Workshops), 2017, pp. 618–623.
[34] R. Li, T. Song, B. Mei, H. Li, X. Cheng, L. Sun, Blockchain for large-scale internet of things data storage and protection, IEEE Trans. Serv. Comput. (2018) 1.
[35] M. Cebe, E. Erdin, K. Akkaya, H. Aksu, S. Uluagac, Block4Forensic: an integrated lightweight blockchain framework for forensics applications of connected vehicles, IEEE Commun. Mag. 56 (10) (2018) 50–57.
[36] O. Novo, Blockchain meets IoT: an architecture for scalable access management in IoT, IEEE Internet Things J. 5 (2) (2018) 1184–1195.
[37] Y. Rahulamathavan, R.C. Phan, M. Rajarajan, S. Misra, A. Kondoz, Privacy-preserving blockchain based IoT ecosystem using attribute-based encryption, in: 2017 IEEE International Conference on Advanced Networks and Telecommunications Systems (ANTS), 2017, pp. 1–6.

[38] O. Alphand, et al., IoTChain: a blockchain security architecture for the Internet of Things, in: IEEE Wireless Communications and Networking Conference, WCNC, vol. 2018, 2018, pp. 1–6.
[39] L. Seitz, G. Selander, E. Wahlstroem, S. Erdtman, H. Tschofenig, Authentication and Authorization for Constrained Environments (ACE), IETF, 2017.
[40] M.T. Hammi, P. Bellot, A. Serhrouchni, BCTrust: a decentralized authentication blockchain-based mechanism, in: 2018 IEEE Wireless Communications and Networking Conference (WCNC), IEEE, 2018, pp. 1–6.
[41] C. Enterprise, Blockchain of Things, [Online]. Available, https://blockchainofthings.com/, 2014.
[42] J. Song, C. He, F. Yang, H. Zhang, A privacy-preserving distance-based incentive scheme in opportunistic VANETs, Secur. Commun. Netw. 9 (15) (2016) 2789–2801.
[43] Z.W. Sun, W.X. Yan, A privacy preserving scheme for vehicle to grid networks based on homomorphic cryptography, Adv. Mat. Res. 1014 (2014) 516–519.
[44] W. Jiang, D. Lin, F. Li, E. Bertino, Randomized and Efficient Authentication in Mobile Environments, Cyber Center Publication, 2014, pp. 1–15.
[45] K. Rabieh, M.M. Mahmoud, M. Younis, Privacy-preserving route reporting schemes for traffic management systems, IEEE Trans. Veh. Technol. 66 (3) (2017) 2703–2713.
[46] Y. Zhang, Q. Pei, F. Dai, L. Zhang, Efficient secure and privacy-preserving route reporting scheme for VANETs, J. Phys. Conf. Ser. 910 (1) (2017).
[47] Y. Song, et al., Toward a secure drone system: flying with real-time homomorphic authenticated encryption, IEEE Access 6 (2018) 24325–24339.
[48] G. Zyskind, O. Nathan, A. Pentland, Enigma: decentralized computation platform with guaranteed privacy, Distrib. Parallel Clust. Comput. (2015) 1–14, arXiv:1506.03471.
[49] D. Grishin, et al., Accelerating genomic data generation and facilitating genomic data access using decentralization, privacy-preserving technologies and equitable compensation, BHTY 1 (2018) 1–23.
[50] Blockstack, Blockstack, [Online]. Available, https://blockstack.org/. Accessed 28 May 2019.
[51] L. Zhou, L. Wang, Y. Sun, P. Lv, BeeKeeper: a Blockchain-based IoT system with secure storage and homomorphic computation, IEEE Access 6 (2018) 43472–43488.
[52] L. Zhou, L. Wang, T. Ai, Y. Sun, BeeKeeper 2.0: confidential blockchain-enabled IoT system with fully homomorphic computation, Sensors 18 (11) (2018).
[53] Q. Lin, H. Yan, Z. Huang, W. Chen, J. Shen, Y. Tang, An ID-based linearly homomorphic signature scheme and its application in blockchain, IEEE Access 6 (2018) 20632–20640.
[54] P.K. Sharma, M. Chen, J.H. Park, A software defined fog node based distributed blockchain cloud architecture for IoT, IEEE Access 6 (2018) 115–124.
[55] A. Reyna, C. Martín, J. Chen, E. Soler, M. Díaz, On blockchain and its integration with IoT. challenges and opportunities, Futur. Gener. Comput. Syst. 88 (2018) 173–190.
[56] N.M. Kumar, P.K. Mallick, Blockchain technology for security issues and challenges in IoT, Procedia Comput. Sci. 132 (2018) 1815–1823.

Further Reading/References for Advance

[57] H. Wang, Homomorphic Encryption on the IoT, Thesis, Department of Information Systems and Technology, Mid Sweden University, IPFS, 2018, pp. 1–44.
[58] M.S. Ali, K. Dolui, F. Antonelli, IoT data privacy via blockchains and IPFS, in: Proceedings of the Seventh International Conference on the Internet of Things (IoT'17), ACM, New York, NY, 2017, pp. 1–7.

[59] IOTW, "A Disruptive Technology to Bring Blockchain to Every Household", white-paper, [Online]. Available: https://icosbull.com/whitepapers/3484/IOTW_whitepaper.pdf. [Accessed: 28-May-2019], 2018.
[60] J. Lee, BIDaaS: blockchain based ID as a service, IEEE Access 6 (2018) 2274–2278.
[61] W. Song, B. Hu, X. Zhao, Privacy protection of IoT based on fully homomorphic encryption, Wirel. Commun. Mob. Comput. (2018) 1–7, 2018. Article ID 5787930.
[62] Q. Santos, Cryptography for Pragmatic Distributed Trust and the Role of Blockchain. Cryptography and Security, Thesis, PSL Research University; École Normale Supérieure, 2018.

About the authors

Rakesh Shrestha received his B.E in Electronics and Communication Engineering from Tribhuvan University (TU), Nepal in 2006. He received his M.E in Information and Communication Engineering from Chosun University, in 2010 and PhD degree in Information and Communication Engineering from Yeungnam University in 2018, respectively. From Feb. 2010 to May 2011, he worked for Honeywell Security System as a Security Engineer. From Jun. 2010 to Sep. 2012, he worked as a Core Network Engineer at Huawei Technologies Co. Ltd, Nepal. From 2018 Feb to 2019 Feb, he worked as a Postdoctoral Researcher at Department of Information and Communication Engineering, Yeungnam University, Korea. He is currently working as a Postdoctoral Researcher in Yonsei University of Convergence Technology, Yonsei University. He was invited as a keynote speaker in Third IEEE ICCCS conference and worked as a reviewer in several renowned journal and conferences. He is currently an IEEE member and his main research interests include wireless communications, Mobile ad-hoc networks, Vehicular ad-hoc network, blockchain, IoT, homomorphic encryption, deep learning, wireless security, etc.

Shiho Kim is a professor in the school of integrated technology at Yonsei University, Seoul, Korea. His previous assignments include, being a System on chip design engineer, at LG Semicon Ltd. (currently SK Hynix), Korea, Seoul [1995–1996], Director of RAVERS (Research center for Advanced hybrid electric Vehicle Energy Recovery System), a government-supported IT research center. Associate Director of the Yonsei Institute of Convergence Technology (YICT) performing Korean National ICT consilience program, which is a Korea National program for cultivating talented engineers in the field of information and communication Technology, Korea [2011 – 2012], Director of Seamless Transportation Lab, at Yonsei University, Korea [since 2011 to present].

His main research interest includes the development of software and hardware technologies for intelligent vehicles, blockchain technology for intelligent transportation systems, and reinforcement learning for autonomous vehicles. He is a member of the editorial board and reviewer for various Journals and International conferences. So far he has organized two International Conference as Technical Chair/General Chair. He is a member of IEIE (Institute of Electronics and Information Engineers of Korea), KSAE (Korean Society of Automotive Engineers), vice president of KINGC (Korean Institute of Next Generation Computing), and a senior member of IEEE. He is the coauthor for over 100 papers and holding more than 50 patents in the field of information and communication technology.

The authors thank for IITP (Institute for Information & communications Technology Planning & Evaluation), this works was support by MSIT (Ministry of Science and ICT), Korea, the under the "ICT Consilience Creative Program" (IITP-2019-2017-0-01015) supervised by IITP.

Printed in the United States
By Bookmasters